DEEPAK VASUDEVA

Basic Statistical Methods

FOR ENGINEERS AND SCIENTISTS

The authors of the first edition of this book were listed as Neville and Kennedy, the order of the names having been decided by chance and therefore not being significant. Since chance is a fickle jade, the order of names in this edition is Kennedy and Neville.

Basic Statistical Methods

FOR ENGINEERS AND SCIENTISTS

2nd Edition

John B. Kennedy
University of Windsor

Adam M. Neville
University of Leeds

iep *A Dun-Donnelley Publisher* *New York*

Library of Congress Cataloging in Publication Data

Kennedy, John B.
Basic statistical methods for engineers and
scientists.

Authors' names appear in reverse order on 1964 ed.
Includes bibliographical references and index.
1. Engineering—Statistical methods. I. Neville,
Adam M., joint author. II. Title.
TA340.N4 1976 620'.001'82 75-33267
ISBN 0-7002-2480-7

IEP A DUN-DONNELLEY PUBLISHER
666 Fifth Avenue
New York, New York 10019

Typography by Chris Simon

Wall-hanging by Elizabeth Ginsberg,
courtesy Soho Tapestries Gallery, New York

To N. K. and L. N.

Contents

Notation

${}_nC_r$	number of combinations of r items from n items
d	range coefficient
d_m	mean deviation
$E(x)$	expected value of a variable x
f	class frequency
$f(z)$	normalized probability density of a variable z
F	cumulative frequency
$F(z)$	area under normal probability curve between $z = 0$ and $z = z$
n	sample size
N	population size
$\binom{n}{r}$	number of combinations of r items from n items
p	probability of success of an event
$p(x)$	probability density of a variable x
$P(x)$	cumulative distribution function of a variable x
${}_nP_r$	number of permutations of r items from n items
q	$= (1 - p) =$ probability of failure of an event

r	correlation coefficient
R	range
$\bar{R}$	mean range
$R(t)$	reliability function
s	estimate of standard deviation
s^2	estimate of variance
S_t	cumulative sum of deviations at any instant t
V	coefficient of variation
w	class width
x	value of an independent variable
$\bar{x}$	estimate of population mean or sample mean
$\bar{\bar{x}}$	mean of means
X	$= (x - \bar{x})$
y	frequency of a continuous variable
z	$= \dfrac{x - \mu}{\sigma}$
$Z(t)$	hazard rate or instantaneous failure rate
α	level of significance, probability of committing a Type I error
β	probability of committing a Type II error
μ	true mean
σ	true standard deviation
$\sigma_{\bar{x}}$	standard error of the mean
σ^2	true variance
ν	number of degrees of freedom

Preface to the Second Edition

The continued demand for the first edition indicated, we believe, that our simple exposition of statistical methods has been found useful in the training of engineers and scientists. However, since the first edition was printed eleven years ago, we decided it was appropriate to update and revise the book.

We added sections on the cumulative sum technique and on the cumulative terms of the binomial distribution, together with an appropriate table on nonparametric tests. We wrote a new chapter on the distribution of extremes and specifically of failure time; this is of interest in the estimated life of a structure or of manufactured components. We also inserted additional material on sampling, degrees of freedom, the null hypothesis, Type I and II errors, and regression. We have removed most of the errors (the probability of removing them all is very small!), clarified ambiguities and improved explanations in a number of places. We have refrained, however, from inserting too much new material, which would change the character of the book from its simple format, because we continue to believe that this is what the majority of engineers and scientists need. In keeping with the modern trend, the Système International d'Unités (SI Units) has been used throughout the book.

We are grateful to Dr. June Adam and Mr. T. Lewis for help with some aspects of Solved Problem 18-4, and to Professor D. B. DeLury for help on some aspects of the power curve in Chapter 21. We wish also to acknowledge our obvious debt to the authors of numerous books on statistics from which we ourselves learned the art. Further, we are indebted to those who have suggested alterations to the first edition, especially Professors Clayborn, Hall, Stratton, and Stolte.

Reproduced material is acknowledged wherever appropriate. We are indebted to the Imperial Chemical Industries Ltd., London, and to Messrs. Oliver and Boyd Ltd., Edinburgh, for permission to reprint Table G from their book by Davies: *Statistical Methods in Research and Production*, and to the Literary Executor of the late Sir R. A. Fisher, F.R.S., Cambridge, and Dr. F. Yates, F.R.S., and the same publishers for permission to reprint Tables II, III, IV, and XXXIII from their book *Statistical Tables for Biological, Agricultural and Medical Research.*

We wish to express our thanks to Mrs. M. Christopherson and Mrs. E. L. Wittig who arranged the typing of the manuscript; to Messrs. D. J. Martin and F. W. Unger who checked the typescript; and to Miss S. M. J. Porter who read the proofs. The organization of the preparation of the second edition was very ably carried out by Mrs. Jenny Bennett with the help of Miss Sue Chapman and Miss Elizabeth Kelly. Last, as well as least, we are grateful to Elizabeth Neville who performed the experiment discussed on page 70.

Preface to the First Edition

We should explain at the outset that we are not mathematicians but engineers. This should not, however, be construed as an apology because this book is written to satisfy, we hope, the needs of engineers and scientists who, as is always the case, need statistics but not necessarily the full treatment of the underlying mathematical theory.

We believe that in order to better understand engineering and scientific problems and to be able to draw correct inferences from experiment, a knowledge of statistics is essential in the very early stages of a university program. This book makes the acquisition of such knowledge possible since it does not build upon rigorous mathematics but rather uses an intuitive approach and common sense. Clearly then this is *not* a book on mathematical statistics, but neither is it a cookbook of ready-to-apply formulae. Jargon is down to a minimum; explanation, though full, is kept simple and there is a large number of illustrative solved problems both in the text and at the end of each chapter.

The book should thus prove acceptable as a textbook either for a one-semester course, possibly with some of the more advanced chapters omitted, or as a basis for a one-year junior course taken at a comfortable rate. We have endeavored to make the book suitable also for the engineer or scientist who wants to teach himself the basic statistical methods, as well as for occasional use by the man who is presented with a specific problem requiring statistical treatment, be it in research, in laboratory, or in production.

We hope to have satisfied these aims since the book is written largely from the practical point of view. The manuscript was understood by our wives which makes us confident that it can be understood by anybody.

Chapter 1

Introduction

It is a truism to say that an improvement in the knowledge of the world around us requires an ever-increasing use of statistical methods and inferences. Almost everyone needs some knowledge of statistics. However, because of the width and depth of the subject, we have to select the field of knowledge and methods relevant to a particular purpose. This book is primarily concerned with basic applications in engineering and in science, although much of the knowledge of basic applied statistics is of use to workers and students in other fields of study.

There are several definitions of statistics. It is defined as the science concerned with problems involving *chance* variations that result from a large number of small and independent influences operating on each measured result which we obtain; or, it can be defined as the science of making decisions from data in such a way that the reliability of conclusions based on the data is evaluated by means of probability. More generally, statistics is a science that deals with the collection, tabulation, analysis, and interpretation of quantitative and qualitative data; this process includes determining the actual attributes or qualities as well as making estimates and the testing of hypotheses by which *probable* or expected values are determined.

Although statistics is a branch of mathematics, it is different from pure mathematics in the following sense. In pure mathematics, values are exact, i.e., either a variable has a particular value (the probability of this being so is unity, since we are *certain*) or it does not have that particular value (the probability in this case being zero, since we are certain that the variable does *not* have that value). In statistics, however, the variable can take on many possible values, and there is a definite probability that it acquires each such value. This probability can have any value between 0 and 1. By means of statistics we attempt to define and control the extent of the uncertainty that is due to the inevitable variability of the data.

Statistics is concerned with two basic types of problems: *descriptive problems* and *inference problems.* The former include presentation of sets of observations in such a manner that they can be comprehended and interpreted. The numerical characteristics used to describe the set are called statistics.[1] *Inference problems* are those that involve inductive generalizations, e.g., from a sample actually tested to the whole from which the sample was drawn. Statistical inference enables us to obtain the maximum amount of accurate information from a given effort of testing; in other words, the use of statistics makes testing more efficient.

In the fields of engineering and experimental sciences the use of statistics is almost invariably required in routine testing in the laboratory, in research work, and in production and construction.

In the laboratory, we may want to know whether our testing is "precise," or whether the variability of our results is greater than expected, or greater than in some other test.

In research, we may want to know whether a change in an ingredient affects the properties of the resulting material; to compare the efficiency of processes or of testing machines; to determine whether the results fit a suspected or postulated form; or to design an experiment that will enable us to separate out the variation due to different causes.

The latter problem also arises in production, as the knowledge of variation in observations caused by a certain factor enables us to decide whether it is economic to control this factor more closely. We may want to know the probability of obtaining a strength above or below a certain value; to check whether the production has altered so as to change this probability; to determine the proportion of items that have a certain attribute; or to know the size of the sample that we have to use in order that our conclusions will have a specified reliability.

1. Singular: a statistic.

There are two basic types of variables with which we are concerned: *continuous variables*, which may differ by infinitesimal amounts, and *discrete variables*, which can have only specified values but not intermediate values between the specified values. These concepts are familiar from a study of mathematics and are of interest as the two types of variables generally follow different *distributions*, or laws of behavior. By distribution we mean the frequency with which different observed values occur.

The "different values" occur in two ways. We may measure a certain property, e.g., a dimension of one particular thing, a number of times. Because of errors in measurement we shall not record exactly the same value every time. The second case occurs when we manufacture items all of which are to have a certain property, e.g., a dimension, the same. Because of variations in the manufacture, as well as errors in measurement, the recorded values vary. In either case, if we take a number of observations, we obtain results that vary among themselves, and it is one of the main functions of statistics to evaluate information of this type so that we can estimate the "best" value of the quantity being measured and assess the precision of our estimate.

The distribution of discrete variables is of interest primarily in problems involving items that have or have not a certain attribute: balls that are black or not black, manufactured items that are defective or not defective, specimens that have or have not a strength in excess of an expected value, and so on.

It may be relevant to mention that for the purpose of statistical analysis, the discrete and continuous variables are not irrevocably separated from one another. If values of a variable that is continuously distributed are grouped in intervals and further treated in the grouped form, the problem becomes essentially one of discrete variables. Conversely, when a discrete variable consists of a large number of classes and is determined a large number of times, its distribution approximates that of a continuous variable, and the use of such an approximation is often convenient.

In statistical analysis we refer to the quantity that varies as the *variate*: It may be the original variable or a derived quantity such as the mean of samples,[2] their standard deviation,[3] and so on.

In a great many practical problems we cannot test or observe all of the items involved (all of them constitute a *population* or universe), and therefore we have to resort to *sampling*. We measure then the properties of a *sample* for the purpose of estimating the properties of all the items (population) from which the sample was drawn. Inference from samples is of tremendous value in many fields, varying from assessing whether a consignment of goods is up

2. See Chapter 3.
3. See Chapter 4.

to specification to the prediction of election results. Experience with the latter type of problem makes us realize that not only must the sample be properly taken so as to be representative of the underlying population, but also that our conclusion is only *probably* correct; certainty on the basis of sampling is not possible.

This is so because samples from the same population or collection of items vary among themselves, and variation is inherent in all natural phenomena and in all manufacturing operations. For this reason, all statistical inference is presented in terms of *probability* statements.

Although the greater part of this book is concerned with the use of statistics in extracting information from the results of experiments that have already been carried out, we must not forget the importance of statistics in planning experiments. With an appropriate program we can obtain more information from a given experimental effort than if the tests are made in a haphazard manner and the use of statistics is brought in only a posteriori. For this reason, we should view statistics not merely as an aid in the interpretation of experimental results but as an integral part of the design of experiments.

STATISTICS AND PROBABILITY

It is clear from what has been mentioned already that the subjects of statistics and probability are fundamentally interrelated. While statistics is largely concerned with drawing conclusions from samples affected by random variations or uncertainties, it is only through the probability theory that we can define or express and control these uncertainties in our results. Variations are said to be *random* when they have no pattern of behavior or regularity.

The relation between a sample and the population might elucidate further the distinction between statistics and probability. Such a relation poses two general problems: the testing of a statistical hypothesis and the estimation of a parameter or parameters characteristic of the population. In the former problem, we are concerned with testing whether it is reasonable to conclude that an observed sample belongs to a particular population (the hypothesis) or whether it is not reasonable to reach such a conclusion. Because of the inherent chance variations in the sample, we cannot be 100 percent sure of our conclusion and hence we must couple our conclusion with a probability statement.

In the problem of estimation, we are attempting to estimate a parameter(s) of the population from a sample by means of some "best" single value(s); again, because of the inherent variation from sample to sample we cannot be certain of our estimate and hence we must assign a probability band about our estimate. Such a band will give us a specified degree of confidence that the true value of the population parameter lies between the confidence limits.

In certain problems a simplified distinction between statistics and probability is possible. For example, if the parameter(s) of the population is known from an earlier record, we can deduce the behavior of the component, or samples, assumed to be a part of the population; this then is a probability problem. However, if the parameter(s) of the population is unknown, and it has to be estimated from the sample, then we have a statistical problem. It should be mentioned that the theory of probability is based on laws of chance or randomness; hence samples must be random in composition. A *random sample* is one selected in such a way that every element in the population has an equal chance of being selected. Obviously, if we are to judge the population (whole) from a sample (part), the sample must be as representative of the population as possible.

Chapter 2

Frequency Distribution

We mentioned in Chapter 1 the variability of engineering and scientific measurements arising from inaccuracies known as errors. Errors associated with the accuracy and precision of measurements are dealt with later on in this chapter, and at this stage we are concerned only with the fact that if we measure something repeatedly we obtain different observations or results, even if our determination of the measured value is made under as closely similar conditions as possible. This is due to variation, however small, in temperature, pressure, potential, instrument setting, etc. By a similar argument it can be seen that there are differences between supposedly similar items manufactured by the same process. If we measure some property of the various items of the same type, we obtain a collection of data, but without an intelligent treatment, interpretation of such data is well-nigh impossible.

We should stress that we have ignored the possibility of a mistake in our measurements and are concerned only with a number of "equally good" observations that are truly representative of the measured quantity.

TABULATION OF DATA

Imagine, then, that we have a collection of measurements, each represented by a number, which gives us information about the quantity being determined. Some of the numbers occur only once; others are repeated several times. If we write down the results in the order in which they occur, they are said to be in the form of *ungrouped data*. This form enables us to study the sequence of the values, e.g., of " high " or " low " values, and, hence, possibly to discover some of the causes of variation. However, arithmetical processing of ungrouped data is cumbersome, and we usually resort to tabulation. A convenient form is to write the numbers in an increasing order of magnitude, i.e., in *rank order*; such a form is sometimes called *ungrouped frequency distribution.*

As an example, let us consider the results of tests on the transverse strength of 270 bricks from one works. Table 2-1 gives the ungrouped data

TABLE 2-1

Transverse Strength of 270 Bricks from One Source, MN/m^2

5.93	9.10	5.66	7.17	6.89	6.96	8.20	8.14	7.45	7.58	7.79
6.34	7.58	8.62	10.20	7.93	5.10	7.45	5.93	6.89	5.58	6.89
8.27	5.66	7.58	6.14	1.86	7.38	5.72	9.51	6.62	9.38	5.03
5.86	6.34	6.48	9.03	9.17	7.03	9.58	5.72	5.66	6.76	9.17
6.34	7.38	11.24	4.62	7.93	8.07	6.34	7.72	8.07	8.00	7.52
7.52	4.83	6.27	8.07	5.52	6.62	7.03	7.52	13.86	6.14	6.41
5.72	6.07	6.00	9.24	5.79	8.14	5.10	6.07	5.45	7.58	8.69
7.17	7.45	7.17	6.76	8.55	5.52	5.93	6.96	7.79	6.69	7.86
10.41	7.31	5.79	6.48	7.65	8.55	8.89	6.00	8.69	7.24	6.21
5.10	8.48	7.03	7.31	6.83	7.03	5.66	7.03	5.93	5.86	6.14
7.93	5.93	7.58	5.79	7.31	7.10	6.83	7.58	7.45	7.45	6.69
6.89	4.96	5.52	8.07	6.69	4.76	7.10	6.14	4.83	6.07	7.93
7.86	7.45	6.83	3.93	5.45	7.38	5.66	4.00	5.66	7.31	6.76
7.10	6.62	6.00	5.52	7.17	5.66	8.14	9.31	8.14	6.55	7.58
4.83	5.93	4.55	8.14	5.38	8.48	6.55	6.21	5.24	9.51	6.21
6.34	7.58	7.45	6.76	5.24	5.72	8.41	7.58	7.52	9.51	8.75
5.93	6.83	6.14	6.48	6.27	7.65	7.03	9.51	6.96	7.10	6.55
6.55	6.07	6.69	6.89	6.83	5.72	5.86	4.34	4.90	6.21	6.14
7.03	5.17	7.38	6.34	6.00	6.96	8.48	5.38	6.89	7.93	9.38
8.96	6.69	5.52	4.48	8.14	5.93	7.93	9.65	6.07	5.03	5.72
6.14	7.10	7.31	11.10	8.20	9.65	5.86	6.96	6.96	8.55	
7.38	6.69	6.62	8.14	7.24	6.27	7.65	5.38	5.38	8.20	
6.27	7.58	6.00	6.76	5.03	5.52	5.52	7.86	6.48	6.76	
6.00	6.69	6.27	5.72	7.10	7.24	4.90	6.14	6.96	7.72	
5.58	7.38	7.58	3.17	5.93	7.38	6.07	8.55	6.48	5.93	

Source: Report of Committee on Manual on Presentation of Data, Proc. ASTM, vol. 33, 1933, Part 1, p. 454. The original data was in psi units.

as they were obtained in the order of testing, and Table 2-2 shows the same data ranked in an increasing order, but even this form is rather difficult to take in at a glance.

TABLE 2-2

Data of Table 2-1 Arranged in Rank (Ascending) Order

1.86	5.38	5.72	6.00	6.34	6.69	7.03	7.38	7.58	8.14	9.03
3.17	5.38	5.72	6.07	6.34	6.76	7.03	7.38	7.58	8.14	9.10
3.93	5.38	5.72	6.07	6.34	6.76	7.03	7.38	7.58	8.14	9.17
4.00	5.45	5.79	6.07	6.34	6.76	7.03	7.38	7.58	8.14	9.17
4.34	5.45	5.79	6.07	6.34	6.76	7.03	7.38	7.65	8.14	9.24
4.48	5.52	5.79	6.07	6.41	6.76	7.03	7.38	7.65	8.14	9.31
4.55	5.52	5.86	6.07	6.48	6.76	7.03	7.38	7.65	8.14	9.38
4.62	5.52	5.86	6.14	6.48	6.83	7.10	7.45	7.72	8.20	9.38
4.76	5.52	5.86	6.14	6.48	6.83	7.10	7.45	7.72	8.20	9.51
4.83	5.52	5.86	6.14	6.48	6.83	7.10	7.45	7.79	8.20	9.51
4.83	5.52	5.93	6.14	6.48	6.83	7.10	7.45	7.79	8.27	9.51
4.83	5.52	5.93	6.14	6.55	6.83	7.10	7.45	7.86	8.41	9.51
4.90	5.58	5.93	6.14	6.55	6.89	7.10	7.45	7.86	8.48	9.58
4.90	5.58	5.93	6.14	6.55	6.89	7.17	7.45	7.86	8.48	9.65
4.96	5.66	5.93	6.14	6.55	6.89	7.17	7.52	7.93	8.48	9.65
5.03	5.66	5.93	6.21	6.62	6.89	7.17	7.52	7.93	8.55	10.20
5.03	5.66	5.93	6.21	6.62	6.89	7.17	7.52	7.93	8.55	10.41
5.03	5.66	5.93	6.21	6.62	6.89	7.24	7.52	7.93	8.55	11.10
5.10	5.66	5.93	6.21	6.62	6.96	7.24	7.58	7.93	8.55	11.24
5.10	5.66	5.93	6.27	6.69	6.96	7.24	7.58	7.93	8.62	13.86
5.10	5.66	6.00	6.27	6.69	6.96	7.31	7.58	8.00	8.69	
5.17	5.72	6.00	6.27	6.69	6.96	7.31	7.58	8.07	8.69	
5.24	5.72	6.00	6.27	6.69	6.96	7.31	7.58	8.07	8.75	
5.24	5.72	6.00	6.27	6.69	6.96	7.31	7.58	8.07	8.89	
5.38	5.72	6.00	6.34	6.69	6.96	7.31	7.58	8.07	8.96	

FREQUENCY GROUPING

For this reason, if there are more than about 40 observations, a more compact representation is advantageous. This is obtained by arranging the results into *class intervals*, usually of equal width, and recording the number of items in each interval, this number being called the *class frequency*. For example, we can choose a class width of 1.00 MN/m^2, with the midpoint of the lowest interval at 2.00 MN/m^2. Such a choice is quite arbitrary, but it is usually convenient to have 10 to 25 intervals. If too many class intervals are used, the class frequencies are low and the saving in computational effort is small. Conversely, with too few class intervals the true character of the distribution may be obscured and information may be lost.

It is generally preferable to choose class intervals in such a way that no result falls on the *class boundary*. Since in the present case the results are given to the nearest 0.01 MN/m^2, the class intervals may be chosen as 1.50–2.49, 2.50–3.49, etc. For simplicity, class intervals such as 1.50–2.50, 2.50–3.50, and 3.50–4.50, are sometimes written down, it being understood that the upper boundary is exclusive. Other ways of dealing with the boundary between class intervals exist, but no fundamental point is involved; what is essential is that there is no gap and no overlap between classes. In the numerical examples given in the book, the class boundaries are in all cases assumed to be such that the interval to either side contains one-half of the values falling on the boundary. This method has the advantage of leading to simple numerical values, desirable in a textbook.

Arrangement of observations in class intervals, with the class frequencies tallied, produces a *grouped frequency distribution*. The concept of frequency distribution is of utmost importance. Table 2-3 gives the grouped frequency

TABLE 2-3

Grouped Frequency Table for Data of Table 2-1

Class interval	*Class midpoint* x_i	*Class frequency* f_i	*Cumulative frequency* F	*Relative cumulative frequency* F/n	$f_i x_i$
1.5–2.5	2.00	1	1	0.00370	2.00
2.5–3.5	3.00	1	2	0.00741	3.00
3.5–4.5	4.00	4	6	0.0222	16.00
4.5–5.5	5.00	24	30	0.111	126.00
5.5–6.5	6.00	81	111	0.411	486.00
6.5–7.5	7.00	78	189	0.700	546.00
7.5–8.5	8.00	51	240	0.889	408.00
8.5–9.5	9.00	18	258	0.956	162.00
9.5–10.5	10.00	9	267	0.989	90.00
10.5–11.5	11.00	2	269	0.996	22.00
11.5–12.5	12.00	0	269	0.996	0
12.5–13.5	13.00	0	269	0.996	0
13.5–14.5	14.00	1	270	1.000	14.00
Totals		$\sum f_i = 270 = n$			$\sum f_i x_i = 1869.00$

distribution for the data of Table 2-1. The advantages of this presentation are clear: We can see whether values near the extremes occur frequently or whether the observations cluster near some central value, and specifically which class intervals contain the most values.

If instead of actual frequency we consider the frequency of each interval divided by the total number of observations, we obtain results in terms of relative frequency; we then deal with a *relative frequency distribution.*

CUMULATIVE FREQUENCY

Furthermore, we may be interested (to continue with our example) in the number of bricks whose strength is higher or lower than a specified value. This is given by the *cumulative frequency* F column of Table 2-3. The "lower than" F is a sum of the frequencies of all class intervals below the specified value. For example, $1 + 1 + 4 + 24 = 30$ bricks have a strength lower than 5.50 MN/m^2. A "not lower than" F is similarly obtained by summing frequencies of intervals from the highest; for example, $1 + 0 + 0 + 2 + 9 = 12$ bricks have a strength not lower than 9.50 MN/m^2. It is not possible to say how many bricks have a strength higher than 9.50 MN/m^2 because bricks whose strength is exactly 9.50 MN/m^2 are included in the interval 9.50–10.50 MN/m^2.

In many cases the proportion of results rather than their number is of interest; to obtain this, the cumulative frequency is simply divided by the total number of results n, the quotient being known as relative or *fractional cumulative frequency.* The calculated values are given in Table 2-3, and it can be seen, for example, that the proportion of bricks whose strength is lower than 5.50 MN/m^2 is $\frac{30}{270} = 0.11$.

GRAPHICAL REPRESENTATION

A grouped frequency distribution may be represented diagrammatically in several ways. The most common of these is the *histogram.* Here, the class intervals are set out on a horizontal axis, and the frequency in a given interval is measured in the vertical direction and marked by means of a horizontal line across the width of the class interval; the scale chosen is quite arbitrary. Figure 2-1 shows a histogram for the data of Table 2-3. Strictly speaking, it is the area of each rectangle that represents the frequency in that interval, and the total area under the histogram represents the total number of results to an appropriate scale; but when the width of all class intervals is the same, the heights and areas are proportional to one another. However, when the class intervals are not all of the same width (and this should be avoided whenever possible), the ordinates should be plotted so that the area of each rectangle is proportional to the class frequency. Thus the ordinate no longer represents the frequency but a ratio of frequency to the class width, i.e., density. This concept is of importance and will be encountered again in Chapter 11.

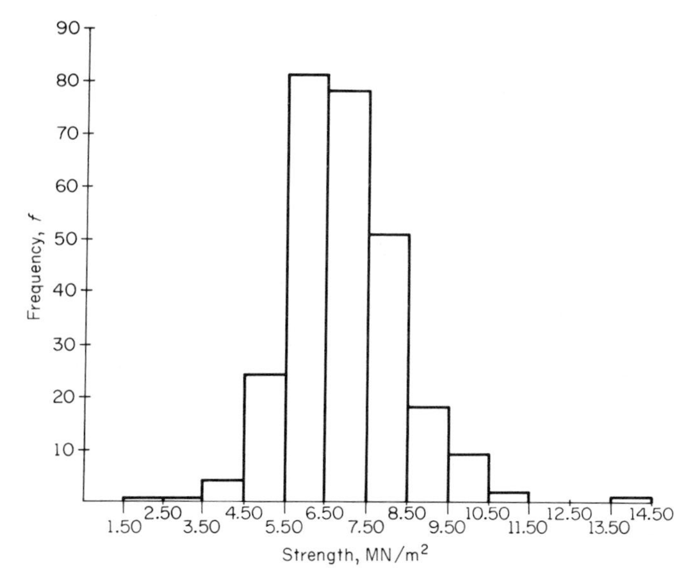

Fig. 2-1. Histogram for the data of Table 2-3.

The choice of the class width is governed by the same considerations as in the case of grouping for computational purposes only. If too many intervals are used, the histogram becomes irregular, and no clear pattern of frequency distribution can be seen. This is illustrated in Fig. 2-2a; by contrast, Fig. 2-2b shows a histogram for the same data but with a class width twice that in case (a).

Sometimes the frequency within a class interval is plotted as a point whose abscissa is that of the class midpoint. If adjacent points are connected, a *frequency polygon* is obtained, as shown in Fig. 2-3 for the data of Fig. 2-1. When plotting a frequency polygon, it is usual to indicate the values of the class midpoints rather than class boundaries as in the case of a histogram. The line between the points has no significance, and intermediate values must not be read off the line. Frequency polygons are particularly applicable to observations on discrete variables, especially when the class width corresponds to the smallest increment in the variable—e.g., as in plotting the number of accidents with a class width of one. With continuous variables, frequency polygons are useful when we want to compare visually two or more frequency distributions. Superimposed frequency polygons give a clearer picture than histograms.

If we are interested in proportions of results within various class intervals, we plot the relative frequency—i.e., frequency divided by the total number of results. *Percentage frequency,* which is simply relative frequency multiplied

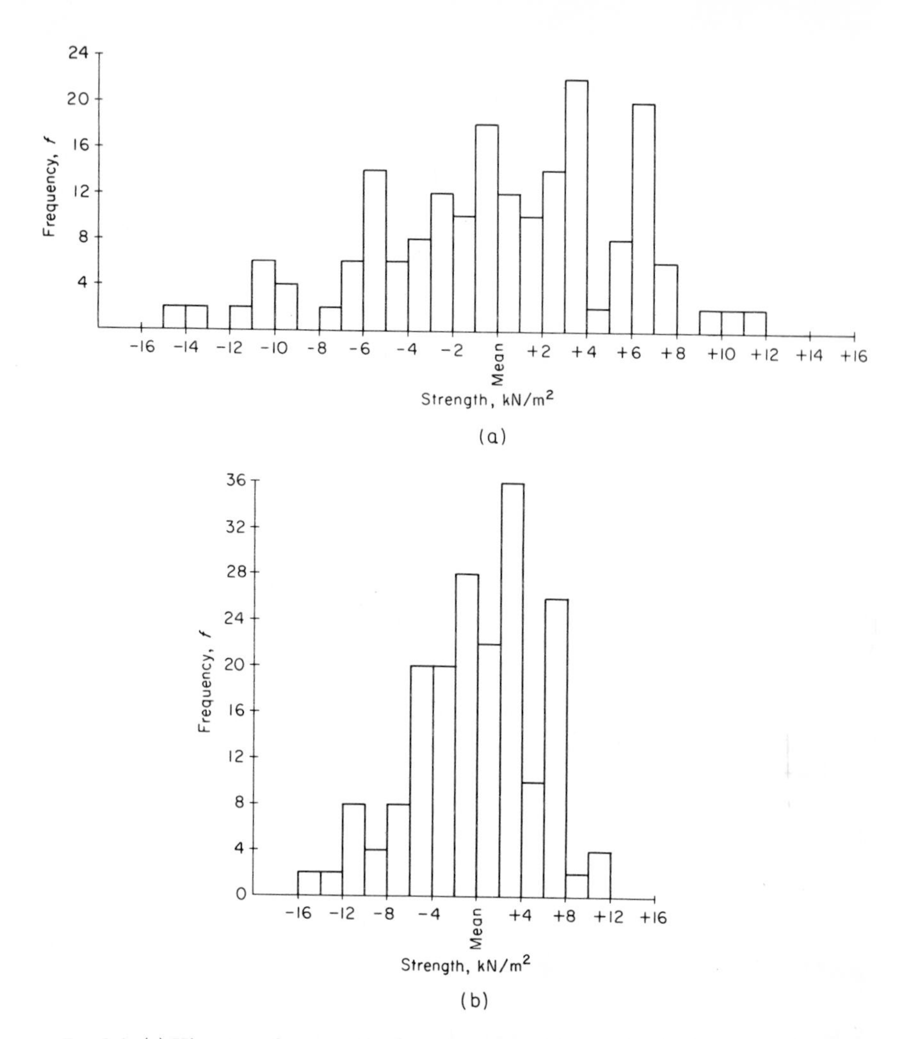

Fig. 2-2. (a) Histogram for strength of mortar cubes with grouping in class intervals of too small a width; (b) histogram for the same data as (a) but with a class width of twice the size.

by 100, is particularly convenient. The difference between histograms showing the frequency and those showing the relative or percentage frequency lies in the scale of the ordinates only; the form of the diagram is the same in either case.

If the total number of results is increased indefinitely so that the class width can be correspondingly decreased, the histogram becomes transformed, in the limiting case, into a *frequency distribution curve*. The ordinate should then properly be regarded to represent the frequency density.

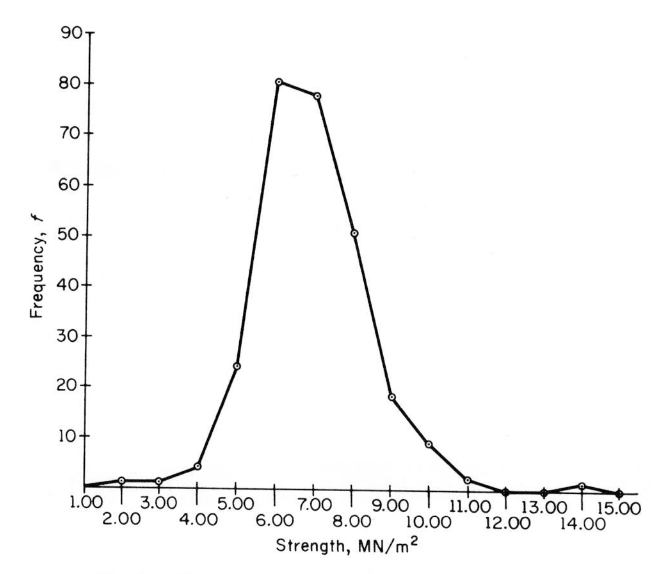

Fig. 2-3. Frequency polygon for the data of Fig. 2-1.

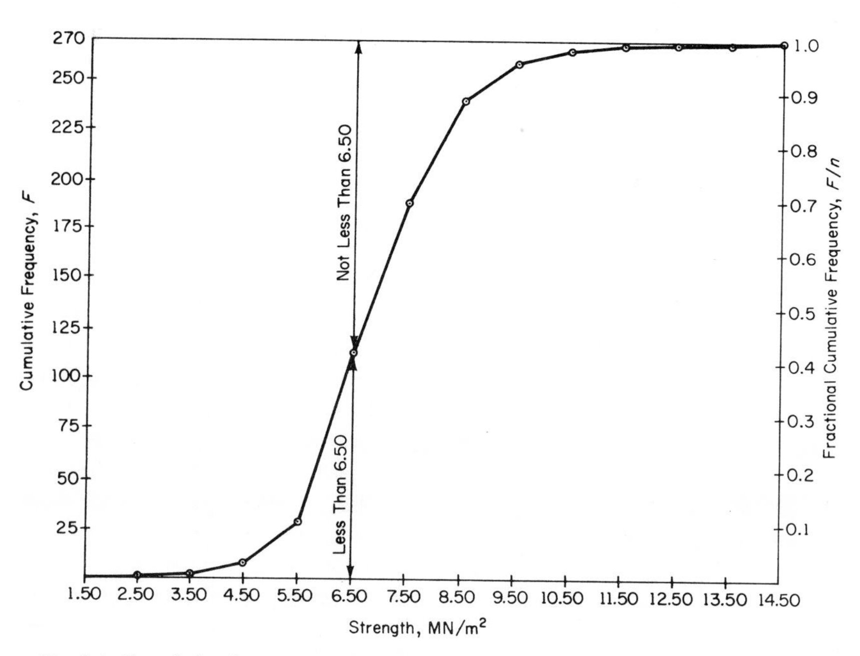

Fig. 2-4. Cumulative frequency and fractional cumulative frequency curve for the data of Table 2-3.

Cumulative frequency can also be represented graphically. The class intervals are set out horizontally as before, and the cumulative frequency is plotted as the ordinate at the right-hand end of each interval. This ordinate thus represents the area to the left of the corresponding ordinate in a histogram. The cumulative frequency curve is also called an *ogive*.[1] Figure 2-4 shows the "less than" curve for the data of Table 2-3, the ordinates giving both the cumulative frequency and the fractional cumulative frequency. The number or the fraction of results below or above any strength can easily be read off such a diagram, but care is required in distinguishing between fractions "less than" and "not greater than" a specified value. The cumulative frequency curve for the so-called normal distribution is considered in Chapter 10.

ACCURACY AND PRECISION OF MEASUREMENTS

When one is dealing with measurements, a distinction should be drawn between accuracy and precision. *Accuracy* is the closeness or nearness of the measurements to the "true" or "actual" value of the quantity being measured. The term *precision* (or repeatability) refers to the closeness with which the measurements agree with each other. To make this distinction clear, we refer to Fig. 2-5. Here we have four tachometers being used to measure the speed of a constant-speed motor. Five measurements are made with each tachometer. It is observed that tachometer I has good precision (the five speeds are closely grouped) as well as good accuracy [the representative value for all five speeds (arithmetic mean[2]) is close to the true value]. Tachometer II is precise but not very accurate (the representative value for the five speeds deviates from the true speed). Following the same reasoning, neither good precision nor good accuracy is exhibited by tachometer III, whereas tachometer IV shows good accuracy but poor precision (wide scatter among the five speeds). Furthermore, comparing tachometer I with tachometer IV we note that both are accurate but the former is more precise than the latter; therefore, we can conclude that tachometer I is more *reliable* than tachometer IV because it possesses good precision as well as good accuracy.

Precision errors are sometimes called *random* or accidental errors, which are usually assessed by applying certain statistical concepts and techniques. Accuracy errors are referred to as *systematic* errors and are usually reduced through *calibration*, which will improve the performance of a measuring device having poor accuracy and good precision. In cases where a systematic error is not completely eliminated, it then becomes part of the random error

1. Pronounced ō-jīv′.
2. See Chapter 3.

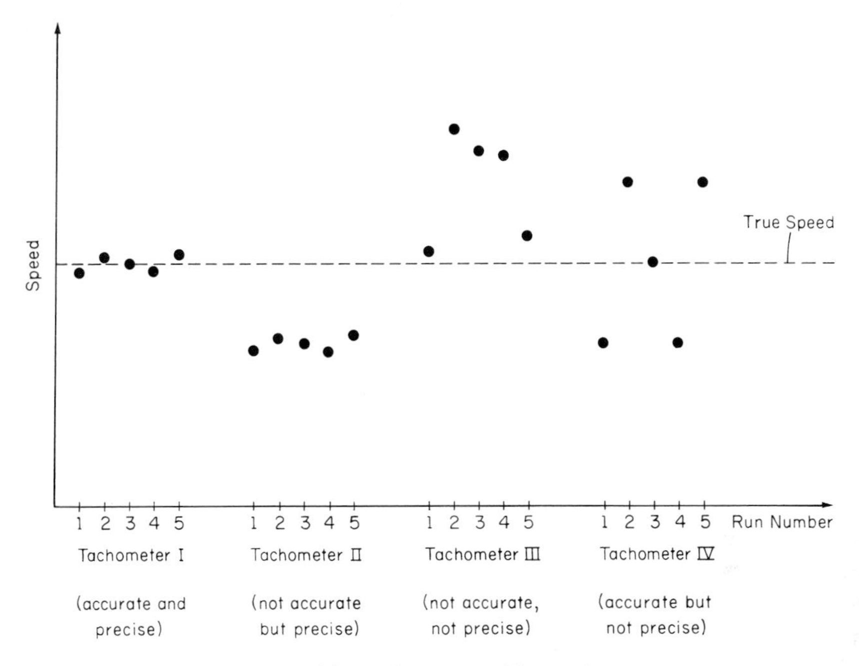

Fig. 2-5. Precision and accuracy of four tachometers.

or uncertainty and is therefore assessed as such; it is measured by means of the standard deviation, a quantity of immense importance in statistical methods, discussed in Chapter 4.

For example, a new rifle cannot be used for tasks requiring precision if test-firing indicates that the standard deviation in distance from a fixed target (precision) is large. On the other hand, if the standard deviation is small, but the average distance of shots from the target (accuracy) is large, it is possible to incorporate an aiming correction that will center the shots on the target. If this correction can be made without increasing the deviation, the rifle will be both accurate and precise. A further discussion on errors can be found in Chapter 9.

Solved Problem

2-1. In a test on 35 glue-laminated beams the following values of the spring constant (in MN/m) were found (Table 2-4).

a. Obtain a frequency table.
b. Draw a histogram and a frequency polygon.
c. Draw a cumulative frequency diagram.

TABLE 2-4

Spring constant/100						
6.72	6.77	6.82	6.70	6.78	6.70	6.62
6.75	6.66	6.66	6.64	6.76	6.73	6.80
6.72	6.76	6.76	6.68	6.66	6.62	6.72
6.76	6.70	6.78	6.76	6.67	6.70	6.72
6.74	6.81	6.79	6.78	6.66	6.76	6.72

d. Estimate the fraction of beams that will have a constant of less than 6.71×100 MN/m. Estimate also the spring constant which is not exceeded by 80 percent of the beams tested.

SOLUTION

a. We shall assign the observations to 6 classes. The lowest and highest values in Table 2-4 are 6.62 and 6.82. The difference is 0.20, which gives, when divided by 6, approximately 0.04. We shall, therefore, adopt 0.04 as class width, the lowest boundary being 6.61.
b. The histogram and the frequency polygon are shown in Fig. 2-6.
c. The cumulative frequency diagram is shown in Fig. 2-7.
d. From the cumulative frequency diagram it is estimated that 13 beams have a constant of less than 6.71×100 MN/m, which corresponds to a fractional cumulative frequency of $\frac{13}{35} = 0.37$. Eighty percent of the beams (28 beams) are estimated to have a spring constant of less than 6.776×100 MN/m.

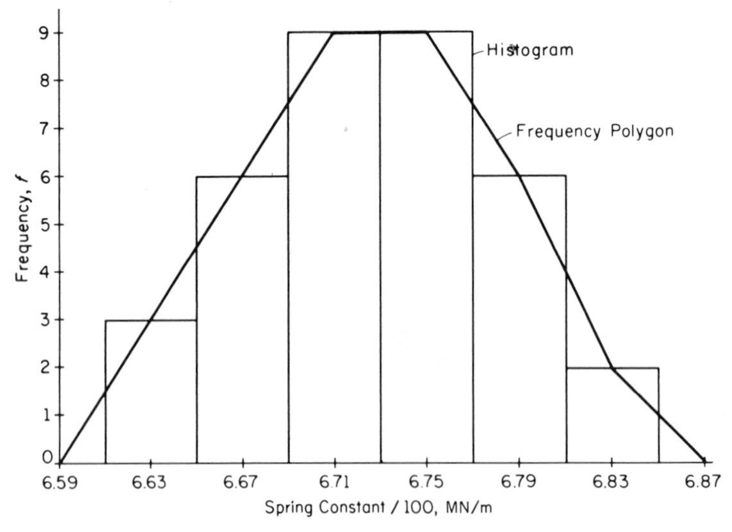

Fig. 2-6

TABLE 2-5

Class interval	*Frequency,* f_i	*Cumulative frequency,* F
6.61–6.65	3	3
6.65–6.69	6	9
6.69–6.73	9	18
6.73–6.77	9	27
6.77–6.81	6	33
6.81–6.85	2	35
Total	$\sum f_i = 35$	

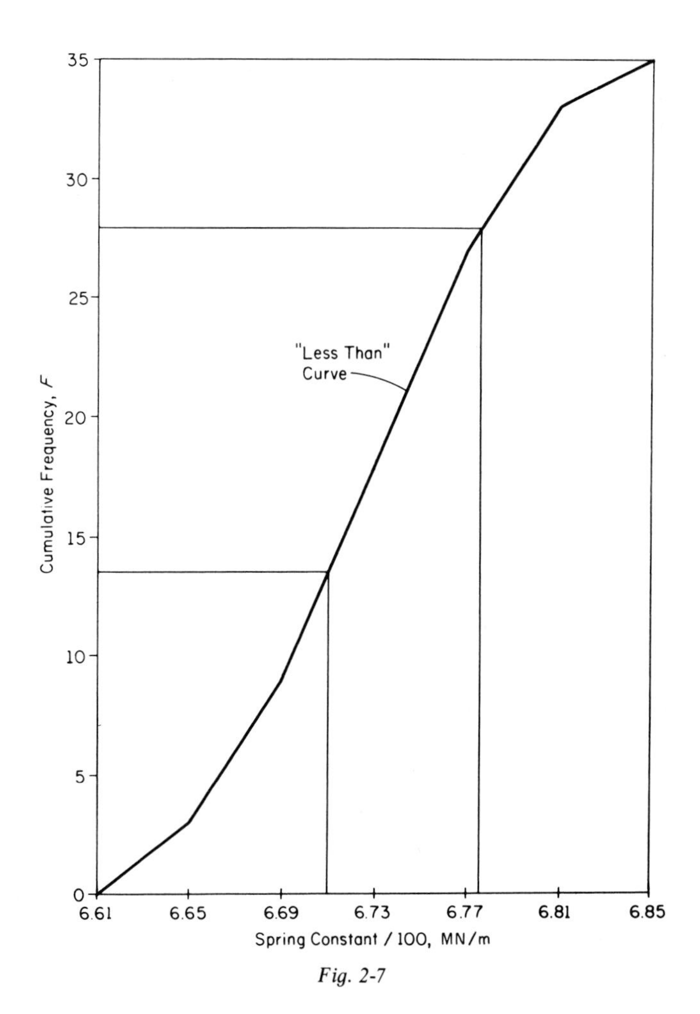

Fig. 2-7

Problems

2-1. The annual precipitation (in centimeters) at some city is listed in Table 2-6.

TABLE 2-6

Year	*Precipitation*	*Year*	*Precipitation*	*Year*	*Precipitation*
1904	19.5	1923	18.8	1942	21.0
1905	10.9	1924	9.3	1943	13.5
1906	12.6	1925	15.9	1944	14.0
1907	10.4	1926	13.6	1945	13.3
1908	14.1	1927	17.0	1946	16.2
1909	15.9	1928	12.9	1947	13.6
1910	11.1	1929	9.0	1948	10.2
1911	19.4	1930	11.5	1949	15.5
1912	16.7	1931	11.6	1950	16.3
1913	13.5	1932	10.1	1951	16.4
1914	12.7	1933	9.8	1952	10.7
1915	10.5	1934	9.9	1953	13.4
1916	17.4	1935	17.8	1954	20.4
1917	10.3	1936	11.3	1955	15.4
1918	12.7	1937	10.7	1956	13.7
1919	13.4	1938	18.0	1957	10.1
1920	15.2	1939	15.7	1958	11.7
1921	21.0	1940	12.9	1959	13.4
1922	11.4	1941	10.3	1960	10.2

a. Arrange the data in (ascending) rank order.
b. Prepare a frequency table. Take a class width of 1.0 cm and commence with the lower boundary of the first class interval of 9.0 cm.
c. Draw a histogram, a frequency polygon, and a cumulative frequency diagram.
d. From the cumulative frequency diagram, estimate the number of years in which the annual precipitation exceeded 14.5 cm and the number of years in which the annual precipitation was lower than 11.0 cm.

2-2. The following marks were obtained in an examination taken by 100 students:

Mark	0–30	31–40	41–45	46–50	51–55	56–65	66–75	76–80	81–90	91–100
Number of students	2	10	7	8	25	18	12	10	5	3

a. Draw a histogram for the above data.
b. Draw a cumulative frequency diagram, and hence estimate the mark exceeded by the top 25 percent of students.
c. Suggest a passing mark if 15 percent of the students are to fail.

2-3. The frequency distribution of 500 radio tubes tested at a certain company is as follows:

Lifetime (hours)	400–499	500–599	600–699	700–799	800–899	900–999	1000–1099	1100–1199
Number of tubes	25	65	79	108	92	76	34	21

a. Construct a histogram, a frequency polygon, and a cumulative frequency diagram.
b. Find the percentage of tubes whose lifetimes do not exceed 700 h.
c. Find the percentage of tubes whose lifetimes are at least 600 but not more than 1000 h.

2-4. The moisture content in percent of 44 samples of clay was measured as follows: 10.3, 10.7, 9.2, 11.4, 11.5, 8.4, 10.3, 10.2, 9.8, 10.2, 11.4, 10.4, 11.6, 10.3, 11.4, 10.3, 7.8, 8.7, 11.7, 9.3, 9.8, 11.4, 10.4, 9.3, 10.4, 11.2, 10.6, 9.8, 10.1, 13.1, 10.5, 12.2, 9.6, 7.0, 12.3, 12.3, 12.3, 9.7, 10.7, 11.5, 10.6, 13.0, 13.0, 9.0.

a. Compute a grouped frequency distribution for these results.
b. Draw a histogram.
c. Draw a frequency polygon and cumulative frequency diagram.

2-5. The distribution of the number of defective bolts found in 400 lots of manufactured bolts is shown in Table 2-7.

TABLE 2-7

Number of defective bolts	0	1	2	3	4	5	6	7	8	9	10	11	12
Number of lots	55	108	80	56	34	21	19	11	8	4	1	2	1

a. Determine the percentage of the lots containing more than 10 defective bolts.
b. Determine the relative frequency of the lots containing 5 or fewer defective bolts.
c. What are the class boundaries?
d. Construct the histogram and the cumulative frequency diagram.

2-6. The diameters in centimeters of a sample of 60 ball bearings manufactured by a company are given in Table 2-8.

a. Using appropriate class intervals, construct a frequency distribution of the diameters.

TABLE 2.8

1.740	1.731	1.740	1.730	1.741	1.735	1.732	1.736	1.738	1.737
1.746	1.742	1.744	1.727	1.734	1.742	1.735	1.724	1.732	1.729
1.740	1.735	1.735	1.736	1.731	1.729	1.741	1.735	1.733	1.736
1.727	1.732	1.732	1.734	1.736	1.736	1.738	1.739	1.728	1.732
1.730	1.743	1.735	1.739	1.733	1.738	1.734	1.728	1.735	1.737
1.735	1.736	1.737	1.726	1.739	1.725	1.733	1.734	1.745	1.730

b. Draw a histogram, a frequency polygon, and a cumulative frequency diagram.
c. Find the percentage of bearings whose diameters do not exceed 1.735 cm.

Chapter 3

Characteristics of Distributions: Central Tendency

In Chapter 2 we explained how to arrange a collection of observations in a frequency distribution form. We shall now proceed with statistical analysis in order to learn how to present information about the distribution in a clear and concise form.

There are two obvious features of the data that can be characterized in a simple form and yet give a very meaningful description of a set of observations: *central tendency* and *dispersion*. The central tendency is measured by averages; these describe the point about which the various observed values cluster. The measure of dispersion is concerned with scatter about the average and is dealt with in Chapter 4.

AVERAGES

Averages are commonly used in everyday life to give a *typical* representation of a group as a whole, possibly as a basis for comparison with other groups. In many cases the data in hand refer to a sample drawn from a larger body of data—for example, we may take a number of rock specimens from some stratum and measure their density. The tested specimens are referred to collectively as a *sample*, and the whole body of rock as the parent *population*. We measure the average density of the sample for the purpose of obtaining information about the average density of the population. This type of problem is dealt with in Chapter 6.

It may be relevant to note that if we take repeated measurements of a quantity, the average does not represent a "true" value of the quantity. In many cases the term "true value" has no meaning (e.g., the time taken by jet planes to fly between New York and San Francisco); in other cases, we may never be able to determine the true value but can determine only the most probable value (e.g., the difference in level between two points on the ground).

There are several types of averages, which will now be discussed; the appropriate one to use depends on the problem in hand.

ARITHMETIC MEAN

This is the most common type of average, often referred to simply as the *average* or *mean*. The latter term will be used here.

The Mean is a value such that the sum of deviations of observations from it is zero; it is thus the sum of the observations divided by their number:

$$\bar{x} = \frac{1}{n} \sum_{i=1}^{n} x_i \tag{3-1}$$

where x_i is an observation or measurement, n the total number of observations, and $\bar{x}$ the mean.

The above definition can also be expressed mathematically as:

$$\sum_{i=1}^{n} (x_i - \bar{x}) = 0 \tag{3-1a}$$

The concept of mean is so well known that no further discussion is needed, but at this stage a brief note on the notation for the mean may be of help. The true mean of a population is usually denoted by μ. The mean of a sample is written as $\bar{x}$. The values of $\bar{x}$ and μ become identical when the sample is, in fact, the total finite population; in such a case either symbol may be used.

A simple example will be given to illustrate some shortcuts in computation.

Example. The solids content of water, in parts per million (ppm), was measured on 11 samples, the following results being obtained: 4520, 4570, 4520, 4490, 4540, 4570, 4500, 4520, 4520, 4500, and 4590. Hence

$$\text{mean} = \bar{x} = \frac{\sum \text{above values}}{11} = 4530.9 \text{ ppm}$$

When the frequency of some of the observations is greater than 1, computation may be simplified by setting the data in a tabular form, such as in Table 3-1, and using frequency grouping. If f_i is the frequency of any value x_i, then the mean can be written as

$$\bar{x} = \frac{\sum f_i x_i}{\sum f_i} \tag{3-2}$$

TABLE 3-1

(1)	*(2)*	*(3)*	*(4)*	*(5)*
Solids content ppm x_i	(-4400)	$(\div 10)$ x'_i	*Frequency* f_i	$x'_i f_i$
4490	90	9	1	9
4500	100	10	2	20
4520	120	12	4	48
4540	140	14	1	14
4570	170	17	2	34
4590	190	19	1	19
Totals			$\sum f_i = 11$	$\sum x'_i f_i = 144$

A further saving in effort is effected by reducing all the observations by a constant value, and possibly also by dividing them by a factor such as 10. This transformation of the original variable is known as *coding*. In our case we can thus subtract 4400 (column 2) and divide by 10 (column 3).

Hence, from Eq. (3-2),

$$\bar{x}' = \frac{144}{11} = 13.09$$

and $\bar{x} = 10\bar{x}' + 4400 = 4530.9$ ppm, as before.

When dealing with a large number of observations, or a large sample, the computation of the mean can be shortened considerably, with only a small loss of accuracy, by using the class-interval method explained in Chapter 2.

Instead of considering each individual observation, we treat all observations within a class interval as a group and assume that, in any class, the observations are uniformly distributed throughout the interval so that the class frequency may be assumed to be concentrated at the class midpoint x_i'. The procedure is then as follows:

1. Take the first class midpoint x_0 as an arbitrary origin for the purpose of calculating the fictitious mean.
2. Calculate deviations X_i' from this origin, expressed in terms of the class width w, i.e., $X_i' = (x_i' - x_0)/w$.
3. Find the product of class frequency f_i and X_i'.
4. Obtain the fictitious mean $\bar{X}' = \dfrac{\sum f_i X_i'}{\sum f_i}$
5. Convert $\bar{X}'$ to the true mean $\bar{x}$:

$$\bar{x} = \text{arbitrary origin} + (\text{fictitious mean}) \times (\text{class width})$$

That is,

$$\bar{x} = x_0 + \frac{\sum f_i X_i'}{\sum f_i} \times w = x_0 + \bar{X}'w \tag{3-3}$$

As an example, the short method of computing the mean will be used for the data of Table 2-1. The class width, as in Table 2-3, is $w = 1.00$ and the

TABLE 3-2

Calculation of the Mean for Data of Table 2-3 Using the Class-Interval Method

Class midpoint	*Class frequency* f_i	*Deviation from origin in terms of class width* X_i'	$f_i X_i'$	$f_i X_i'^2$
2.00	1	0	0	0
3.00	1	1	1	1
4.00	4	2	8	16
5.00	24	3	72	216
6.00	81	4	324	1296
7.00	78	5	390	1950
8.00	51	6	306	1836
9.00	18	7	126	882
10.00	9	8	72	576
11.00	2	9	18	162
12.00	0	10	0	0
13.00	0	11	0	0
14.00	1	12	12	144
Totals	$\sum f_i = 270$		$\sum f_i X_i' = 1329$	$\sum f_i X_i'^2 = 7079$

origin is taken at $x_0 = 2.00$ MN/m^2. From Table 3-2 the fictitious mean (in terms of class width) is

$$\bar{X}' = \frac{\sum f_i X'_i}{\sum f_i} = \frac{1329}{270} = 4.92$$

Using Eq. (3-3), we find that the true mean is

$$\bar{x} = 2.00 + 4.92 \times 1.00$$
$$= 6.92 \text{ MN/m}^2$$

The exact mean of the measurements in Table 2-1 is 6.89 MN/m^2. For many purposes the difference between the two is not significant.

The value that we have found is the mean of the values comprising our sample of bricks, but if we took another sample from the same source, it would, in all likelihood, have a different mean. Thus, so far as the population of all bricks manufactured by the given works is concerned, the mean as determined from samples is a variable quantity, but much less variable than the strength of the individual bricks. This applies, of course, to measurements in all problems of this type. In general terms we can state that statistics derived from a random sample are also random variables.

As we saw earlier, the arithmetic mean is the average that is most commonly used, but it is not the only one of importance; we shall now consider the other averages.

MEDIAN

The median of a set of observations is the middle observation when the observations are ranked or arranged in order of magnitude. The term *middle observation* refers to the distance from the extremes and not to the numerical value. More precisely, if the number of observations is odd, say $2m + 1$, the median is the $(m + 1)$th value; if the number of observations is even, say $2m$, the "middle" values of the set are the mth and $(m + 1)$th, and their arithmetic mean is taken as the median. For example, the numbers 21, 22, 31, 34, 31, 22, 17, and 26 when arranged in ascending order of magnitude are 17, 21, 22, 22, 26, 31, 31, and 34, with 22 and 26 being the two "middle" values. The median in this case is $(22 + 26)/2 = 24$. If, however, the last value 34 were not present, the median would be 22.

Since the area of a histogram is proportional to the number of observations, it follows from the above definition that the median divides a histogram (or a frequency polygon or curve) into two equal areas.

Because the median is a positional value (in contrast with the arithmetic character of the mean), it is less affected by extreme values within the group than the mean. This property of the median makes it in some cases a useful measure of the central tendency. For example, a median of 2, 3, 6, 8, 9, 9, 12 is 8. If the extreme values change so that the set is now 3, 3, 6, 8, 9, 9, 18, the median is still 8 but the mean has increased from 7 to 8.

We can now find the median value for the data on the solids content of water, given previously. Arranging the results in ascending order, we have 4490, 4500, 4500, 4520, 4520, 4520, 4520, 4540, 4570, 4570, 4590, and the median value is 4520 ppm.

MODE

Mode is the value of the observation that occurs most frequently if the variable is discrete or the class interval (often quoted as a class midpoint) that has the highest frequency if the distribution is continuous. The mode thus represents a peak value in a frequency distribution. Like the median, the mode is less affected by extreme values than the mean.

In the example concerning the solids content of water, the value of 4520 has the highest frequency, namely 4, and is, therefore, the modal value. We have thus a mode and a median of 4520, and a mean of 4531 ppm.

Some distributions have more than one mode, but in experimental work this is rare, although bimodal distribution is encountered in microscope counts and particle-size gradings. In many other cases the appearance of a bimodal distribution means that the data contain values from two different distributions. For example, if two machines manufacturing an item are set at different averages sufficiently far apart and all the items are pooled, the resulting frequency distribution will be bimodal.

SKEWNESS

We can now view the three measures of central tendency: mean, median, and mode on a general frequency distribution curve, shown in Fig. 3-1. The mode is the value corresponding to the highest point on the curve; the median divides the area under the curve into two halves; and the mean passes through the centroid of the area. (The latter arises from the fact that the sum of deviations of all observations from the mean is zero.) The median lies between or coincides with the mean and the mode.

When the three averages do not coincide, the frequency distribution curve is said to be *skew* or *skewed*. It is skewed to the right when the median is to the right of the mode, i.e., when the tail to the right (the direction of increasing values) is longer than the tail to the left. Such a curve is also said to be *positively skewed*.

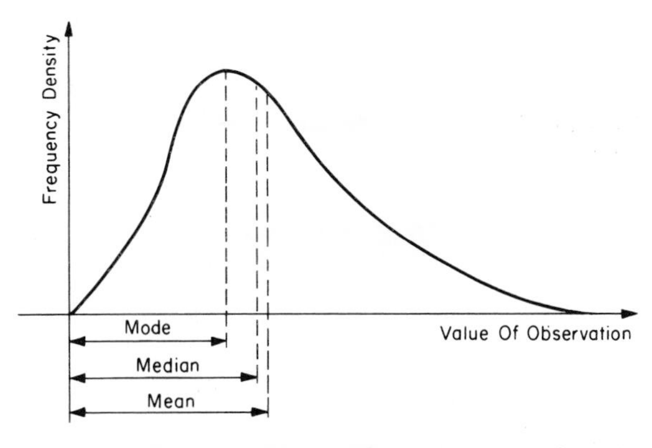

Fig. 3-1. Mean, median, and mode.

There is no accepted method of measuring skewness. To compare the skewness of different distributions the ratio,

$$\frac{\text{mean} - \text{mode}}{\text{standard deviation}}$$

may be used. The term in the denominator, standard deviation (see Chapter 4), makes a comparison of distributions with different widths of scatter possible.

For moderately skewed distributions there is an approximate relation between the various averages:

$$\text{mean} - \text{mode} = 3(\text{mean} - \text{median}) \tag{3-4}$$

It is interesting to note that in skew distributions that are sharply peaked, the median is often a particularly useful measure of central tendency. For example, if we are interested in the distribution of periods of time at which atoms of a radioactive element have disintegrated, the median represents the time at which half the atoms have disintegrated, and is a common measure of radioactivity.

Skewness arises usually from natural causes and is characteristic of many distributions, such as those dealt with in Chapters 7 and 8, and also of some continuous distributions. For example, the frequency of the occurrence of a very large defect in a material (e.g., a blowhole) is smaller than that of a very small defect. Thus the frequency curve for defects of various sizes is skewed to the left. However, skewness can also be caused by selection. For instance, in testing concrete specimens the very poorly made ones may well be discarded prior to testing, with the result that a frequency curve for the strength of the concrete specimens would be skewed to the right. This behavior has, in fact, been observed in some tests.

A great many random variables are, of course, distributed symmetrically, i.e., the departure of observations from the mean by any given amount occurs with a sensibly equal frequency in the up and down directions. Such a distribution is simply said to be *symmetrical*, and the mean, mode, and median all coincide.

QUANTILES

In a manner similar to that of the median, which divides a set of observations so that 50 percent of them fall above and 50 percent below the median, we can introduce other points, which divide the observations into a number of equal parts, known as *quantiles*. An example of commonly used quantiles are *quartiles* which divide the set into four equal parts; e.g., the upper quartile is the value above which 25 percent of the set falls. The *interquartile range* contains the middle 50 percent of the set, with 25 percent falling above and 25 percent below the range, and is sometimes used as a measure of dispersion. A *decile* divides the set of observations into 10 groups, the lowest decile, for example, being a value below which 10 percent of the set falls. Other quantiles are described as *percentiles*, e.g., a 5th percentile.

GEOMETRIC MEAN

There is one more type of average that is of interest in engineering calculations. This is the *geometric mean*, defined as the nth root of the product of n observations. Thus the geometric mean $\bar{x}_g$, of n observations $x_1, x_2, \ldots, x_n$ is

$$\bar{x}_g = \sqrt[n]{x_1 \times x_2 \times \cdots \times x_n} \tag{3-5}$$

This average is used when dealing with observations each of which bears an approximately constant ratio to the preceding one, e.g., in averaging rates of growth (increase or decrease) of a statistical population, as illustrated in the following example.

Example. The number of degrees *cum laude* awarded at a university during six consecutive years is given in Table 3-3. What is the average percentage increase in the number of such degrees per annum?

To find the answer, we calculate the geometric mean of the ratios given in the last column. This is

$$\sqrt[5]{1.2 \times 1.5 \times 1.67 \times 2.0 \times 1.67} = 1.585$$

i.e., an average increase per year of 58.5 percent.

It might be asked why an arithmetic mean cannot be used. This is $\frac{1}{5}(1.2 + 1.5 + 1.67 + 2.0 + 1.67) = 1.61$ or an increase of 61 percent, which is higher

TABLE 3-3

Year	Number of degrees	Ratio to previous year's value
1959	5	—
1960	6	1.20
1961	9	1.50
1962	15	1.67
1963	30	2.00
1964	50	1.67

than that given by the geometric mean; the arithmetic mean is always higher than the geometric mean. The bias in the answer given by the arithmetic mean arises from the influence of the absolute magnitude of the ratios. For example, doubling a value represents a ratio of 2, while halving means a ratio of $\frac{1}{2}$. Thus if we consider a value of 100 which falls to 50 and subsequently rises to 100, the ratios are $\frac{1}{2}$ and 2, respectively. The geometric mean is $\sqrt{\frac{1}{2} \times 2} = 1$, and this is the average rate of increase. This answer is intuitively correct as the overall change is zero. However, the arithmetic mean of the ratios is $\frac{1}{2}(\frac{1}{2} + 2) = 1.25$. If the ratios were 3 and $\frac{1}{3}$, the geometric mean would still be 1 but the arithmetic mean would be $1\frac{2}{3}$.

The use of the geometric mean can be avoided by transforming the original variate x into $\log x$: the arithmetic mean of the new variate will then give the right answer, since from Eq. (3-5),

$$\log \bar{x}_g = \frac{\sum (\log x_i)}{n}$$

Solved Problems

3-1. Three hundred and three tensile pieces of a certain new brittle lacquer (used for experimental stress analysis) gave the tensile strengths in Table 3-4 at the age of 7 days.

TABLE 3-4

Strength interval kN/m^2	200–230	230–260	260–290	290–320	320–350	350–380	380–410	410–440	440–470	470–500
Number of test pieces	7	30	50	77	53	40	35	6	3	2

a. Complete the frequency table.
b. Draw the histogram and frequency polygon.
c. Draw the cumulative frequency diagram.
d. Calculate the mean tensile strength and indicate this on the histogram.

TABLE 3-5

Class interval	*Class midpoint*	*Class frequency* f_i	*Cumulative frequency* F	*Deviation from origin in terms of class width* X'_i	$f_i X'_i$	$f_i X'^2_i$
200–230	215	7	7	0	0	0
230–260	245	30	37	1	30	30
260–290	275	50	87	2	100	200
290–320	305	77	164	3	231	693
320–350	335	53	217	4	212	848
350–380	365	40	257	5	200	1000
380–410	395	35	292	6	210	1260
410–440	425	6	298	7	42	294
440–470	455	3	301	8	24	192
470–500	485	2	303	9	18	162
Totals					$\sum f_i X'_i = 1067$	$\sum f_i X'^2_i = 4679$

SOLUTION

a. In view of the form in which the data are presented we shall adopt a class width of 30 kN/m², with the first class midpoint at 215 kN/m². See Table 3-5.
b. The histogram and frequency polygon are shown in Fig. 3-2.
c. The cumulative frequency diagram is shown in Fig. 3-3.
d. From the Table in a,

$$\begin{array}{c}\text{mean from first class midpoint} \\ \text{in terms of class width}\end{array} = \frac{1067}{303} = 3.52$$

$$\text{mean} = 215 + 3.52 \times 30 = 321 \text{ kN/m}^2$$

The mean tensile strength of the brittle lacquer test pieces is marked in Fig. 3-2.

3-2. Mid-block passenger-car spot speeds under urban conditions were found to be as listed in Table 3-6.

a. Complete the frequency table.
b. Draw the cumulative frequency diagram.
c. Find the mode, median, and the mean speed.
d. If the 85th and 98th percentiles are employed in speed regulation and design respectively, calculate the speed limit and the design speed for the locality.

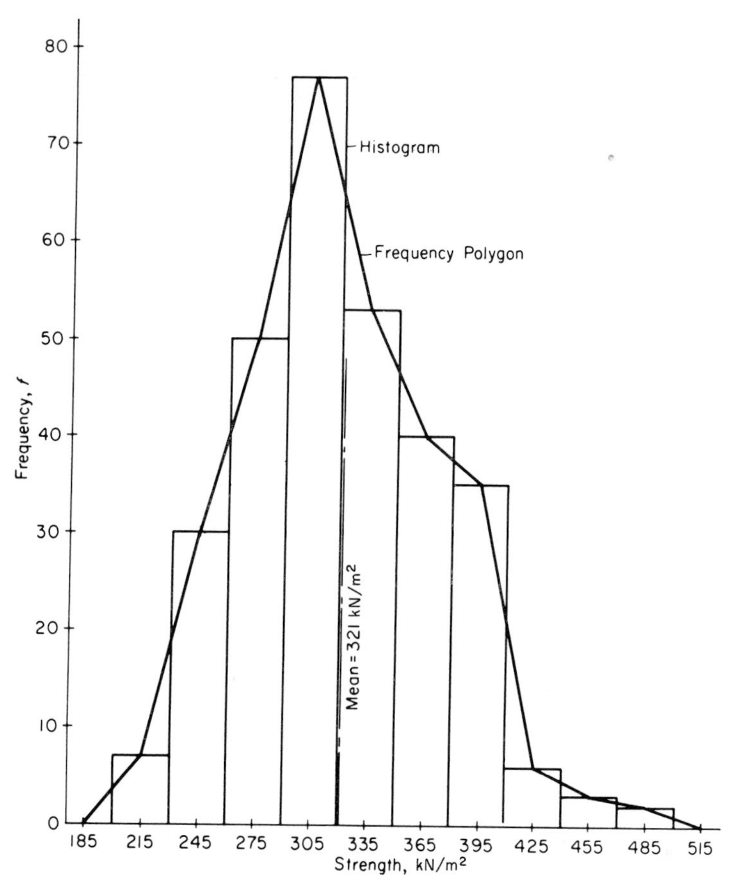

Fig. 3-2

TABLE 3-6

Speed interval, km/h	8–16	16–24	24–32	32–40	40–48	48–56	56–64	64–72	72–80	80–88	88–96
Number of vehicles	0	1	34	146	178	130	31	16	3	2	0

Source: Data from T. M. Matson, W. S. Smith, and F. W. Hurd, *Traffic Engineering* (New York: McGraw-Hill Book Company, 1941), p. 50. The original data was in mph units.

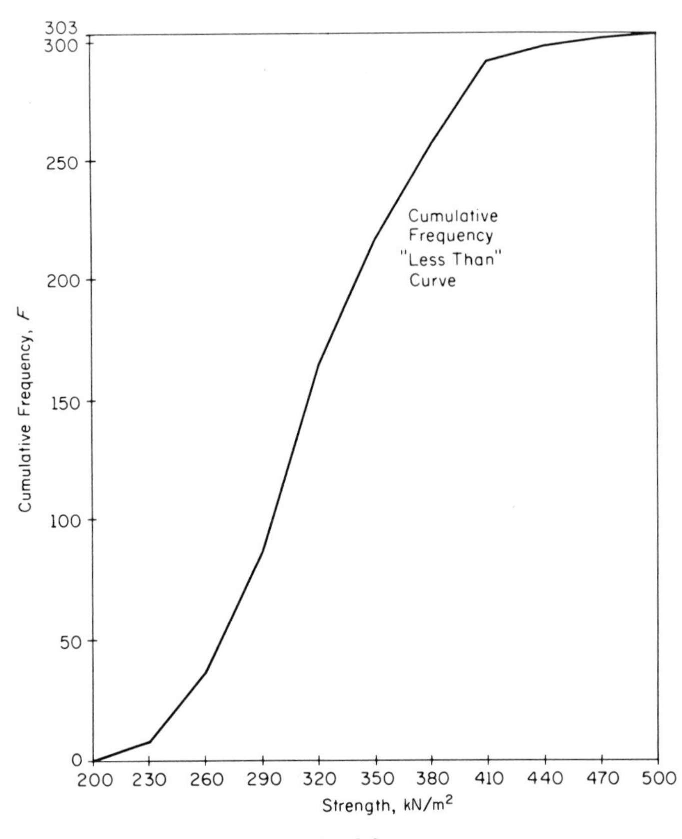

Fig. 3-3

SOLUTION

a. Take the origin at the first class midpoint, i.e., at 12 km/h. With a class width $w = 8$ km/h we tabulate the data in Table 3-7.
b. The cumulative frequency diagram is shown in Fig. 3-4.
c. By inspection of Table 3-7 the mode = 44 km/h. The median divides the histogram into two equal areas.

Now, one-half the area of the histogram = $(w \times 541)/2 = 270.5 \times w$. The median will fall to the right of the class interval with the highest cumulative frequency below 270.5, i.e., to the right of the 32 to 40 km/h class interval. Hence

$$181 \times w + 178 \times x = \frac{541 \times w}{2}$$

or

$$x = \frac{89.5}{178} \times w = \frac{89.5}{178} \times 8$$

$$= 4.02 \text{ km/h to the right of 40 km/h}$$

TABLE 3-7

Class interval	Class midpoint	Class frequency f_i	Cumulative frequency F	Deviation from origin in terms of class width X'_i	$f_i X'_i$	$f_i X_i^2$
8–16	12	0	0	0	0	0
16–24	20	1	1	1	1	1
24–32	28	34	35	2	68	136
32–40	36	146	181	3	438	1314
40–48	44	178	359	4	712	2848
48–56	52	130	489	5	650	3250
56–64	60	31	520	6	186	1116
64–72	68	16	536	7	112	784
72–80	76	3	539	8	24	192
80–88	84	2	541	9	18	162
88–96	92	0	541	10	0	0
Totals		$\sum f_i = 541$			$\sum f_i X'_i = 2209$	$\sum f_i X_i^2 = 9803$

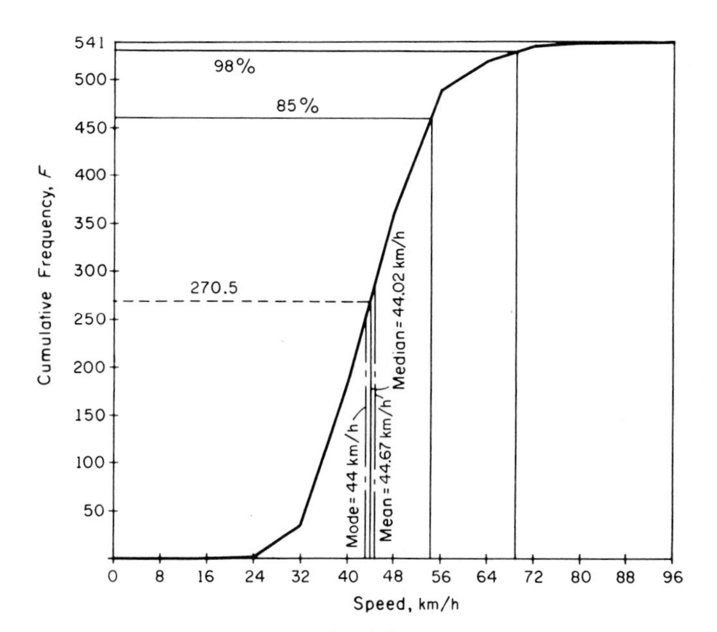

Fig. 3-4

Therefore,

$$\text{median} = 40 + 4.02 = 44.02 \text{ km/h}$$

$$\text{mean} = 12 + \frac{2209}{541} \times 8 = 44.67 \text{ km/h}$$

d. 98 percent of 541 cars = 530 cars, and 85 percent of 541 cars = 460 cars.

From the cumulative frequency diagram the speed not exceeded by 530 cars is 67 to 69 km/h; i.e., the design speed is, say, 68 km/h. For 460 cars the corresponding speed is 54 km/h—i.e., the speed limit for the locality is 54 km/h.

Problems

3-1. The following was the number of vehicles passing a certain point on different days:

1310 1207 760 983 260 618 1262 1152 598 1218 1391

Determine the mean, median, and mode for the above values.

3-2. The diameters in centimeters of a sample of 50 ball bearings manufactured by one company are as shown in Table 3-8.

TABLE 3-8

0.529	0.538	0.532	0.529	0.535
0.536	0.534	0.542	0.537	0.530
0.538	0.536	0.536	0.536	0.526
0.525	0.524	0.543	0.530	0.539
0.542	0.528	0.546	0.532	0.534
0.535	0.539	0.527	0.544	0.527
0.535	0.534	0.540	0.540	0.536
0.532	0.535	0.535	0.535	0.528
0.541	0.531	0.540	0.532	0.535
0.533	0.535	0.537	0.537	0.545

a. Prepare a frequency table.
b. Draw a histogram, a frequency polygon, and a cumulative frequency diagram.
c. Calculate the mean diameter from the grouped data in the frequency table and indicate the mean on the histogram.

3-3. Calculate the mean of data for the annual precipitation given in Problem 2-1 from

a. The ungrouped data.
b. The frequency table.

Calculate and then comment on the percentage error between the values of the mean found from (a) and (b) above.

3-4. Calculate the mean, median, and mode for the data given in Problem 2-2.

3-5. The number of students entering engineering increased in 5 years from 3000 to 4880. Calculate the average annual rate of increase.

3-6. For the data of Problem 2-4 find the mean, mode, and median. Also find the mean using a grouped frequency table.

3-7. The distribution for times between failures of electronic components are shown in Table 3-9.

TABLE 3-9

Time in hours	0–100	100–200	200–300	300–400	400–500	500–600	600–700	700–800	800–900	900–1000	1000–1100	1100–1200	1200–1300
Number of failures	280	200	170	100	75	45	10	8	6	4	4	2	1

a. Draw a histogram, a frequency polygon, and a cumulative frequency diagram.
b. Calculate the mean, median, and mode of failure time.

3-8. Find the mean, mode, and median for the number of defective bolts in Problem 2-5.

Chapter 4

Characteristics of Distributions: Dispersion

In the preceding chapter we considered the central tendency of sets of observations and the calculation of averages. The average is a single value that typifies the whole group but does not generally give adequate information about the distribution of observations within the group. To choose a simple example,

mean of 99.9 cm, 100.0 cm, and 100.1 cm is 100.0 cm

mean of 99.0 cm, 100.0 cm, and 101.0 cm is also 100.0 cm

but it is clear that the two sets of observations differ appreciably in the scatter of the values about their mean. And yet the scatter may be of considerable importance. Let us assume, for example, that tests show the mean strength of a structural material to be 30,000 kN/m^2, but it is only the knowledge of scatter of the results that will tell us the proportion of specimens whose strength is, say, below 20,000 kN/m^2, the latter value being critical for safety.

There are several measures of scatter or dispersion, which will now be considered.

VARIANCE

If our set of values (a finite population) consists of n observations x_i, whose mean is μ, we can write for each observation the *deviation* $(x_i - \mu)$, known also as the *residual*. The mean square deviation is known as variance, which is given by

$$\sigma^2 = \frac{\sum_{i=1}^{n} (x_i - \mu)^2}{n} \tag{4-1}$$

For reasons to be discussed later, it is important to remember that the deviations must be calculated from the true mean of the set of values.

STANDARD DEVIATION

While variance is a fundamental measure of dispersion, it is not a convenient practical measure as its units are the square of the units of the variate.[1] Furthermore, many numerical characteristics of distributions are expressed directly in terms of the square root of variance. It is, therefore, preferable to give this square root the name *standard deviation* σ. Standard deviation is thus the root-mean-square (rms) deviation and is always positive. Its units are the same as those of the variate. The standard deviation is thus

$$\sigma = \sqrt{\frac{\sum_{i=1}^{n} (x_i - \mu)^2}{n}} \tag{4-2}$$

For the simple case of the groups of observations given at the beginning of this chapter, the values of the standard deviation in the two cases are, respectively,

$$\sigma_1 = \sqrt{\frac{(0.1)^2 + (0.1)^2}{3}} = \sqrt{0.0067} = 0.08 \text{ cm}$$

and

$$\sigma_2 = \sqrt{\frac{1^2 + 1^2}{3}} = \sqrt{\frac{2}{3}} = 0.8 \text{ cm}$$

Equation (4-2) is applicable when we are interested in the mean of a set of n items (specimens or observations), all of which have been determined; μ is

1. The quantity that varies and is being studied.

then the true mean. In many cases, however, we actually test only a limited number of specimens out of a *population* (or a *universe*, the two terms being synonymous) of all those that could be tested. The same applies to taking a limited number of observations (e.g., temperature at intervals of time) when we are really interested in all the observations that could be made (in this case, an infinite number of determinations of temperature during the given time interval). Under such circumstances, the true or population mean μ is unknown, and we have only the mean of the actual observations $\bar{x}$, i.e., the sample mean. We calculate the deviations from $\bar{x}$ and not from μ, and therefore put $(n - 1)$ instead of n in the denominator of the expression for the estimate of σ; the estimate is denoted by s to distinguish it from the true standard deviation σ. The reason for this correction of $\sqrt{n/(n-1)}$, known as *Bessel's correction,*[2] is that the sum of squares of deviations has a minimum value when taken about the sample mean $\bar{x}$, and is therefore smaller than it would be if taken about the population mean (which is presumably different from $\bar{x}$). Thus the estimate of σ is

$$s = \sqrt{\frac{\sum_{i=1}^{n} (x_i - \bar{x})^2}{n - 1}} \tag{4-3}$$

Appendix A gives a proof of the above equation. Bessel's correction can, of course, be neglected when n is large.

It is important to appreciate the difference between Eq. (4-2) and Eq. (4-3), and to use the correct one, depending on whether we are interested in the standard deviation of the observations in hand (considered as a finite population) or in an estimate of the standard deviation of the population.

From Eq. (4-3) we can see that if only one observation is made, nothing can be said about its precision. Assume that the observation is x_1. Then the best estimate of the population mean is given by $\bar{x} = x_1$. Hence

$$s^2 = \frac{(\bar{x} - x_1)^2}{n - 1} = \frac{0}{0}$$

and the estimate of the population standard deviation is, therefore, indeterminate.

The calculation of the standard deviation by Eqs. (4-2) and (4-3) is laborious, especially when the mean involves more significant places than the variate and thus introduces fractional values. It is more convenient, therefore, to use another form of Eq. (4-2), namely,

$$\sigma = \sqrt{\frac{\sum x_i^2}{n} - \mu^2} = \frac{1}{n}\sqrt{n \sum x_i^2 - \left(\sum x_i\right)^2} \tag{4-4}$$

2. Strictly speaking, Bessel's correction is applied to variance and is equal to $n/(n-1)$.

Here σ^2 denotes the variance of the set of n items and we require s^2, the unbiased estimate of the population from which the n items are drawn.

It may be noted that we have now omitted the symbols indicating that the summation proceeds from $i = 1$ to $i = n$, as such a notation is cumbersome. We must remember, however, that these are the limits of our summation unless otherwise indicated. Equation (4-4) arises from the algebraic identity (which we now apply to a finite population with a mean $\bar{x}$):

$$\sum (x_i - \bar{x})^2 = \sum x_i^2 - 2\bar{x} \sum x_i + n\bar{x}^2$$

But

$$\bar{x} = \frac{\sum x_i}{n} \qquad \text{(by definition)}$$

Therefore,

$$2\bar{x} \sum x_i = 2\bar{x}n\bar{x} = 2n\bar{x}^2$$

Hence

$$\sum (x_i - \bar{x})^2 = \sum x_i^2 - n\bar{x}^2$$

or

$$\sum (x_i - \bar{x})^2 = \sum x_i^2 - \frac{(\sum x_i)^2}{n}$$

The advantage of Eq. (4-4) is that we do not need to find the deviations $(x_i - \bar{x})$. The squares of the variate x_i^2 can be obtained rapidly from tables or by means of a calculating machine. The fundamental difference between $\sum x_i^2$ and $(\sum x_i)^2$ must not be overlooked.

To obtain s, Bessel's correction is applied:

$$s = \sigma \sqrt{\frac{n}{n-1}} \qquad \text{(4-5)}$$

DEGREES OF FREEDOM

At this stage it may be convenient to introduce the concept of the number of *degrees of freedom*. The term *degrees of freedom* plays an important role in statistical methods and its meaning, therefore, must be understood. A degree of freedom is defined as a comparison between the data, independent of the other comparisons in the analysis. Each observation in a random sample of size n can be compared with $(n - 1)$ other observations and hence there are

$(n - 1)$ degrees of freedom. In the case where an estimate s^2 [see Eq. (4-3)] is made of the population variance σ^2, the true population mean μ is not known. Thus, when the observations are compared with the sample mean $\bar{x}$, there is a constraint on the values of $(x_i - \bar{x})$ imposed by the sum of the deviations about the sample mean $\bar{x}$ being zero, i.e.,

$$\sum_{i=1}^{n} (x_i - \bar{x}) = 0 \qquad [3\text{-}1a]$$

Hence one degree of freedom is lost, leaving $(n - 1)$ comparisons or degrees of freedom. Therefore, to obtain an estimate of the variance, i.e., mean square of deviations, we divide the sum of the squares of deviations by the number of comparisons, i.e., degrees of freedom $(n - 1)$

$$s^2 = \frac{\sum (x_i - \bar{x})^2}{n - 1}$$

In the case when the true mean μ is known and we want to compare it to n observations of a sample, then each of the n observations can be compared independently to μ; hence, when we calculate the mean square deviation, σ^2, we divide the term $\sum (x_i - \mu)^2$ by n, the degrees of freedom in this case [see Eq. (4-2)].

The concept of degrees of freedom can also be visualized by referring to a point that can move freely in three-dimensional space. The point can be located in space by three variable coordinates x, y, and z. If we constrain the point to move in a plane, such as $ax + by + cz = d$, then it will just have 2 degrees of freedom, which is the number of independent variables (x, y and z, i.e., 3) minus the number of constraints (the equation $ax + by + cz = d$, i.e., 1). Hence, in general, the degrees of freedom represent the number of independent variables minus the number of constraints. If we were dealing with, say, n independent variables related by m equations (restrictions), then the number of degrees of freedom would be $(n - m)$.

The analogy between this geometric concept and the degrees of freedom in statistical language is quite clear. For example, for a sample of size n from a population whose mean μ is unknown, the term

$$\sum_{i=1}^{n} (x_i - \bar{x})^2$$

will have $(n - 1)$ degrees of freedom. This is obtained from the number of variables n, minus the one constraint imposed, i.e.,

$$\sum_{i=1}^{n} (x_i - \bar{x}) = 0.$$

In other words, when we use the sample to determine $\bar{x}$ as representing the population mean μ, we force a certain degree of agreement among the variables, e.g., the quantity

$$\sum_{i=1}^{n} (x_i - \bar{x})^2.$$

However, if μ is known, then the degrees of freedom will be simply n, since there is no longer a constraint. We shall see that the sampling distributions of some statistics depend on the number of degrees of freedom, such as χ^2 (Chapter 12), t (Chapter 13) and F (Chapter 14). The concept of degrees of freedom will be illustrated further in Chapters 12–15, 17, 19, and 21.

SIMPLIFIED COMPUTATION OF STANDARD DEVIATION

Computations can be simplified without any loss in accuracy by a suitable reduction of data. Two rules are useful:

a. If a constant number is added to or subtracted from a set of numerical variates, their mean will increase or decrease by the same constant number, the standard deviation remaining unaffected.
b. If a set of numerical variates is multiplied or divided by a constant number, their mean and standard deviation are multiplied or divided by the same constant number.

Example. Let us consider the data on the strength of bricks, given in Table 2-2. To compute the standard deviation by Eq. (4-2), we first require the mean; this was quoted previously to be $\bar{x} = 6.89$ MN/m^2. We now compute the deviations $(x_i - \bar{x})$ and find their squares; a tabular form is convenient (Table 4-1).

TABLE 4-1

Observation	x_i	$\lvert x_i - \bar{x} \rvert^a$	$(x_i - \bar{x})^2$
1	1.86	5.03	25.3009
2	3.17	3.72	13.8384
3	3.93	2.96	8.7616
⋮	⋮	⋮	⋮
269	11.24	4.35	18.9225
270	13.86	6.97	48.5809
$n = 270$			$\sum (x_i - \bar{x})^2 = 518.9963$

[a] The absolute value of the deviation is entered, since its sign is immaterial.

Thus, from Eq. (4-2),

$$\sigma = \sqrt{\frac{518.9963}{270}} = 1.39 \text{ MN/m}^2$$

Let us now find the solution using Eq. (4-4). See Table 4-2.

TABLE 4-2

Observation	x_i	x_i^2
1	1.86	3.4596
2	3.17	10.0489
3	3.93	15.4449
⋮	⋮	⋮
269	11.24	126.3376
270	13.86	192.0996
$n = 270$	$\sum x_i = 1861.30$	$\sum x_i^2 = 13351.7562$

Thus

$$\sum x_i = 1861.30$$

$$(\sum x_i)^2 = 3{,}464{,}437.69$$

and

$$\sum x_i^2 = 13{,}351.7562$$

Hence

$$\sigma = \frac{1}{270} \times \sqrt{270 \times 13{,}351.7562 - 3{,}464{,}437.69}$$

$$= 1.39 \text{ MN/m}^2\text{, as before}$$

We can now use the simplifications of rules (a) and (b). Since all values have two decimal places, we can easily multiply them by 100. To reduce the numerical values further, we can subtract 186 from all the values. (It is generally preferable to avoid negative values, such as would result from subtracting, say, 700, but there is nothing incorrect in doing so.) We can tabulate the computation as shown in Table 4-3.

Thus $(\sum X_i')^2 = 18{,}471{,}528{,}100$. Hence, in terms of X',

$$\sigma_{X'} = \frac{1}{270} \times \sqrt{270 \times 73{,}618{,}121 - 18{,}471{,}528{,}100}$$

$$= 139$$

TABLE 4-3

Observation	$X'_i = (x_i \times 100) - 186$	X'^2_i
1	0	0
2	131	17 161
3	207	42 849
⋮	⋮	⋮
269	938	879 844
270	1200	1440 000
$n = 270$	$\sum X'_i = 135\,910$	$\sum X'^2_i = 73\,618\,121$

To allow for the multiplication by 100 we have to divide $\sigma_{X'}$ by 100; hence

$$\sigma = \frac{139}{100}$$

$$= 1.39 \text{ MN/m}^2\text{, as before}$$

The decision on the amount of simplification is a matter for individual preference, but the advantage of Eq. (4-4) over Eq. (4-2) is great, especially when a calculating machine is used.

With data in a grouped-frequency form, a shorter, though approximate, computation of the standard deviation can be made. We assume that each observation is replaced by an observation at the class midpoint so that

$$\sigma = \sqrt{\frac{\sum f_i(x'_i - \bar{x})^2}{n}} \tag{4-6}$$

where f_i is the frequency in the class interval and x'_i is the class midpoint. Following Eq. (4-4), the expression becomes

$$\sigma = \sqrt{\frac{\sum f_i x'^2_i - [(\sum f_i x'_i)^2/n]}{n}} \tag{4-7}$$

We can further replace x'_i by X'_i = deviation from an arbitrary origin, measured in terms of class width w. Then

$$\sigma = w\sqrt{\frac{\sum f_i X'^2_i - [(\sum f_i X'_i)^2/n]}{n}} \tag{4-8}$$

As an example, let us apply this method of calculation of the standard deviation to the data of Table 3-2. Then

$$\sigma = w\sqrt{\frac{7079 - [(1329)^2/270]}{270}} = 1.41w$$

Since the class width w is 1.00,

$$\sigma = 1.41 \times 1.00 = 1.41 \text{ MN/m}^2$$

The difference between this value and the accurate value of σ from Eq. (4-4) is small and for most purposes not significant.

MEAN DEVIATION

In some cases, instead of standard deviation, mean (absolute) deviation d_m is used; this is the mean of the absolute values of deviations:

$$d_m = \frac{\sum |x_i - \bar{x}|}{n} \tag{4-9}$$

The use of absolute values is necessary because the algebraic sum of deviations from the mean is, by definition of the mean, always equal to zero.

The usefulness of the mean deviation in statistical calculations is small, and practically no statistical methods of analysis involve its use. However, in the case of normal distribution there is a simple relation between the mean deviation and standard deviation; this is discussed in Chapter 10.

COEFFICIENT OF VARIATION

As mentioned before, the standard deviation is expressed in the same units as the original variate x_i, but for many purposes it is convenient to express the dispersion of results on a percentage basis, i.e., in relative rather than absolute terms. To achieve this, we take the ratio of the standard deviation to the mean and define the coefficient of variation V as:

$$V = \frac{\sigma}{\bar{x}} \times 100 \tag{4-10}$$

It is a dimensionless quantity.

For the observations on the strength of bricks used in the preceding example, the coefficient of variation is

$$V = \frac{1.39}{6.89} \times 100 = 20.17 \text{ percent}$$

While the coefficient of variation is extremely useful in giving a value that is independent of the units employed, it may sometimes be meaningless. This is the case when the origin of measurement is not uniquely fixed; for example, if we measure temperature and find the mean to be 10°C and the standard deviation 1°C, we could report the coefficient of variation as

(1/10) × 100 = 10 percent. If, however, the measurements were converted to degrees Fahrenheit, we would have a mean of 50°F and a standard deviation of 1.8°F. One could thus report a coefficient of variation of (1.8/50) × 100 = 3.6 percent. The absurdity of these calculations is obvious, and we would have done well to have reported our results in terms of standard deviation.

RANGE

Range is a simple measure of dispersion, very rapid to compute as it is merely the difference between the highest and the lowest observations. However, because of this dependence on two values only, range is a rather crude measure of dispersion, and is an efficient statistic[3] when we deal with small samples only.

It is important to note that the larger the number of observations in a sample the more likely it is that values remote from the mean will be encountered. Thus range increases with the sample size. If this is not remembered and the ranges of samples of different sizes are compared indiscriminately, misleading results are obtained.

The range and the standard deviation are related to one another so that for any given number of observations n, an estimate of the standard deviation s of the underlying population can be obtained from the mean value of sample range $\overline{R}$:

$$s = \overline{R} \times d \tag{4-11}$$

This expression is valid when the variate is normally distributed,[4] and the estimate s becomes less accurate the more the distribution departs from normality.

The values of d are given in Table A-1. It may be observed that for n between 3 and 12, d varies approximately as $1/\sqrt{n}$.

We must remember that the estimate of the standard deviation given by Eq. (4-11) is no more than an approximation and should be used only when the average range $\overline{R}$ is obtained from a reasonably large number of samples (say, not less than 10) all of the same size. For the same total number of observations the estimate is considerably more accurate when the samples are many and small, rather than few and large.

If a large number of observations has been made without subdivision into samples, a breakdown of the observations into equal subgroups can be achieved by random sampling. The range R of each subgroup is then calculated, and, hence, the mean value of range $\overline{R}$. This is then multiplied by the

3. A value that characterizes the data.
4. For a definition of the normal distribution, see Chapter 9.

coefficient d from Table A-1, corresponding to the number of observations in a subgroup, and hence an estimate of the standard deviation is obtained. This procedure is illustrated by the following example.

Example. The modulus of rupture was determined on 52 concrete test beams. By random selection the observations were arranged in 13 subgroups, as shown in Table 4-4.

TABLE 4-4

Subgroup	*Observations, MN/m^2*	*Range, R*
1	3.90—4.00—4.03—3.74	0.29
2	3.62—4.05—3.93—3.96	0.43
3	4.09—4.19—4.50—3.76	0.74
4	4.02—4.01—3.45—4.05	0.60
5	4.27—4.13—3.96—3.99	0.31
6	4.71—4.32—4.14—4.43	0.57
7	4.07—3.96—4.52—4.08	0.56
8	4.50—4.32—3.99—4.05	0.51
9	4.09—4.12—3.64—4.14	0.50
10	3.96—4.05—4.07—3.90	0.17
11	3.95—3.94—4.05—3.83	0.22
12	4.16—3.92—4.25—3.95	0.33
13	4.36—4.01—4.10—3.98	0.38
Total		$\sum R = 5.61$

Source: K. E. C. Nielsen, "Calculation of Statistical Data," *Beton-Teknik* (Copenhagen), vol. 3, no. 24, 1958, p. 170. The original data was in psi units.

The value of range for each subgroup is shown in the right-hand column. Hence the mean value of range is

$$\overline{R} = \frac{5.61}{13} = 0.4315$$

For $n = 4$, Table A-1 gives $d = 0.4857$. Using Eq. (4-11), we estimate the standard deviation to be $s = 0.4315 \times 0.4857 = 0.21$ MN/m^2.

A direct calculation of s using Eq. (4-3) gives $s = 0.23$ MN/m^2. The difference may for some purposes not be important, while the saving in computational effort is considerable. A comparison of experimental results with the theoretical relation between range and standard deviation is shown in

Fig. 4-1 for tests on the compressive strength of concrete. (It has been shown that the compressive strength of concrete sensibly follows a normal distribution.[5])

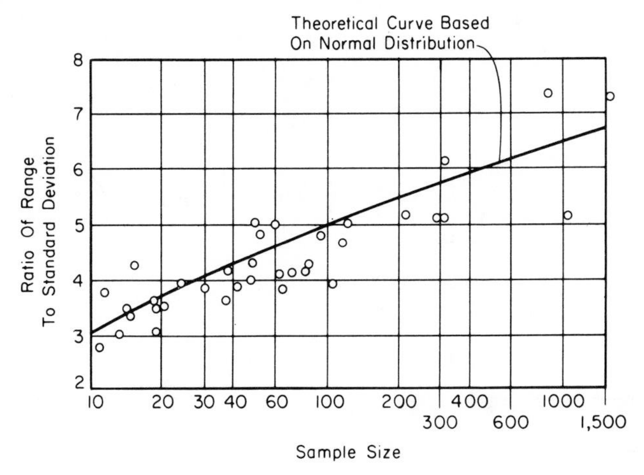

Fig. 4-1. Ratio of range to standard deviation for samples of different sizes; tests on concrete compression cubes. (From P. J. F. Wright, "Variations in the Strength of Portland Cement," *Magazine of Concrete Research*, vol. 10, no. 30, Nov. 1958, pp. 123–132.)

Solved Problems

4-1. To obtain the shrinkage limit of a particular clay, the moisture content in percent of dry weight was determined on 20 specimens with the following results:

15.2	16.7	15.8	14.6	18.1	17.2	18.0	15.9	16.1	16.9
14.8	17.6	18.2	16.9	17.3	16.5	15.6	16.7	15.8	18.2

a. Calculate the mean and standard deviation by the long method.
b. Calculate the mean and standard deviation using an arbitrary origin.

SOLUTION

a. $\text{mean} = \bar{x} = \dfrac{\sum x}{n}$

$$= \frac{15.2 + 16.7 + \cdots + 16.9 + 14.8 + 17.6 + \cdots + 18.2}{20}$$

$$= \frac{332.1}{20}$$

$$= 16.605$$

5. A. M. Neville, "Some Aspects of the Strength of Concrete," *Civil Engineering* (London), vol. 54, Oct.–Dec. 1959.

The standard deviation of the population of clay samples is estimated by

$$s = \sqrt{\frac{\sum x^2 - [(\sum x)^2/n]}{n-1}}$$

Now

$$\sum x^2 = 15.2^2 + 16.7^2 + \cdots + 18.2^2 = 5538.13$$

and

$$(\sum x)^2 = (332.1)^2 = 110{,}290.41$$

Hence

$$s = \sqrt{\frac{5538.13 - (110{,}290.41/20)}{19}}$$

$$= \sqrt{\frac{23.61}{19}}$$

$$= 1.11$$

b. Let the arbitrary origin = 16. Then, as shown in Table 4-5,

TABLE 4-5

	Deviation from arbitrary origin X'_i		
x_i	*Negative*	*Positive*	X'^2_i
15.2	0.8		0.64
16.7		0.7	0.49
15.8	0.2		0.04
14.6	1.4		1.96
18.1		2.1	4.41
17.2		1.2	1.44
18.0		2.0	4.00
15.9	0.1		0.01
16.1		0.1	0.01
16.9		0.9	0.81
14.8	1.2		1.44
17.6		1.6	2.56
18.2		2.2	4.84
16.9		0.9	0.81
17.3		1.3	1.69
16.5		0.5	0.25
15.6	0.4		0.16
16.7		0.7	0.49
15.8	0.2		0.04
18.2		2.2	4.84
Totals	$\sum X'_i = -4.3$ +	16.4 = 12.1	$\sum X'^2_i = 30.93$

Therefore,

$$\text{mean} = \bar{x} = 16 + \frac{\sum X_i'}{n}$$

$$= 16 + \frac{12.1}{20}$$

$$= 16.605$$

$$\text{standard deviation} = s = \sqrt{\frac{\sum X'^2 - [(\sum X')^2/n]}{n-1}}$$

$$= \sqrt{\frac{30.93 - [(12.1)^2/20]}{19}}$$

$$= \sqrt{1.243}$$

$$s = 1.11$$

4-2. Calculate the range and the standard deviation of the data given in Solved Problem 3-1. If the permissible tensile stress allowed in design is equal to the mean less 2.33 times the standard deviation, calculate this allowable stress and indicate whether any of the 303 brittle lacquer test pieces fell below this stress. If this criterion for allowable stress is used, show that when the coefficient of variation $V = 10$ percent the maximum allowable design tensile stress = 0.77 × mean stress.

SOLUTION

Range = 500 − 200 = 300 kN/m^2. For grouped data,

$$s = w\sqrt{\frac{\sum f_i X_i'^2 - [(\sum f_i X_i')^2/n]}{n-1}}$$

From the values in Table 3-5, we have

$$s = w\sqrt{\frac{4679 - [(1067)^2/303]}{302}} = 1.75 \times \text{class width}$$

$$= 1.75 \times 30 = 52.5 \text{ kN/m}^2$$

Allowable tensile stress in design = 321 − 2.33 × 52.5 = 199 kN/m^2. According to Table 3-5, no brittle lacquer test piece had a tensile strength below 199 kN/m^2.

Now the maximum allowable design tensile stress = $\bar{x} - 2.33s$. But

$$V = \frac{s}{\bar{x}}$$

Thus

$$s = V\bar{x}$$

Hence maximum allowable design tensile stress

$$= \bar{x} - 2.33V\bar{x}$$

$$= (1 - 2.33 \times 0.10)\bar{x}$$

$$= 0.77\bar{x} \text{ (approximately)}, \quad \text{for } V = 0.10$$

Problems

4-1. The following measurements (in newtons) were obtained from a test on the tensile strength of rubber samples:

1419	1410	1410
1403	1396	1389
1400	1380	1422

Calculate the mean and the estimated standard deviation of the tensile strength of the rubber from which the samples were drawn. What is the range?

4-2. Determine the variance, standard deviation, and coefficient of variation of the distribution in Problem 3-1. Can you estimate the standard deviation from the range in this case? State the reason.

4-3. Determine the range of the data in Problem 3-2. Also find the standard deviation of the diameters of the ball bearings from

a. The ungrouped data.
b. The frequency table.

What is the coefficient of variation in each case?

By random selection, arrange the data in 10 subgroups, and hence estimate the standard deviation from the mean range. Compare this value with the values obtained in (a) and (b).

4-4. Calculate the estimate of the standard deviation of the population of radio tubes in Problem 2-3. Determine the variance and the coefficient of variation of the tubes.

4-5. In order to determine the hardness of concrete in a shell roof the roof area was divided into 64 equal parts.[6] The average of 10 measurements of hardness (sclerometer readings) in each part is given in Table 4-6 for all 64 parts.

6. Détermination de la Dispersion des Valeurs de la Résistance du Béton en Place au Moyen du Scléromètre," *Bulletin du Ciment*, no. 10 (Switzerland), Oct. 1962.

TABLE 4-6

	1	*2*	*3*	*4*	*5*	*6*	*7*	*8*
A	49.7	52.2	55.9	57.4	59.5	56.5	55.4	55.2
B	52.0	56.2	56.2	56.8	54.8	55.6	53.5	51.2
C	58.1	52.3	55.4	49.6	51.2	50.6	51.0	50.9
D	53.8	56.6	54.1	55.7	54.2	53.4	50.9	54.6
E	53.0	54.6	55.3	54.3	56.8	50.6	55.7	55.2
F	57.7	55.6	52.7	53.8	53.7	56.4	53.5	52.3
G	52.4	55.2	51.4	49.4	50.5	56.3	52.4	55.2
H	54.9	52.3	53.9	55.4	51.6	57.1	52.4	59.4

The individual values of hardness readings in eight parts were as follows:

TABLE 4-7

Point	*Individual values of hardness*										*Mean value*	*Standard deviation*
*A*1	48	52	50	45	45	44	49	54	55	55	49.7	4.21
*B*1	56	54	52	51	52	52	52	50	50	51	52.0	1.82
*C*1	55	61	61	57	56	56	57	60	60	58	58.1	2.23
*D*1	54	56	57	58	55	54	52	50	50	52	53.8	2.78
*E*1	52	50	48	57	50	53	59	58	49	54	53.0	3.92
*F*1	61	59	56	57	57	57	55	58	58	59	57.7	1.70
*G*1	55	53	56	51	52	53	51	50	55	48	52.4	2.50
*H*1	52	56	52	58	59	60	54	55	51	52	54.9	3.25

a. Find the root-mean-square value of the standard deviation in the eight parts (no. 1). Hence estimate the testing error (the rms value divided by the square root of the number of values, namely, eight). [HINT: The individual readings need not be used.]

b. Find the standard deviation of the hardness readings for the entire roof, exclusive of the testing error (the square root of the difference between the variance of the 64 average values and the square of the testing error).

4-6. On a construction job it was required to make concrete with a specified minimum compressive strength of 17.2 MN/m^2. The minimum was understood to be a value exceeded by not less than 96 percent of test results. The values of strength of 50 test cubes are given in Table 4-8.

Calculate the mean strength, range, mean deviation, standard deviation, and coefficient of variation. Find whether the specification requirements are satisfied.

TABLE 4-8

Cube Number	*Strength, MN/m²*	*Cube Number*	*Strength, MN/m²*
1	22.3	26	22.8
2	18.2	27	22.1
3	24.8	28	21.2
4	23.5	29	23.4
5	19.8	30	18.1
6	27.0	31	23.4
7	25.9	32	23.0
8	29.2	33	23.6
9	24.1	34	25.4
10	24.3	35	22.2
11	21.7	36	17.8
12	23.4	37	28.0
13	23.3	38	28.8
14	28.3	39	22.1
15	23.4	40	20.1
16	20.0	41	20.8
17	26.1	42	23.0
18	15.2	43	28.2
19	20.0	44	23.0
20	22.8	45	30.4
21	32.0	46	25.1
22	25.6	47	24.8
23	20.8	48	19.9
24	23.2	49	18.2
25	26.9	50	25.1

Source: Cement and Concrete Association: *Technical Memorandum*, No. 8, (London), April 1960. *The original data was in psi units.*

4-7. The strengths of concrete specimens, made two at a time over a period of 2 months on a construction site, are given in Table 4-9.

Find the mean strength, standard deviation, and coefficient of variation, working in terms of mean test values.

4-8. For the data of Problem 4-7, estimate the standard deviation from the mean range.

4-9. For the data of Problem 4-7, determine the value of strength exceeded by 90 percent of tests (a "test" is the average of two specimens).

TABLE 4-9

Test Number	28-day strength, MN/m^2 Cylinder Number 1	28-day strength, MN/m^2 Cylinder Number 2	Test Number	28-day strength, MN/m^2 Cylinder Number 1	28-day strength, MN/m^2 Cylinder Number 2
1	24.8	22.0	24	23.5	24.1
2	24.5	24.5	25	19.2	21.9
3	23.6	25.3	26	21.3	22.8
4	20.1	22.8	27	22.1	22.8
5	23.0	22.0	28	24.1	22.8
6	25.9	23.4	29	19.9	21.0
7	27.8	27.0	30	26.3	24.1
8	27.6	24.4	31	26.4	26.2
9	25.4	25.2	32	31.6	30.1
10	19.5	21.5	33	24.8	25.5
11	20.3	23.8	34	24.2	22.5
12	21.0	22.1	35	23.1	26.5
13	22.1	20.0	36	20.8	24.0
14	26.3	22.0	37	21.7	20.3
15	27.2	25.7	38	17.3	19.4
16	21.7	20.4	39	23.2	24.8
17	20.2	18.6	40	29.0	26.6
18	24.9	22.1	41	26.2	22.1
19	30.6	27.6	42	20.7	21.0
20	26.7	26.0	43	24.3	23.0
21	30.3	24.8	44	27.2	22.4
22	24.5	26.1	45	24.8	23.7
23	21.0	23.4	46	27.0	25.9

Source: "Evaluation of Compression Test Results of Field Concrete," *Journal of American Concrete Institute*, Vol. 52, Nov. 1955, p. 255. *The original data was in psi units.*

4-10. Readings on a filament temperature were taken by several persons using an optical pyrometer. The results are given in Table 4-10.

TABLE 4-10

Temperature, °C	1000	1025	1050	1075	1100	1125	1150
Number of readings	1	10	14	20	10	8	2

a. Calculate the "best" representative value for the filament temperature.
b. Estimate the variance and standard deviation of the optical pyrometer. (Assume there is no reading error due to personnel.) What is the coefficient of variation?

Chapter 5

Probability, Probability Distributions, and Expectation

The purpose of this chapter is to present briefly information on some simple concepts extensively used in statistical computations. The computation of probabilities in some complicated problems can be more readily handled if one uses the ideas of permutations and combinations. These two topics are undoubtedly familiar to the majority of readers, but for the sake of presenting a complete picture, brief notes on permutations and combinations will be given.

PERMUTATIONS

The number of permutations of n different items is the number of different arrangements in which these items can be placed. We can take all n items every time, or r ($r < n$) items at a time. Not only their identity, but also the order in which the items are arranged is significant.

We write the number of permutations as ${}_nP_r$, n being the number of items available and r the number chosen at a time. Thus, if all items are chosen every time, the number of permutations is ${}_nP_n$.

In the general case we have r places to fill with n items to choose from so that

the first place can be filled in n ways,

the second place can be filled in $n - 1$ ways,

the third place can be filled in $n - 2$ ways,

$\vdots$

the rth place can be filled in $n - r + 1$ ways.

We now apply a fundamental principle which states that if one selection can be made in p ways and if, after this selection has been made, the second selection can be made in q ways, the two selections together can be made in pq ways. Using this principle in succession, the number of permutations is

$$
\begin{aligned}
{}_nP_r &= n(n-1)(n-2)\cdots(n-r+1) \\
&= \frac{n!}{(n-r)!}
\end{aligned}
\tag{5-1}
$$

in which $n!$ is the symbol for factorial $n = 1 \times 2 \times 3 \times \cdots (n-1)n$.

When $r = n$,

$$
{}_nP_n = n! \tag{5-2}
$$

since by definition $0! = 1$.

Example. How many three-digit numbers can be formed from the numbers 1, 2, 3, 4, and 5 if each number can be repeated?

The first place can be filled in five ways; the second place can also be filled in five ways; and so can the third place. Hence we can form

$$
5 \times 5 \times 5 = 125 \text{ different numbers}
$$

Example. How many four-digit numbers can be formed from the numbers 1 to 9? There is no provision for repeating any number within any four-digit number. We can thus use the standard expression with $n = 9$ and $r = 4$:

$$
{}_nP_r = \frac{n!}{(n-r)!} = \frac{9!}{5!} = 9 \times 8 \times 7 \times 6 = 3024 \text{ numbers}
$$

Example. A product is coded by three letters and two numbers, the letters preceding the numbers. Only the letters A and B and the numbers 1 to 6 can be used. How many different code "numbers" are possible?

Consider the letters:

each can appear as A or B, i.e., in 2 ways

Therefore,

3 letters can be arranged in $2 \times 2 \times 2 = 8$ ways

Consider the numbers:

each can appear as 1 or 2 or $\cdots$ or 6, i.e., in 6 ways

Therefore,

2 numbers can be arranged in $6 \times 6 = 36$ ways

Thus the total number of code "numbers" is $8 \times 36 = 288$.

Consider now the permutation of n items taken all at a time, when the n items consist of r_1 alike, r_2 alike, ..., r_k alike, so that $r_1 + r_2 + \cdots + r_k = n$. The number of permutations is then

$$P = \frac{n!}{r_1!r_2! \cdots r_k!} \tag{5-3}$$

Example. How many different patterns (in a single row) can be made with 3 yellow tabs, 2 red ones, and 7 green ones?

We have

$$n = 3 + 2 + 7 = 12$$

From Eq. (5-3),

$$P = \frac{12!}{3! \times 2! \times 7!} = 7920$$

COMBINATIONS

The number of combinations of n different items is the number of different selections of r items each, without reference to the order or arrangement of the items in the group. This disregard of arrangement distinguishes combinations from permutations.

The reason that r items can be arranged in $r!$ ways is that:

the first place can be filled in r ways,

the second place can be filled in $r - 1$ ways,

the third place can be filled in $r - 2$ ways,

$\vdots$

the last can be filled in 1 way,

so that r places can be filled in $r(r-1)(r-2) \cdots 1 = r!$ ways.

Thus the number of combinations of r items from n items, denoted by

$$ {}_nC_r \quad \text{or} \quad \binom{n}{r} $$

is $r!$ times smaller than the number of permutations. Hence

$$ {}_nC_r = \frac{{}_nP_r}{r!} = \frac{n!}{(n-r)!\,r!} \tag{5-4} $$

It follows from symmetry that

$$ {}_nC_r = {}_nC_{n-r} \tag{5-5} $$

The use of this identity may save time in computations.

Example. In how many ways can a team of 9 be selected from 12 people?

This is evidently a problem of selection and not of arrangement, as the assignment of positions is not considered. We use, therefore, Eq. (5-4) with $n = 12$ and $r = 9$

$$ {}_nC_r = \frac{n!}{r!\,(n-r)!} = \frac{12!}{9! \times 3!} = \frac{12 \times 11 \times 10}{2 \times 3} = 220 $$

Example. From 5 men and 4 women, in how many ways can we select a group of 3 men and 2 women?

a. We can select 3 men from 5 men in ${}_5C_3$ ways.
b. We can select 2 women from 4 women in ${}_4C_2$ ways.

By the fundamental principle of selections we can do (a) and (b) in ${}_5C_3 \times {}_4C_2$ ways. Thus the number of possible selections is

$$ \frac{5 \times 4}{2} \times \frac{4 \times 3}{2} = 60 $$

Example. From 6 men and 5 women, how many committees of 8 members can be formed when each committee is to contain at least 3 women?

The conditions of the problem are satisfied if a committee contains

5 men and 3 women	selected in ${}_6C_5 \times {}_5C_3$ ways
4 men and 4 women	selected in ${}_6C_4 \times {}_5C_4$ ways
3 men and 5 women	selected in ${}_6C_3 \times {}_5C_5$ ways

The number of possible committees is thus:

$$ {}_6C_5 \times {}_5C_3 + {}_6C_4 \times {}_5C_4 + {}_6C_3 \times {}_5C_5 = 155 $$

It frequently happens that a problem involves both a selection and an arrangement with a limitation upon either or both. A safe procedure is to deal first with the selections (combinations) and then with the arrangements (permutations).

Example. How many lineups are possible in choosing a hockey team composed of 4 seniors and 2 juniors from 8 seniors and 7 juniors if any man can be used in any position?

4 seniors can be selected in ${}_8C_4$ ways
2 juniors can be selected in ${}_7C_2$ ways

Hence a team of players can be selected in ${}_8C_4 \times {}_7C_2$ ways. Any one set of 6 men can be arranged in 6! ways. Thus the total number of possible lineups is ${}_8C_4 \times {}_7C_2 \times 6! = 1{,}058{,}400$.

Example. Suppose that in the last example any chosen team had to include a certain senior player for center and a certain junior player as a goal keeper. How many lineups are then possible?

When a particular senior player is always included in a team, the problem is to find the number of combinations of 3 seniors from the remaining 7 seniors, and with regard to juniors, to find the number of combinations of 1 junior from the remaining 6 juniors. Thus

3 seniors can be selected in ${}_7C_3$ ways
1 junior can be selected in ${}_6C_1$ ways

Hence a team of players can be selected in ${}_7C_3 \times {}_6C_1$ ways.

Since the positions of goal and center are already assigned, the remaining four positions in any one team can be filled in 4! ways. Thus the total number of possible lineups is ${}_7C_3 \times {}_6C_1 \times 4! = 5040$.

It is important to observe that ${}_nC_r$ is the coefficient of the $(r + 1)$th term in the binomial expression $(a + b)^n$, whose expansion is

$$(a+b)^n = a^n + na^{n-1}b + \frac{n(n-1)}{2!}a^{n-2}b^2 + \cdots + \frac{n(n-1)\cdots(n-r+1)}{r!}a^{n-r}b^r + \cdots + b^n \qquad (5\text{-}6)$$

and can be conveniently written as:

$$(a+b)^n = a^n + {}_nC_1a^{n-1}b + {}_nC_2a^{n-2}b^2 + \cdots + {}_nC_ra^{n-r}b^r + \cdots + b^n$$

or

$$(a+b)^n = \sum_{r=0}^{n} {}_nC_ra^{n-r}b^r \qquad (5\text{-}7)$$

provided that we define ${}_nC_0 = 1$.

The use of binomial coefficients is considered again in Chapter 7.

PROBABILITY

In Chapter 1 we discussed the relationship between statistics and probability. We noted that in statistical problems we estimate parameters of a distribution from sample data, whereas in probability problems the parameters of the mathematical model of the system are known and it is the behavior of parts of the system, or samples, that is deduced. Thus, broadly speaking, the two approaches are the inverse of one another.

Let us denote the occurrence of a certain event as success and its nonoccurrence as failure. If we consider all the possible (imaginable) arrangements (or trials), and also the arrangements corresponding to success, then the ratio of the latter to the former defines the probability of success. All the arrangements considered must be mutually exclusive and equally likely, and, as stated before, exhaustive. If there are n possible arrangements, and success occurs in p cases, the probability of success is p/n, and the probability of failure is $1 - (p/n)$. Probability is thus expressed as a number not greater than 1. A value of unity denotes a certainty of success, and a value of zero means an impossibility of success.

Strictly speaking, we should distinguish two types of probability. In the first, the probability of an event is established solely by the definition of the system, e.g., the probability of obtaining a given number on rolling a die is $\frac{1}{6}$. This is an *a priori* probability. In many other cases we are concerned with an *empirical* probability, which is based solely on experience, e.g., on the past records of deaths, accidents, and so on. Engineering statistics are frequently concerned with a combination of both types of probabilities.

There are two rules of probability that are of fundamental importance:

a. The probability of occurrence of several of a number of *mutually exclusive* events is the sum of the probabilities of the separate events. Events are said to be mutually exclusive if only one of them can occur at a time (e.g., a die shows a 1 or a 6 but not both).

For example, in tossing a die, the probability of throwing a 3 *or* 6 is equal to the probability of throwing 3 plus the probability of throwing 6, i.e., it is equal to

$$\tfrac{1}{6} + \tfrac{1}{6} = \tfrac{1}{3}$$

We may note that the sum of the probabilities of *all* the possible events is always 1; for example, the probability of not throwing a 3 or 6 is $\frac{2}{3}$, and hence the sum of the probabilities of all possible outcomes is $\frac{1}{3} + \frac{2}{3} = 1$.

Proof of the addition rule. Let there be n equally likely cases such that event A occurs in r of them and event B occurs in s of them. Since the occurrence of A does not coincide with that of B (mutually exclusive), there is no overlapping between the $(r + s)$ cases; i.e., the event A or B occurs in only the $(r + s)$ cases. Hence

$$P(A \text{ or } B) = P(A + B) = \frac{r + s}{n} = \frac{r}{n} + \frac{s}{n} = P(A) + P(B) \tag{5-8}$$

This rule can be readily generalized to several mutually exclusive events, $A, B, C, \ldots$; thus we can write

$$P(A \text{ or } B \text{ or } C \text{ or } \cdots) = P(A) + P(B) + P(C) + \cdots$$

or

$$P(A + B + C + \cdots) = P(A) + P(B) + P(C) + \cdots \tag{5-9}$$

b. The probability of a *simultaneous* occurrence of a number of *independent* events is the product of the separate probabilities. An event is considered independent if its occurrence does not affect the probability of the occurrence of other events.

For example, in tossing two dice, the probability of throwing a double 6 is equal to the probability of throwing a 6 times the probability of throwing a 6; i.e., it is equal to

$$\tfrac{1}{6} \times \tfrac{1}{6} = \tfrac{1}{36}$$

Proof of the multiplication rule. Let us assume that there are m favorable cases for event A in a total of n cases, and M favorable cases for event B in a total of N cases. Then $P(A) = m/n$ and $P(B) = M/N$. When we consider the joint event "A and B," we have altogether nN possible cases all equally likely, since A and B are independent. Each favorable case for A combines with each favorable case for B to yield a favorable case for the joint event A and B. Thus there are mM favorable cases for the event A and B; hence

$$P(A \text{ and } B) = P(AB) = \frac{mM}{nN} = \frac{m}{n} \cdot \frac{M}{N} = P(A) \cdot P(B) \tag{5-10}$$

We can readily generalize this rule to the situation where there are several independent events $A, B, C, \ldots$; thus

$$P(A \text{ and } B \text{ and } C \text{ and } \cdots) = P(A) \cdot P(B) \cdot P(C) \cdots$$

or

$$P(ABC \cdots) = P(A) \cdot P(B) \cdot P(C) \cdots \tag{5-11}$$

Example. Thirty high-strength bolts became mixed by mistake with 25 ordinary bolts and it was not possible to tell them apart from appearance. If two bolts are drawn in succession, what is the probability that one of them is of the high-strength type and the other one is ordinary?

In this problem, the mutually exclusive events are

a. Drawing a high-strength bolt on the first trial and an ordinary one on the second.
b. Drawing an ordinary bolt on the first trial and a high-strength one on the second.

Now the probability of (a) is

$$\tfrac{30}{55} \times \tfrac{25}{54} = \tfrac{25}{99}$$

and the probability of (b) is

$$\tfrac{25}{55} \times \tfrac{30}{54} = \tfrac{25}{99}$$

The probability of either (a) or (b) is the sum of the two probabilities:

$$\tfrac{25}{99} + \tfrac{25}{99} = \tfrac{50}{99}$$

Earlier, we considered events that are mutually exclusive, but this, of course, need not always be the case. For instance, if we are interested in the proportion of people who are left-handed or myopic, we find in addition to those who have either of these attributes also those who have both. Suppose that the probability of being left-handed is $P(A)$ and the probability of being myopic is $P(B)$. Then the probability of being either left-handed or myopic, $P(A + B)$, is given by an addition expression which has to recognize that the group with both attributes must not be counted twice. Thus

$$P(A + B) = P(A) + P(B) - P(AB) \tag{5-12}$$

where $P(AB)$ is the probability of being both left-handed and myopic. A graphical representation of this situation is given in Fig. 5-1.

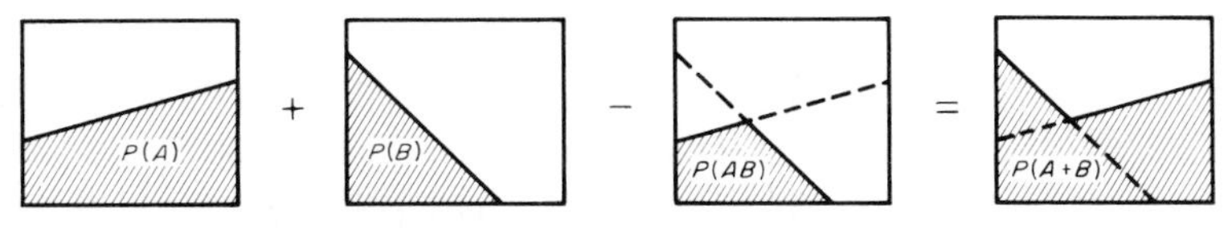

Fig. 5-1. Probability of occurrence of either or both of two independent events.

Example. If one card is selected at random from a deck of 52 cards, what is the probability that the card is a club or a face card or both? (Count an ace as a face card.)

The probability of a club, $P(A)$, is $\tfrac{13}{52}$. The probability of a face card, $P(B)$, is $\tfrac{16}{52}$. Since four of the clubs are also face cards, then $P(AB)$ is $\tfrac{4}{52}$. Thus the required probability is

$$P(A + B) = P(A) + P(B) - P(AB) = \tfrac{13}{52} + \tfrac{16}{52} - \tfrac{4}{52} = \tfrac{25}{52}$$

We should further recognize that events which are not mutually exclusive may be independent or *dependent*. In the latter case, the occurrence of an event affects the probability of other events. For instance, drawing a card from a pack (e.g., an ace) affects the probability of drawing any other given card (e.g., a queen) unless the first card is returned before the second draw. We are dealing, therefore, with *conditional probability*, which is the probability of an event A, given that the event B has occurred; such a conditional probability can be written as $P(A \mid B)$. Thus, in our notation,

$$P(A \mid B) = \frac{P(AB)}{P(B)} \tag{5-13}$$

A diagrammatic representation of the classification of various types of events as shown in Fig. 5-2 may be helpful.

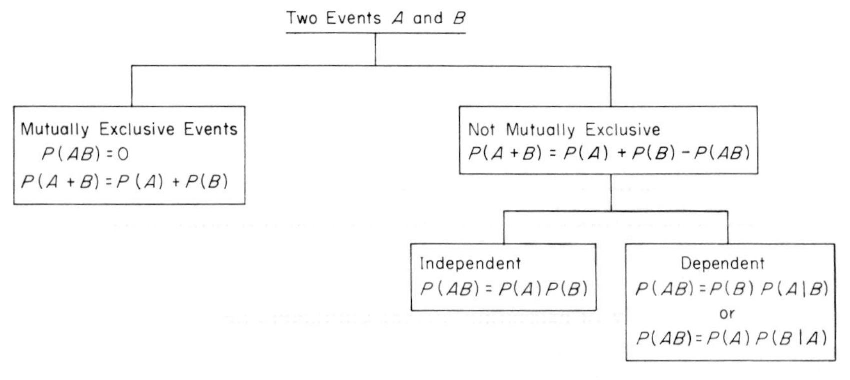

Fig. 5-2

We can use the definition of conditional probability to deduce the definition of probabilistic independence; thus two events A and B are said to be independent if and only if

$$P(A \mid B) = P(A) \qquad \text{or} \qquad P(B \mid A) = P(B) \tag{5-14}$$

Intuitively, if two events are unrelated, then the probability of one event's occurring is unaltered if the other event has occurred.

Example. A roll of two fair dice has shown the number 4 on one of the dice. We are not told what the other die is. What is the probability that the sum of numbers on both dice is 9?

There are 36 sample points. The number of points in which at least one 4 turns up is 11; thus the corresponding probability, $P(B)$, is $\frac{11}{36}$. Of the four points that result in the sum of 9, [(6, 3), (5, 4), (4, 5), and (3, 6)], there are two in which 4 appears. Therefore, $P(AB)$ is $\frac{2}{36}$.

Hence the required probability, $P(A \mid B)$, is

$$P(A \mid B) = \frac{P(AB)}{P(B)} = \frac{\frac{2}{36}}{\frac{11}{36}} = \frac{2}{11}.$$

Example. A milk delivery company operating a fleet of vans driven both by gasoline and by electric motors maintained annual records on motor overhauls. In Table 5-1 is listed the number of kilometers before an overhaul is necessary for each type of van.

TABLE 5-1

	Vans with		
Kilometers	*Gasoline motor*	*Electric motor*	*Total*
40,001 and over	10	20	30
20,001–40,000	60	50	110
0–20,000	30	10	40
Total	100	80	180

a. What is the probability that a motor (of whatever type) will exceed 40,000 kilometers?
b. What influence does the motor type have on this probability?

We find

a. The probability of exceeding 40,000 kilometers before an overhaul is necessary is $\frac{30}{180}$ or $\frac{1}{6}$.
b. Here we deal with conditional probability, since we wish to find the probability that a motor will exceed 40,000 kilometers prior to overhaul, given that the motor is either gasoline or electric.

For gasoline motors, the required value of probability is

$$P = \frac{\frac{10}{180}}{\frac{100}{180}} = 0.1$$

For electric motors, the value is

$$P = \frac{\frac{20}{180}}{\frac{80}{180}} = 0.25$$

Example. A shopping complex is to be designed to accommodate firms with different needs for hydro and water. The engineer is to design these utilities in such a manner that the provisions should not greatly exceed the actual demand; nor should the capacities be inadequate.

There are four classifications: for hydro either 10 or 20 units, namely, H_{10} and H_{20}; and for water 2 or 4 units, namely, W_2 and W_4. The owner of the complex, based on previous experience, has given the engineer the following estimates of the probability of encountering the four classifications (see Table 5-2).

TABLE 5-2

Event	*Probability*
$H_{10} W_2$	0.1
$H_{10} W_4$	0.2
$H_{20} W_2$	0.2
$H_{20} W_4$	0.5

Find the probability of

a. The water demand being 4 units.
b. The hydro demand being 20 units.
c. Either the water demand being 4 units or the hydro demand being 20 units.
d. A firm known to have a hydro demand of 10 units, also having a water demand of 4 units.

We find

a. Probability $= P[H_{10} W_4] + P[H_{20} W_4] = 0.2 + 0.5 = 0.7$.
b. Probability $= P[H_{20} W_2] + P[H_{20} W_4] = 0.2 + 0.5 = 0.7$.
c. Probability $= (a) + (b) - P[H_{20} W_4] = 0.7 + 0.7 - 0.5 = 0.9$.
d. $P[W_4 \mid H_{10}] = P[H_{10} W_4]/P[H_{10}] = 0.2/0.3 = 0.66$, which is the conditional probability that W_4 will occur, given the knowledge that H_{10} has occurred.

Example. A system S consists of components A and B. It functions 0.10 of the time; component A fails 0.15 of the time and component B fails 0.30 of the time. Check whether components A and B function independently of each other.

Let us denote the functioning of component A as (A) and its nonfunctioning as $(\bar{A})$, and similarly for component B. See Table 5-3.

TABLE 5-3

	(B)	*($\bar{B}$)*	*Total*
(A)	0.10	0.30	0.40
$(\bar{A})$	0.15	0.45	0.60
Total	0.25	0.75	1.00

Now

$$P(A) = P(AB) + P(A\bar{B}) = 0.10 + 0.30 = 0.40$$
$$P(B) = P(AB) + P(\bar{A}B) = 0.10 + 0.15 = 0.25$$

Thus

$$P(A) \cdot P(B) = (0.40)(0.25) = 0.10$$

which is equal to the value given for $P(AB)$. Therefore, the two components A and B function independently.

Also, from the definition of independence, we have

$$P(A \mid B) = \frac{P(AB)}{P(B)} = \frac{0.10}{0.25} = 0.40$$

which is equal to $P(A)$, and

$$P(B \mid A) = \frac{P(AB)}{P(A)} = \frac{0.10}{0.40} = 0.25$$

which is equal to $P(B)$, thus satisfying Eq. (5-14).

PROBABILITY DISTRIBUTIONS

In Chapter 1 we mentioned that there are two types of random variables, continuous variables and discrete variables. A variable is continuous if it can theoretically assume any value between two given limits; otherwise it is a discrete variable. In Chapter 2 we dealt with frequency distributions of data from random *samples*. Now we shall deal with probability distributions of *populations*. We should remember that a random sample is one that is chosen in such a manner that every individual in the population has an equal chance of being chosen for the sample.

Discrete Probability Distributions

If a random variable x can assume discrete values $x_0, x_1, x_2, \ldots, x_k$ with respective probabilities $p_0, p_1, p_2, \ldots, p_k$, where $p_i \geq 0$ for all i, and

$$\sum_{i=0}^{i=k} p_i = 1$$

then probability $(x = x_i)$, or simply $p(x_i) = p_i$, characterizes a discrete probability distribution for the variable x; see Fig. 5-3.

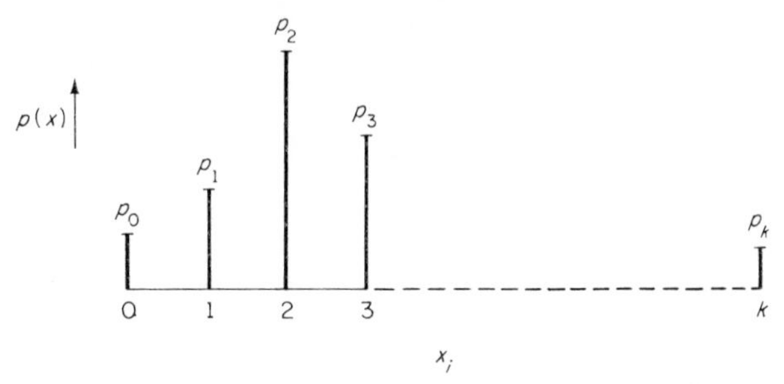

Fig. 5-3. Probability distribution of a discrete random variable x.

The cumulative distribution function of a discrete random variable is defined as

$$P(X) = \sum_{x_i \leq X} p_i \tag{5-15}$$

This function is a step function that is constant over every interval not containing any of the points x_i, as shown in Fig. 5-4.

It is interesting to note that $p(x)$ and $P(X)$ represent probabilities and cumulative probabilities, respectively, for a large parent *population* in contrast to f and F, which represent frequency and cumulative frequency, respectively (see Chapter 2) in a *sample* of grouped data.

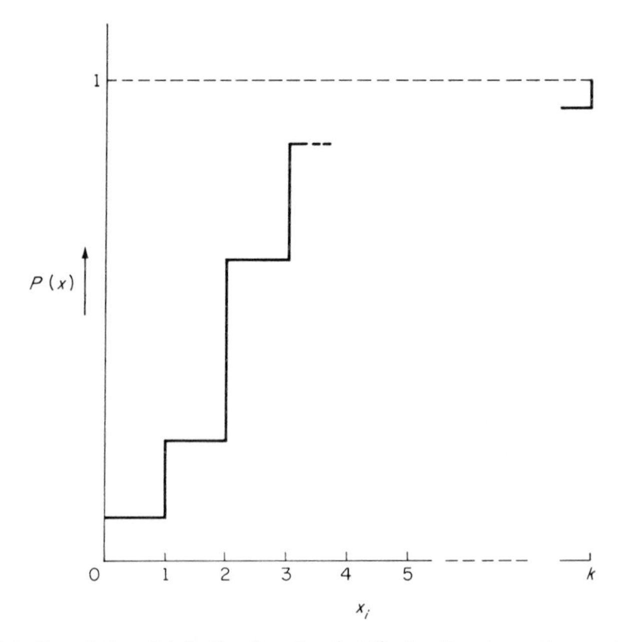

Fig. 5-4. Cumulative distribution function (c.d.f.) of a discrete random variable x.

Thus we can associate probability distributions with *population*, whereas the relative frequency distributions (Chapter 2) are distributions for *samples* drawn from this population; in other words, we can regard a probability distribution as a relative frequency distribution for an indefinitely *large sample* (population).

Example. Draw the probability distribution and cumulative probability distribution for a random variable x, defined as the sum of the upturned faces of two fair dice.

Now x can take the values 2, 3, 4, 5, ... 11, 12.

Probability of obtaining the sum of 2 is $(\frac{1}{6})(\frac{1}{6}) = \frac{1}{36} = p(2)$

Probability of obtaining the sum of 3 is $(\frac{1}{6})(\frac{1}{6}) + (\frac{1}{6})(\frac{1}{6}) = \frac{2}{36} = p(3)$

Probability of obtaining the sum of 4 is $(\frac{1}{6})(\frac{1}{6}) + (\frac{1}{6})(\frac{1}{6}) + (\frac{1}{6})(\frac{1}{6}) = \frac{3}{36} = p(4)$

Probability of obtaining the sum of 5 is $4(\frac{1}{6})(\frac{1}{6}) = \frac{4}{36} = p(5)$

Probability of obtaining the sum of 6 is $5(\frac{1}{6})(\frac{1}{6}) = \frac{5}{36} = p(6)$

Probability of obtaining the sum of 7 is $6(\frac{1}{6})(\frac{1}{6}) = \frac{6}{36} = p(7)$

Probability of obtaining the sum of 8 is $5(\frac{1}{6})(\frac{1}{6}) = \frac{5}{36} = p(8)$

$\vdots$

Probability of obtaining the sum of 12 is $(\frac{1}{6})(\frac{1}{6}) = \frac{1}{36} = p(12)$

These probabilities are plotted in Fig. 5-5.

The cumulative probability distribution follows:

The cumulative probability of obtaining the sum of 2 or less is $p(2) = \frac{1}{36}$.

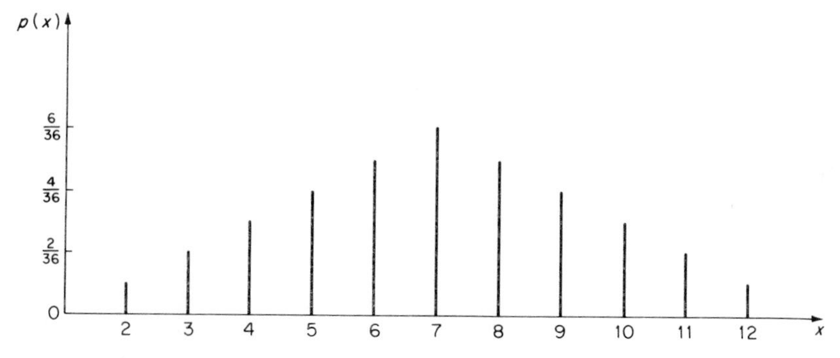

Fig. 5-5. Probability distribution of the sum of the upturned faces of two fair dice.

The cumulative probability of obtaining the sum of 3 or less is

$$p(2) + p(3) = \tfrac{1}{36} + \tfrac{2}{36} = \tfrac{3}{36}$$

$$\vdots$$

The cumulative probability of obtaining the sum of 12 or less is

$$p(2) + p(3) + p(4) + \cdots + p(12) = \tfrac{36}{36} = 1$$

A plot of the cumulative probability is shown in Fig. 5-6. See also Table 5-4.

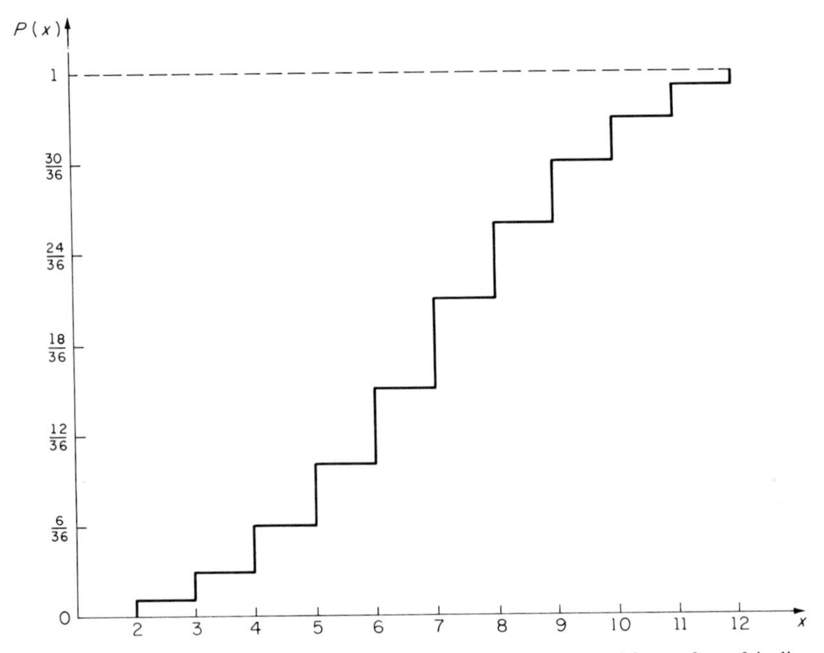

Fig. 5-6. Cumulative distribution function of the sum of the upturned faces of two fair dice.

TABLE 5-4

x	2	3	4	5	6	7	8	9	10	11	12
$p(x)$	$\frac{1}{36}$	$\frac{2}{36}$	$\frac{3}{36}$	$\frac{4}{36}$	$\frac{5}{36}$	$\frac{6}{36}$	$\frac{5}{36}$	$\frac{4}{36}$	$\frac{3}{36}$	$\frac{2}{36}$	$\frac{1}{36}$
$P(x)$	$\frac{1}{36}$	$\frac{3}{36}$	$\frac{6}{36}$	$\frac{10}{36}$	$\frac{15}{36}$	$\frac{21}{36}$	$\frac{26}{36}$	$\frac{30}{36}$	$\frac{33}{36}$	$\frac{35}{36}$	$\frac{36}{36}$ (= 1)

Note that

$$\sum_{x=2}^{x=12} p(x) = 1$$

Now if the two dice are "fair" the actual outcomes of the random variable in question should follow the laws of probability in the long run; in this case the probability is known a priori. However, in most statistical problems, the probability distribution of the random variable is not known and, therefore, we must use the outcomes of the random variable from a sample to make decisions about the unknown probability distribution of the population. The binomial distribution and Poisson's distribution discussed in Chapters 7 and 8 are examples of a discrete distribution.

Continuous Probability Distributions

In Chapter 2 we explained how data describing a continuous variable can be plotted as a histogram and then as a frequency polygon. We also mentioned that if the total number of observations is increased indefinitely and the class width correspondingly approaches zero, the histogram as well as the frequency polygon will approach a continuous curve, namely, a frequency distribution curve (or a relative frequency distribution curve), as shown in Fig. 5-7.

If the height of the frequency curve A is standardized so that the area under curve A is made equal to unity, then a continuous probability distrib-

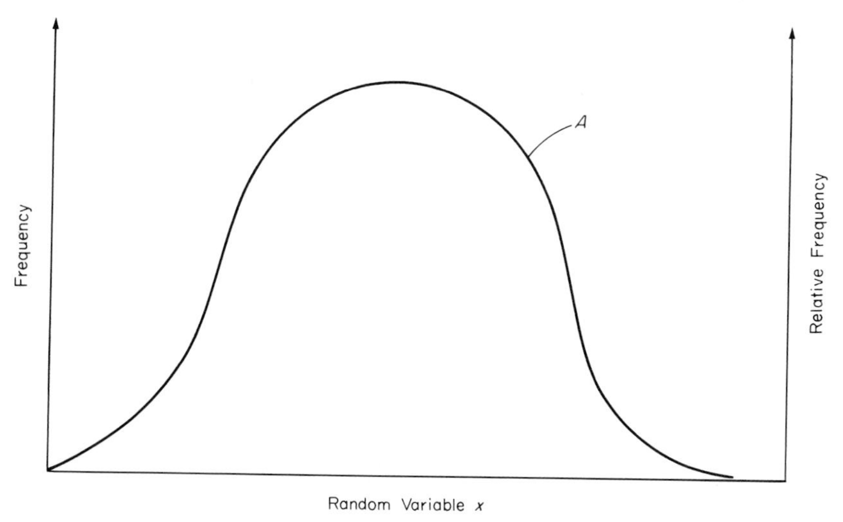

Fig. 5-7. Frequency curve and relative frequency curve of a random variable x.

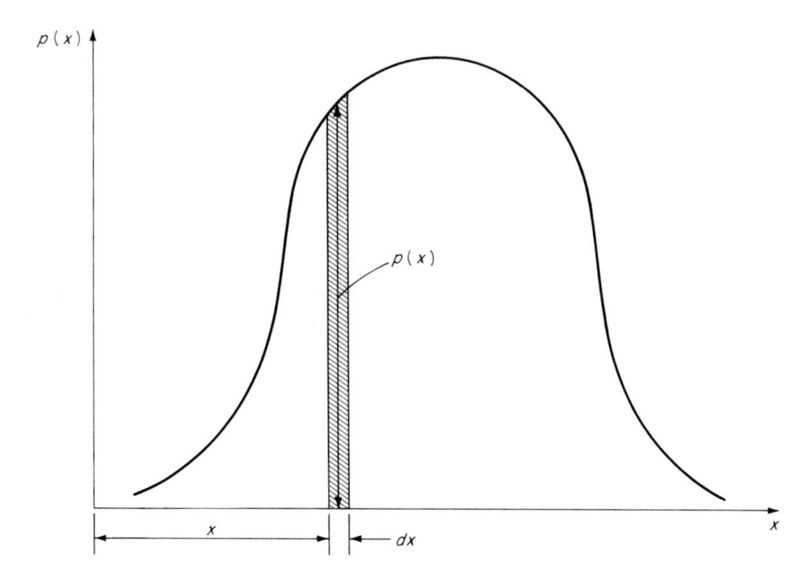

Fig. 5-8. Probability density function.

ution is determined, as shown in Fig. 5-8, where $p(x)$ is the probability density function such that

$$\int_{-\infty}^{+\infty} p(x)\,dx = 1$$

Thus the area under the curve and between lines $x = x_1$, and $x = x_2$ (the shaded area in Fig. 5-9) is the probability that x lies between x_1 and x_2, or

$$\text{probability } (x_1 < x < x_2) = \int_{x=x_1}^{x=x_2} p(x)\,dx \tag{5-16}$$

It is essential to understand that the probability of observing a specific value such as $x = x_3$ is zero, since the calculation of a probability entails the calculation of an area under the curve; in this case, dx for a continuous variable is equal to zero and hence the corresponding probability will also be zero. This is expressed mathematically as

$$\int_{x=x_3}^{x=x_3} p(x)\,dx = 0$$

Thus the proper definition of $p(x)$ is that for a small interval dx, $p(x)\,dx$ is the probability of observing a value between x and $x + dx$; i.e., it is the shaded area in Fig. 5-8.

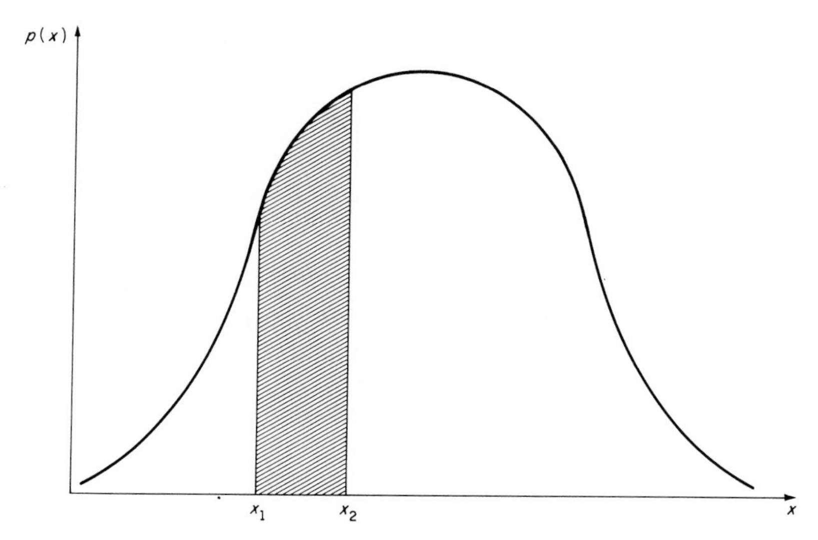

Fig. 5-9. Probability density function.

The cumulative distribution function, $P(x)$, for a continuous random variable x is defined as the probability of observing a value less than or equal to X, i.e., the shaded area in Fig. 5-10. Thus

$$P(X) = \int_{-\infty}^{X} p(x)\, dx \tag{5-17}$$

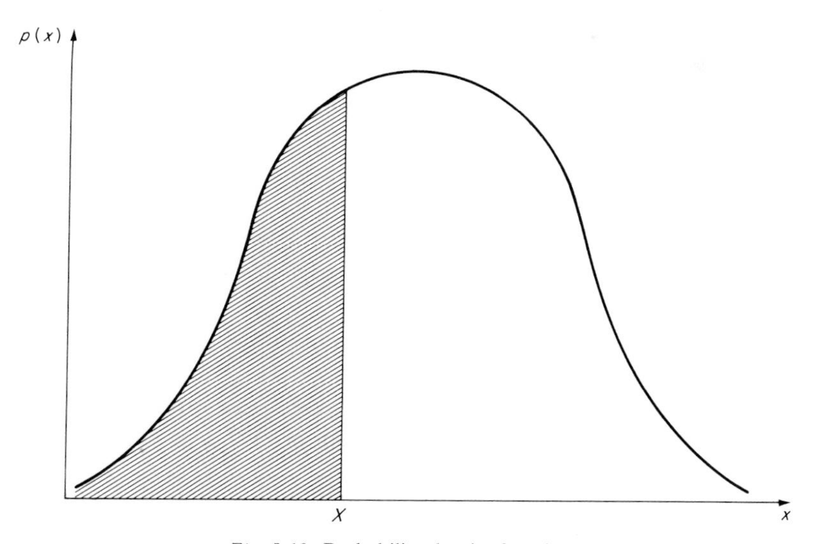

Fig. 5-10. Probability density function.

The probability that x lies between x_1 and x_2, Fig. 5-9, can now be written as

$$\text{probability } (x_1 < x < x_2) = \int_{x_1}^{x_2} p(x)\, dx = P(x_2) - P(x_1) \qquad (5\text{-}18)$$

The cumulative distribution function must increase from zero to one, since we have

$$P(-\infty) = 0 \qquad \text{and} \qquad P(\infty) = 1$$

and it is often an S-shaped curve. as shown in Fig. 5-11. Examples of continuous probability distributions are the normal, χ^2, t, and F distributions, which will be considered later on in the book.

We can calculate means, variances, and other parameters from a probability distribution in the same manner as we did in the case of a frequency distribution (see Chapters 3 and 4). This leads us to the subject of expectation.

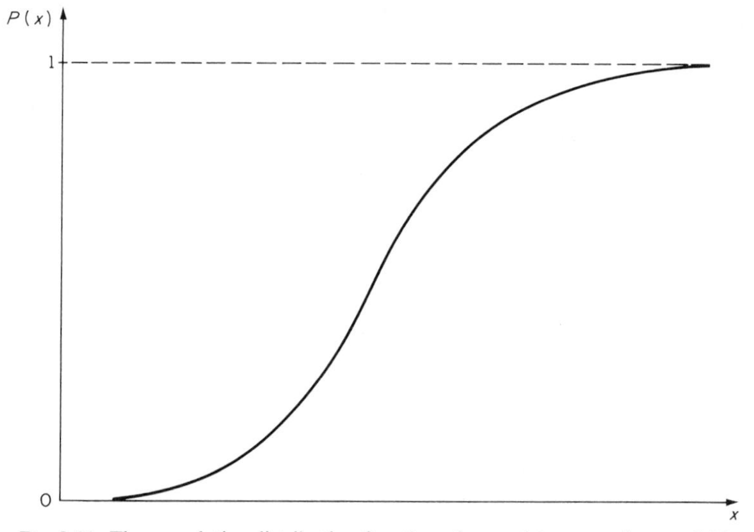

Fig. 5-11. The cumulative distribution function of a continuous random variable.

EXPECTATION

Consider a random variable x. If x is discrete with a probability distribution p, then the *expected value* of x (or mean of x) is defined as

$$E(x) = \sum_i p_i x_i \qquad (5\text{-}19)$$

where x_i is the value of the ith event and p_i is the probability of the ith event.

Now, if x is continuous with $p(x)$, the probability density function, then

$$E(x) = \int_{-\infty}^{+\infty} xp(x)\,dx \tag{5-20}$$

or, we can write

$$\mu = E(x)$$

where μ is the mean of the probability distribution. We must remember that our sample must be random; i.e., every member of the population being sampled has an equal probability of being selected in every trial.

Example. Find the mean μ for the example on the upturned faces of two fair dice.

Using Eq. (5-19), we obtain

$$E(x) = \mu = 2(\tfrac{1}{36}) + 3(\tfrac{2}{36}) + 4(\tfrac{3}{36}) + 5(\tfrac{4}{36}) + 6(\tfrac{5}{36}) + 7(\tfrac{6}{36}) + 8(\tfrac{5}{36})$$
$$+ 9(\tfrac{4}{36}) + 10(\tfrac{3}{36}) + 11(\tfrac{2}{36}) + 12(\tfrac{1}{36}) = \tfrac{252}{36} = 7$$

Example. The probability density function, $p(x)$, of the life of electric light bulbs, random variable x, is given as

$$p(x) = 0 \qquad \text{for } x < 0$$

and

$$p(x) = \frac{1}{800} e^{(-x/800)} \qquad \text{for } x \geq 0$$

as shown in Fig. 5-12. Using Eq. (5-20), we obtain

$$E(x) = \mu = \int_{-\infty}^{+\infty} xp(x)\,dx = \int_{0}^{+\infty} \frac{x}{800} e^{(-x/800)}\,dx$$
$$= 800 \text{ h}$$

(This definite integral can be solved by the use of standard integral tables or by integration by parts.)

The variance σ^2 of x is the *expected* value of the squared deviation from the population mean. Thus

$$\sigma^2 = E(x - \mu)^2 = \sum_i p_i(x_i - \mu)^2 \qquad \text{if } x \text{ is discrete} \tag{5-21}$$

and

$$\sigma^2 = E(x - \mu)^2 = \int_{-\infty}^{+\infty} (x - \mu)^2 p(x)\,dx \qquad \text{if } x \text{ is continuous} \tag{5-22}$$

It should be noted that the expressions for $E(x)$ and $E(x - \mu)^2$ are, in fact, the first moment of the probability distribution about the origin and the second moment about the mean, respectively.

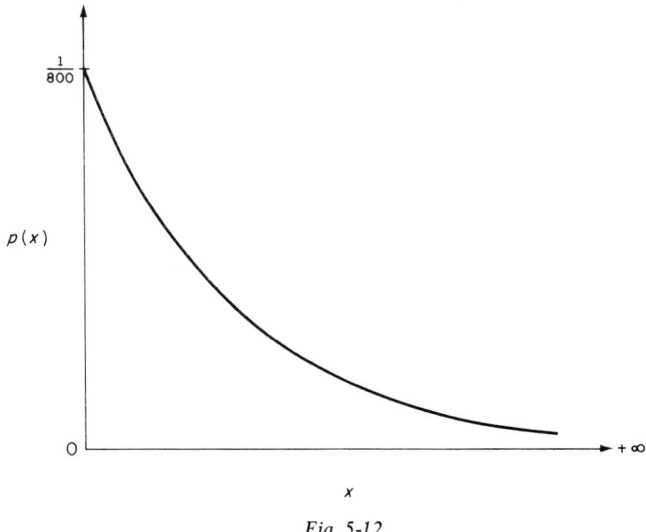

Fig. 5-12

Solved Problems

5-1. How many squads of 6 men can be selected from 60 men?

SOLUTION

This is a problem in combinations. Hence

$$_{60}C_6 = \frac{60!}{6!\,54!} = \frac{60 \times 59 \times 58 \times 57 \times 56 \times 55}{2 \times 3 \times 4 \times 5 \times 6} = 50{,}063{,}860 \text{ squads}$$

5-2. How many straight lines are determined by 10 points, no 3 of which are in the same straight line?

SOLUTION

The number of combinations of 2 points for a straight line from 10 points is

$$_{10}C_2 = \frac{10!}{2!\,8!} = \frac{10 \times 9}{2} = 45 \text{ lines}$$

5-3. From 10 items, in how many ways can a selection of 6 be made:

a. When a specified item is always included?
b. When a specified item is always excluded?

SOLUTION

a. When a specified item is always included in the selection of 6 items, the problem is to find the number of combinations of 5 items from the remaining 9 items. Hence

$$_9C_5 = \frac{9!}{5!4!} = \frac{9 \times 8 \times 7 \times 6}{2 \times 3 \times 4} = 126 \text{ selections of 6 items}$$

b. When a specified item is always excluded, the problem is to find the number of combinations of 6 items from the 9 remaining items. Hence

$$_9C_6 = \frac{9!}{6!3!} = \frac{9 \times 8 \times 7}{2 \times 3} = 84 \text{ selections of 6 items}$$

5-4. A committee of 7 is to be chosen from 8 Canadians and 5 Americans. In how many ways can a committee be chosen if it is to contain:

a. Just four Canadians?
b. At least four Canadians?

SOLUTION

a. To choose just 4 Canadians from 8, we have $_8C_4$ ways. Since there are only 3 positions remaining to be filled by Americans, the number of ways in which 3 Americans can be selected from 5 is $_5C_3$.
Then the number of ways of selecting a committee of 7 is

$$_8C_4 \times {_5C_3} = \frac{8 \times 7 \times 6 \times 5}{2 \times 3 \times 4} \times \frac{5 \times 4}{2} = 700 \text{ ways}$$

b. When the committee is to include at least 4 Canadians, there are several possibilities, as shown in Table 5-5.

TABLE 5-5

Americans	*Canadians*	*Number of ways*
3	4	$_5C_3 \times {_8C_4} = 700$
2	5	$_5C_2 \times {_8C_5} = 560$
1	6	$_5C_1 \times {_8C_6} = 140$
0	7	$_5C_0 \times {_8C_7} = 8$
		Total = 1408 ways

5-5. If $_nP_r = 110$ and $_nC_r = 55$, find n and r.

SOLUTION

$$_nP_r = \frac{n!}{(n-r)!}$$

and

$$_nC_r = \frac{n!}{r!(n-r)!}$$

Therefore,

$$\frac{_nP_r}{_nC_r} = r! = 2$$

Hence $r = 2$. Now

$$_nP_r = \frac{n!}{(n-r)!} = \frac{n!}{(n-2)!} = 110$$

But

$$\frac{n!}{(n-2)!} = n(n-1) = 110$$

Therefore,

$$n^2 - n - 110 = 0 \qquad \text{or} \qquad (n-11)(n+10) = 0$$

Hence

$$n = -10 \text{ (not possible)} \qquad \text{or} \qquad n = 11$$

Thus

$$r = 2 \qquad \text{and} \qquad n = 11$$

5-6.

a. In how many ways can 6 soldiers stand in a line so that two soldiers in particular will not be next to one another?

b. Show that

$$(1) \quad _nC_r + {}_nC_{r-1} = {}_{n+1}C_r$$

and

$$(2) \quad _{n+2}C_{r+1} = {}_nC_{r+1} + 2\,{}_nC_r + {}_nC_{r-1}$$

c. A university senate is composed of 50 staff members of whom 6 are engineers. In how many ways can a committee of 10 be chosen so as to contain at least 4 engineers?

SOLUTION

a. Let us consider 5 soldiers. These can be arranged in $n! = 5!$ ways. For each arrangement of these 5 soldiers, the sixth one can stand in four different locations without being adjacent to the particular soldier in question (see Fig. 5-13). Therefore,

$$\text{total number of ways} = 4 \times 5! = 480$$

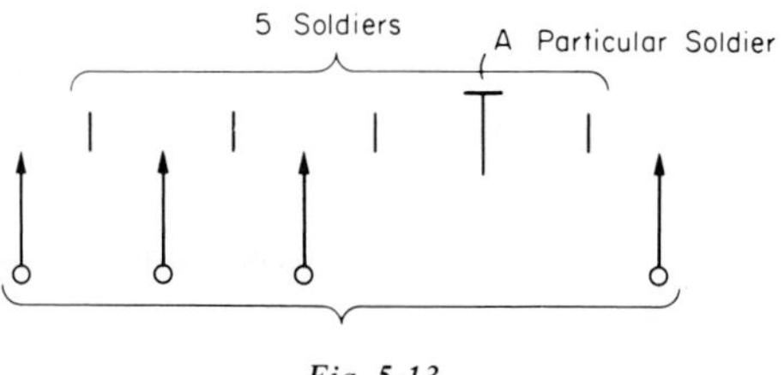

Fig. 5-13

b. (1) Expanding the left-hand side, we have

$$ {}_nC_r + {}_nC_{r-1} = \frac{n!}{r!\,(n-r)!} + \frac{n!}{(r-1)!\,(n-r+1)!} $$
$$ = \frac{n!\,(n-r+1)}{r!\,(n-r+1)!} + \frac{n!\,r}{r!\,(n-r+1)!} $$
$$ = \frac{n!\,(n+1)}{r!\,(n-r+1)!} $$

Therefore,

$$ {}_nC_r + {}_nC_{r-1} = \frac{(n+1)!}{r!\,(n-r+1)!} $$

But

$$ {}_{n+1}C_r = \frac{(n+1)!}{r!\,(n+1-r)!} $$

Thus

$$ {}_nC_r + {}_nC_{r-1} = {}_{n+1}C_r $$

(2) Expanding the right-hand side of the given identity, we have

$$ {}_nC_{r+1} + 2{}_nC_r + {}_nC_{r-1} $$
$$ = \frac{n!}{(r+1)!\,(n-r-1)!} $$
$$ + \frac{2n!}{r!\,(n-r)!} + \frac{n!}{(r-1)!\,(n-r+1)!} $$
$$ = \frac{n!\,[(n-r)(n-r+1) + 2(r+1)(n-r+1) + r(r+1)]}{(r+1)!\,(n-r+1)!} $$
$$ = \frac{n!\,(n+1)(n+2)}{(r+1)!\,(n-r+1)!} $$
$$ = \frac{(n+2)!}{(r+1)!\,(n-r+1)!} $$

But

$$_{n+2}C_{r+1} = \frac{(n+2)!}{(r+1)!\,(n+2-r-1)!} = \frac{(n+2)!}{(r+1)!\,(n-r+1)!}$$

Thus the two sides of the identity are equivalent.

c. A committee of 10 can be chosen so that there are either 4, 5, or 6 engineers on it. Hence the number of possible selections is

$$({}_6C_4 \times {}_{44}C_6) + ({}_6C_5 \times {}_{44}C_5) + ({}_6C_6 \times {}_{44}C_4)$$

$$= \frac{6!}{4!\,2!} \times \frac{44!}{6!\,38!} + \frac{6!}{5!\,1!} \times \frac{44!}{5!\,39!} + \frac{6!}{6!\,0!} \times \frac{44!}{4!\,40!}$$

$$= 112{,}537{,}579 \text{ ways (an adequate number even for academics)}$$

5-7. A machine contains a component A that is vital to its operation. The reliability of component A is 80 percent. To improve the reliability of the machine, a similar component is used in parallel to form system S, as shown in Fig. 5-14. The machine

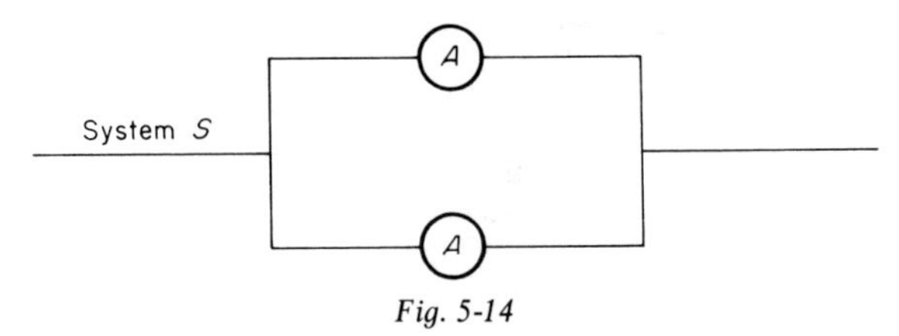

Fig. 5-14

will work, provided that one of these components functions correctly. Calculate the reliability of the system S.

SOLUTION

Denote the probability of success of A as $P(A)$ and the probability of its failure as $P(\bar{A})$. Given $P(A) = 0.8$, we have $P(\bar{A}) = 0.2$. The probability of both components A failing simultaneously is $P(\bar{A}) \cdot P(\bar{A}) = 0.04$. Therefore, the probability of at least one component A functioning properly is $1 - 0.04 = 0.96$, which is the reliability of the system S.

5-8. An experimental fighter aircraft is estimated to have a probability of 0.85 for a successful first flight. In case of failure there is a 0.05 probability of a catastrophic explosion, in which case the abort system cannot be used. The abort system has a reliability of 0.95. Calculate the probability of every possible outcome of the initial flight.

SOLUTION

Let us define the events: flight successes as A; flight failure as $\bar{A}$; noncatastrophic failure as B; catastrophic failure as $\bar{B}$; successful abort as C; abort failure as $\bar{C}$; pilot survives as D; pilot is killed as $\bar{D}$.

From the nature of the problem, we note that A is mutually exclusive with B, $\bar{B}$, C, $\bar{C}$, and $\bar{D}$, since a successful flight cannot be associated with catastrophic or noncatastrophic failure, abort success or failure, or with the pilot's being killed. Similarly, $\bar{B}$ is mutually exclusive with C, $\bar{C}$, and D. Also, C is mutually exclusive with $\bar{D}$.

Now $P(A) = 0.85$ and hence $P(\bar{A}) = 1 - 0.85 = 0.15$, since events A and $\bar{A}$ are complementary.

$$P(\bar{A}\bar{B}) = P(\bar{A})P(\bar{B} \mid \bar{A})$$

But a catastrophic failure ($\bar{B}$) can only occur if a flight failure ($\bar{A}$) occurs. Thus we can write

$$\begin{aligned} P(\bar{B}) = P(\bar{A}\bar{B}) &= P(\bar{A})P(\bar{B} \mid \bar{A}) \\ &= (0.15)(0.05) \\ &= 0.0075 \end{aligned}$$

Similarly,

$$\begin{aligned} P(B) = P(\bar{A}B) &= P(\bar{A})P(B \mid \bar{A}) \\ &= (0.15)(0.95) \\ &= 0.1425 \end{aligned}$$

$$\begin{aligned} P(C) = P(CB) &= P(B)P(C \mid B) \\ &= (0.1425)(0.95) \\ &= 0.1354 \end{aligned}$$

$$\begin{aligned} P(D) &= P(A) + P(C) \\ &= 0.85 + 0.1354 \\ &= 0.9854 \end{aligned}$$

$$P(\bar{D}) = 1 - P(D) = 0.0146$$

(since events D and $\bar{D}$ are complementary)

$$\begin{aligned} P(\bar{C}) = P(\bar{C}B) &= P(B)P(\bar{C} \mid B) \\ &= (0.1425)(0.05) \\ &= 0.007125 \end{aligned}$$

To check our solution, we see that

$$P(\bar{D}) = P(\bar{B}) + P(\bar{C})$$

$$0.0146 = 0.0075 + 0.0071$$

Problems

5-1. From a box containing 20 balls, one-half of them white, one-half black, four balls are drawn at random. What is the probability of obtaining (a) all of them of the same color; (b) all of them black; (c) all of them black if each ball is replaced before the next one is drawn?

5-2. In dealing a pack of 52 cards to four players, what is the probability of one of them obtaining 13 cards of a given suit?

5-3. If one-quarter of the dancers are eliminated after each dance, the elimination being random, what is the mathematical probability of successfully completing 5 consecutive dances?

5-4. A fair die is tossed twice. Find (a) the probability of a 3 turning up at least once, (b) the probability of getting a 3, 4, or 5 on the first toss and 1, 2, 3, or 6 on the second toss.

5-5. In a classroom it was required to seat 7 men students and 6 women students in a row so that the women occupy the even places. How many such arrangements are possible?

5-6. In the "football pools" operated in England one has to predict the result of 14 matches, there being three possible results of each match (home win, away win, and draw). How many entries would you have to submit in order to make sure that the winning entry is included?

5-7. A system has two components, A and B. If the probability of A's failing is 0.7 and the probability of B's failing is 0.8, what is the probability of (a) the system's remaining sound; (b) both components failing; (c) either component's failing?

5-8. Two machines have components A and B arranged in the systems shown in Figs. 5-15 and 5-16. The reliabilities of correct functioning of components A and B, are 0.7 and 0.8, respectively. Assuming that A and B function independently of each other, determine the reliability of each system if

a. Both components must function correctly for system 1 to function correctly (see Fig. 5-15).

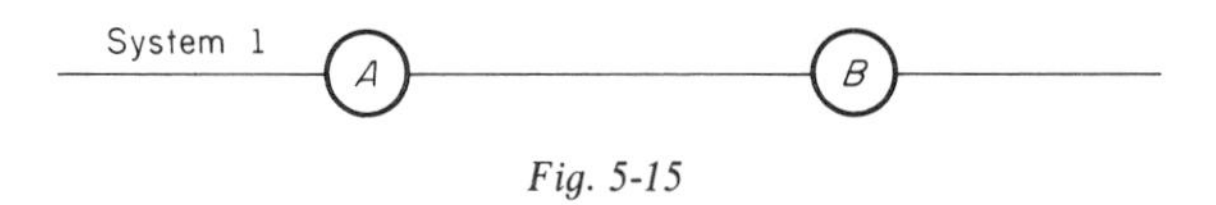

Fig. 5-15

b. The components are connected in parallel in such a way that, if either link A–B functions correctly, system 2 functions correctly (see Fig. 5-16).

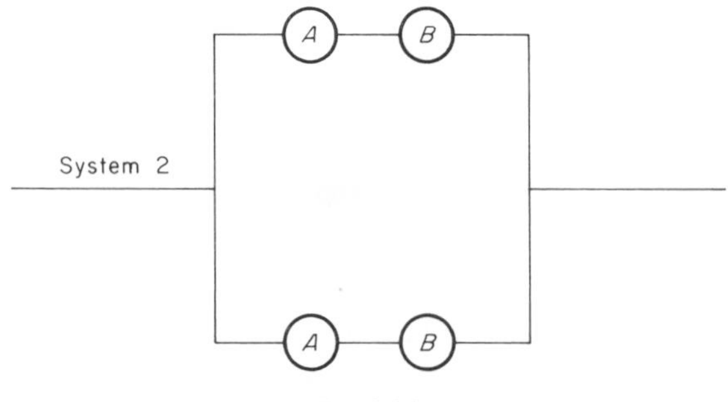

Fig. 5-16

5-9. The probability that a missile propulsion system will function is 0.90 and the probability that its guidance system will work is 0.80. What is the probability of a successful mission?

5-10. The chance that a certain old bridge will fail is 0.002. A new bridge will cost $50,000. If losses for the bridge failure will amount to $5 million, determine whether the new bridge should be built.

5-11. A system consists of components A and B; it functions 0.06 of the time. Experience has shown that component A fails 0.14 of the time and component B fails 0.24 of the time. Determine whether components A and B perform independently of each other.

5-12. A loaded die was tossed. The toss values 1, 2, 3, 4, 5, and 6 occurred with probabilities $\frac{1}{4}$, $\frac{1}{3}$, $\frac{1}{12}$, $\frac{1}{12}$, p, and $\frac{1}{6}$, respectively. (a) Calculate the value p. (b) Find probability $(2 \leq x < 5)$. (c) Compute the cumulative distribution function.

5-13. A machine produces daily either 0, 1, 2, or 3 defective items with probabilities $\frac{1}{6}$, $\frac{1}{2}$, $\frac{2}{3}$, $\frac{1}{6}$, respectively. Calculate the mean value and the variance of the defective items turned out.

5-14. A probability distribution that approximates the number of defective rivets in an airplane is given by:

$$\text{probability } (x = c) = \frac{2^c e^{-2}}{c!} \qquad \text{for } c = 0, 1, 2, 3, \ldots$$

Calculate the probability that the number of defective rivets is less than or equal to 3. (Note that $0! = 1$.)

5-15. A certain tube has a life (in hours) with a probability density $p(x) = c/x^3$ where $x \geq 600$, and $p(x) = 0$, where $x < 600$. Find an expression for the cumulative distribution function of the tube life. Calculate the probability that a tube will last at least 1000 h. Calculate the mean of the tube life in hours.

Chapter 6

Samples: Accuracy of the Mean

The previous chapter has interrupted our considerations of distributions, but the concepts introduced in it are necessary for the calculations to follow.

SAMPLING

We shall now consider more fully some of the properties of samples. Sampling is resorted to because testing the entire population is rarely practicable, especially in destructive tests or in tests involving a large expense or effort. But the sample must be representative of the population from which it is drawn and must, therefore, be drawn in a random manner—which means that every member of the population has an equal chance of being drawn in every trial.[1] How this is done in practice depends on the problem in hand, but the requirement of randomness should always be borne in mind.

1. The qualification "in every trial" precludes the use of methods such as choosing randomly a name in a list, and then taking every tenth name thereafter.

We use the word *sample* in a very broad sense, meaning the observations (of any type) that have been made, while the term *population* refers to all the observations that could be made. Samples are tested for the purpose of making inferences about properties of the population, and the investigator must be clear as to what population he is interested in; e.g., a particular batch of concrete, or all concrete placed in a given day, or all concrete in a given structure.

Tests on a sample can give results that are beyond doubt only so far as the sample itself is concerned. With respect to the underlying population a sampling investigation can give results only in terms of probability, and the confidence that we can have in our answers depends on the size of the sample.

There are two kinds of sampling—with and without replacement. In studying a particular property of a population, we may choose several items for measurement. If each item is chosen, measured, and then returned to the population, the sampling is said to be *with replacement*, since after each measurement the population remains the same as before and an item that has already been measured might be selected again. If each item is chosen one by one to form the sample and then measured, the sampling is said to be *without replacement*. Sampling with replacement is quite frequent; however, sampling without replacement is sometimes necessary, since the chosen items are destroyed in the process of measurement, e.g., the testing of a concrete cylinder for strength. In deriving the equations in this chapter, it is assumed that sampling is without replacement.

MEAN OF SAMPLE MEANS

In Chapter 4 we showed how an estimate of the population variance can be made from the sample variance; we shall now consider the determination of the population mean from the sampling mean.

A problem that sometimes arises in the collection of data in practice is to decide how many observations to make.

Specifically, if we are taking measurements of what is presumed to be the same unique quantity, and if we are satisfied to take the mean of the observations as the appropriate single measure, how is the accuracy increased, if at all, as the number of observations is increased? For example, suppose that a certain length has been measured by four people who have found it to be: 30.45 m, 30.20 m, 30.45 m, and 30.10 m. The mean is 30.30 m. Suppose that five more measurements are made by equally skilled people: 30.25 m, 30.15 m, 30.20 m, 30.40 m, and 30.05 m. The mean of all 9 measurements is 30.25 m instead of 30.30 m for the first four measurements. We can now ask: Is 30.25 m a more accurate result than 30.30 m? If so, how much more accurate can we expect it to be?

We should intuitively expect a larger sample to yield a more accurate result, but in order to have a quantitative appreciation, we have to determine how the representative nature of a sample improves with the sample size.

Suppose that we have a population of N observations from which we draw samples, each of n observations. We find the mean of each sample:

$$\bar{x} = \frac{x_1 + x_2 + \cdots + x_n}{n}$$

The number of different samples that can be drawn is the number of combinations of n elements from a population of N, i.e.,

$${}_N C_n = \frac{N!}{n!(N-n)!}$$

We shall now show that the mean of the means of all the samples is equal to the mean of the original population. Now, since there are $N!/[n!(N-n)!]$ samples, the mean of the means of all the samples is

$$\bar{\bar{x}} = \frac{\bar{x}_1 + \bar{x}_2 + \bar{x}_3 + \cdots}{\dfrac{N!}{n!(N-n)!}}$$

That is,

$$\bar{\bar{x}} = \frac{\dfrac{(x_1 + x_2 + \cdots)}{n} + \dfrac{(x_1 + x_3 + \cdots)}{n} + \dfrac{(x_2 + x_3 + \cdots)}{n} + \cdots}{\dfrac{N!}{n!(N-n)!}} \tag{6-1}$$

Consider all samples containing x_1. We obtain these by removing x_1 from the original N observations and selecting $n-1$ observations out of the remaining $N-1$. There are $(N-1)!/[(n-1)!(N-n)!]$ ways of doing this, and this will be the number of samples that contain x_1. Collecting all the terms of Eq. (6-1) containing x_1, we obtain the coefficient of x_1:

$$\frac{(N-1)!/[(n-1)!(N-n)!] \times 1/n}{N!/[n!(N-n)!]} = \frac{1}{N}$$

Similarly, the coefficient of $x_2 = 1/N$, and the coefficient of $x_3 = 1/N$, and so on. Hence the mean of all the means is

$$\begin{aligned} \bar{\bar{x}} &= \frac{x_1}{N} + \frac{x_2}{N} + \frac{x_3}{N} + \cdots + \frac{x_N}{N} \\ &= \frac{x_1 + x_2 + x_3 + \cdots + x_N}{N} \\ &= \text{mean of the original population} \end{aligned}$$

This relation is also true for sampling with replacement. In this case, the number of samples would be N^n.

Example. Consider observations: 1, 3, 4, 5, 12. Then $N = 5$. Take samples of 2 (i.e., $n = 2$) and verify that the mean of all the sample means = mean of the population. (See Table 6-1.)

TABLE 6-1

Sample	*Mean of sample*	*Number of samples*
1, 3	2	
1, 4	2.5	$N - 1 = 4$
1, 5	3	
1, 12	6.5	
3, 4	3.5	
3, 5	4	$N - 2 = 3$
3, 12	7.5	
4, 5	4.5	$N - 3 = 2$
4, 12	8	
5, 12	8.5	$N - 4 = 1$
Total		$= 10$

$$\text{number of different samples} = \frac{N!}{n!(N-n)!} = \frac{5!}{2!3!} = 10$$

mean of all sample means $= \bar{\bar{x}}$

$$= (2 + 2.5 + 3 + 6.5 + 3.5 + 4 + 7.5 + 4.5 + 8 + 8.5) \times \frac{1}{10} = 5$$

$$\text{mean of original population} = \mu = (1 + 3 + 4 + 5 + 12) \times \frac{1}{5} = 5$$

Thus $\bar{\bar{x}} = \mu$; i.e., mean of all sample means = mean of original population.

DISTRIBUTION OF SAMPLE MEANS

So far we have established the value of the mean of all the sample means, but we still know nothing about how these sample means vary one from the other. We shall, therefore, now consider the distribution of the sample means, and also the relation between the standard deviation of the means and the standard deviation of the original population.

As before, we describe the population mean and standard deviation by μ and σ, respectively. The number of different samples was shown to be

$$\frac{N!}{n!\,(N-n)!}$$

Therefore, using Eq. (4-1), the variance of the means of samples of size n drawn from a population of N is

$$\sigma_{\bar{x}}^2 = \frac{\sum (\bar{x}-\mu)^2}{N!/[n!\,(N-n)!]} \tag{6-2}$$

or, using the form of Eq. (4-4),

$$\sigma_{\bar{x}}^2 = \frac{\sum \bar{x}^2}{N!/[n!\,(N-n)!]} - \mu^2 \tag{6-3}$$

In each case the summation extends over all, i.e., $N!/[n!\,(N-n)!]$ samples. Now, the sum of squares of the sample means is

$$\sum \bar{x}^2 = \left(\frac{x_1 + x_2 + \cdots}{n}\right)^2 + \left(\frac{x_1 + x_3 + \cdots}{n}\right)^2 + \left(\frac{x_2 + x_3 + \cdots}{n}\right)^2 + \cdots$$

Expanding the right-hand side,

$$\sum \bar{x}^2 = A(x_1^2 + x_2^2 + \cdots + x_N^2) + B(x_1 x_2 + x_1 x_3 + x_2 x_3 + \cdots) \tag{6-4}$$

To obtain A, we have to determine the coefficient of x_1^2. The number of samples containing x_1 was shown to be $(N-1)!/[(n-1)!\,(N-n)!]$, and the number of samples containing x_1^2 is the same. Therefore, the coefficient of x_1^2 is

$$A = \frac{1}{n^2} \times \frac{(N-1)!}{(n-1)!\,(N-n)!} = \frac{(N-1)!}{n(n!)(N-n)!}$$

To obtain B in Eq. (6-4), we find the coefficient of $x_1 x_2$. The number of samples containing x_1 and x_2 is $(N-2)!/[(n-2)!\,(N-n)!]$. Hence the coefficient of $x_1 x_2$ is

$$B = \frac{2}{n^2} \times \frac{(N-2)!}{(n-2)!\,(N-n)!} = \frac{2(n-1)(N-2)!}{n(n!)(N-n)!}$$

Therefore, the sum of the squares of sample means is

$$\sum \bar{x}^2 = \frac{(N-1)!}{n(n!)(N-n)!}(x_1^2 + x_2^2 + \cdots + x_N^2) + \frac{2(n-1)(N-2)!}{n(n!)(N-n)!}(x_1 x_2 + x_1 x_3 + x_2 x_3 + \cdots)$$

Also,

$$\mu^2 = \left(\frac{x_1 + x_2 + \cdots + x_N}{N}\right)^2$$

$$= \frac{x_1^2 + x_2^2 + \cdots + x_N^2}{N^2} + \frac{2}{N^2}(x_1 x_2 + x_1 x_3 + x_2 x_3 + \cdots)$$

Substituting in Eq. (6-3), we have

$$\sigma_{\bar{x}}^2 = (x_1^2 + x_2^2 + \cdots + x_N^2)\left(\frac{1}{Nn} - \frac{1}{N^2}\right) + (x_1 x_2 + x_1 x_3 + \cdots)\left(\frac{2(n-1)}{nN(N-1)} - \frac{2}{N^2}\right)$$

or

$$\sigma_{\bar{x}}^2 = (x_1^2 + x_2^2 + \cdots + x_N^2)\left(\frac{N-n}{N^2 n}\right) - 2(x_1 x_2 + x_1 x_3 + \cdots)\left(\frac{N-n}{nN^2(N-1)}\right) \qquad (6\text{-}5)$$

We may remember that our intention is to compare the variance of the sample means with the variance of the original population. Using Eq. (4-4), the latter variance can be written

$$\sigma^2 = \frac{x_1^2 + x_2^2 + \cdots + x_N^2}{N} - \frac{(x_1 + x_2 + \cdots + x_N)^2}{N^2}$$

or

$$\sigma^2 = \frac{N-1}{N^2}(x_1^2 + x_2^2 + \cdots + x_N^2) - \frac{2}{N^2}(x_1 x_2 + x_1 x_3 + \cdots) \qquad (6\text{-}6)$$

STANDARD DEVIATION OF THE MEAN

Comparing Eqs. (6-5) and (6-6), we obtain

$$\sigma_{\bar{x}}^2 = \frac{N-n}{n(N-1)}\sigma^2 \qquad (6\text{-}7)$$

If, as is usually the case, N is very large (i.e., $N \gg n$), then

$$\frac{N-n}{N-1} \to 1$$

Thus Eq. (6-7) becomes

$$\sigma_{\bar{x}}^2 = \frac{\sigma^2}{n}$$

or, in terms of standard deviations,

$$\sigma_{\bar{x}} = \frac{\sigma}{\sqrt{n}} \tag{6-8}$$

This equation is of considerable importance.[2] It should be noted that Eq. (6-8) applies also to sampling *with replacement* for both finite and infinite population.

In many cases, σ is not known but is estimated from the sample. Such an estimate is denoted by s and is more precise the larger the sample. Using this estimate, we can estimate the standard deviation of the mean to be

$$s_{\bar{x}} = \frac{s}{\sqrt{n}} \tag{6-9}$$

This shows that the standard deviation of the sample means varies inversely as the square root of the sample size. Since the standard deviation of the mean is a measure of the scatter of the sample means, it affords a measure of the precision that we can expect of a mean of one sample. For this reason $\sigma_{\bar{x}}$ is often called the *standard error of the mean.* We may observe that a sample of 16 observations is only twice as precise as a sample of 4 so that the gain in precision is small relative to the effort in taking the additional 12 observations. The argument cannot, however, be used too far because a sample of 2 is only $\sqrt{2}/2$ as precise as a sample of 4; here a doubling of the sample size may be well worthwhile.

It is further obvious that a sample of one tells us nothing about the precision of the estimated mean, as s in Eq. (6-9) cannot be estimated.[3] With a number of samples of unit size the standard deviation could be estimated, but $s_{\bar{x}}$ would be no smaller than the standard deviation of the underlying distribution.

Figure 6-1 shows a comparison between the (probability) distribution of individual observations and the (probability) distribution of means of samples drawn from this underlying distribution. The considerably narrower distribution of the sample means is apparent; this indicates that the

2. It may be relevant to note that the standard deviation of the median is $1.25\sigma/\sqrt{n}$. Thus we can see that, for the same precision of the estimate, the sample size in the case of an estimate of the median has to be $(1.25)^2$ times greater than in the case of the mean. We can say, therefore, that the median is less efficient than the mean as a measure of central tendency.

3. The calculation of s [Eq. (4-3)] involves $n - 1$ in the denominator.

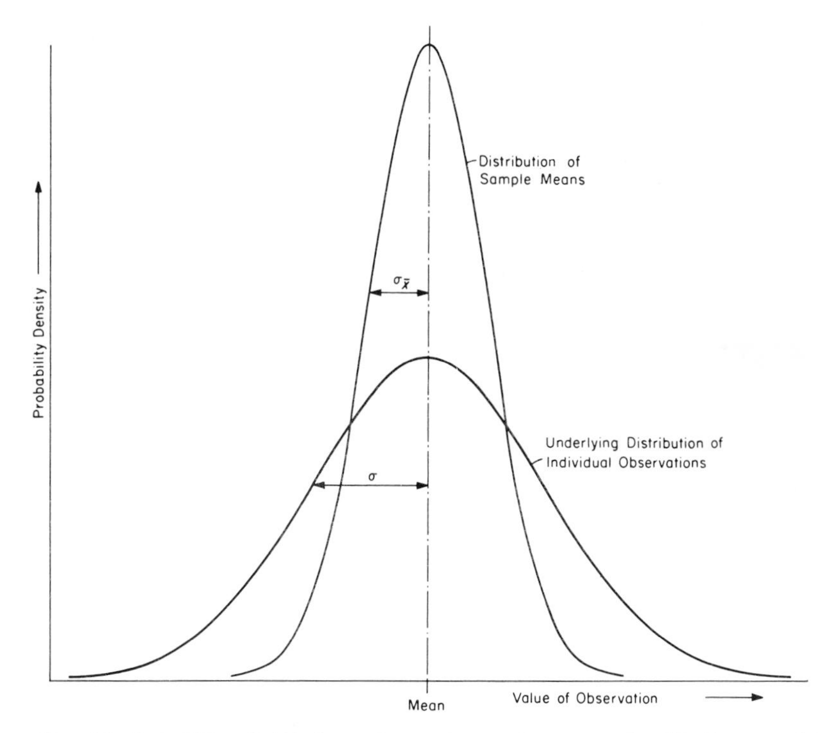

Fig. 6-1. Probability distribution of sample means compared with the underlying distribution; sample size = 4.

mean is an efficient measure of central tendency.[4] The higher peak of the curve for the means is due to the fact that the areas under the two curves are equal to one another and to unity, as each represents the sum of all probabilities.

It may be interesting to note that means of samples are approximately normally distributed even when the underlying distribution is not normal; normal distribution is defined in Chapter 9, and the problem of distribution of means is considered more fully in Chapter 10.

We can now answer, at least in part, the question posed at the beginning of this chapter on the precision of length measurements.

Let us refer to the sample consisting of the first four measurements as sample 1, and to the sample consisting of all nine measurements as sample 2.

We thus have

$$\bar{x}_1 = 30.30 \text{ m} \qquad \text{and} \qquad \bar{x}_2 = 30.25 \text{ m}$$

4. The median is less efficient (see footnote 2).

We calculate

$$s_1 = 0.178 \text{ m} \qquad s_2 = 0.150 \text{ m}$$

From Eq. (6-9),

$$s_{\bar{x}1} = 0.089 \text{ m} \qquad s_{\bar{x}2} = 0.050 \text{ m}$$

The standard error of sample 2 is thus smaller than that of sample 1, and we can write our estimates of the population mean (i.e., "true" length) as: 30.30 ± 0.089 m and 30.25 ± 0.050 m, respectively. The exact interpretation of statements of this type is discussed in Chapter 10.

Example. To verify Eq. (6-7) for a population of observations: 2, 4, 6. Here $N = 3$ and $\mu = 4$. Thus

$$\sigma^2 = \frac{2^2 + 0^2 + 2^2}{3} = \frac{8}{3}$$

Consider $n = 2$. The samples are then 2, 4; 2, 6; and 4, 6. The sample means are 3, 4, and 5.

$$\text{mean of sample means} = \frac{3 + 4 + 5}{3}$$

$$= 4$$

(which checks with the population mean) and

$$\text{variance of sample means} = \frac{1^2 + 0^2 + 1^2}{3} = \frac{2}{3}$$

Substituting in Eq. (6-7), we find that

$$\sigma_{\bar{x}}^2 = \frac{N - n}{n(N - 1)} \sigma^2$$

$$= \frac{1}{2 \times 2} \times \frac{8}{3} = \frac{2}{3}$$

which checks with the value calculated directly from the sample means.

Example. If we consider the results of strength tests on 270 bricks, given in Table 2-2, as a sample of the population consisting of all the bricks made by the given works during the sampling period, then we can say that the standard deviation of the sample mean (calculated in Chapter 3 to be 6.89 MN/m^2) is, from Eq. (6-9):

$$s_{\bar{x}} = \frac{s}{\sqrt{n}} = \frac{1.39}{\sqrt{270}} = 0.0846 \text{ MN/m}^2$$

[1.39 is the value of the *sample* standard deviation (found in Chapter 4), but because n is large, the error involved in ignoring Bessel's correction is negligible].

The interpretation of the standard deviation of the mean in relation to the precision of the calculated mean is considered on p. 159.

Problems

6-1. Calculate the standard deviation of the mean of Problem 3-2.

6-2. Calculate the standard error of the mean tensile strength of rubber of Problem 4-1. By what factor would this standard error be changed if we took 18 measurements? Assume that the estimated standard deviation is the same in both cases.

6-3. The coded length of specimens of a certain type was found to be: 17, 17, 12, 15, and 20. By considering all possible samples of size two, drawn with replacement from the above population, find: (a) the population mean, (b) the population standard deviation, (c) the mean of the sample means, (d) the standard deviation of the sample means.

Check (c) and (d) from (a) and (b), respectively, by using the appropriate formulas.

6-4. Determine the standard error of the mean precipitation found in Problem 3-3.

6-5. Samples of 4 building blocks were taken, and the mean range of their mass was found to be 0.09 kg. Estimate the standard deviation of mass of the blocks. Estimate also the standard deviation of the mean mass of the sample.

6-6. The vibration time of a member was measured 13 times, the following values in seconds being obtained:

59.6, 60.4, 60.2, 60.7, 60.1, 59.8, 59.8, 60.3, 60.0, 59.9, 59.5, 60.2, 60.3

Find the mean and standard deviation for these values. Find also the standard deviation of the mean, mean deviation, and coefficient of variation.

6-7. The wavelength of a spectral line was measured and the results were as follows: 3452, 3458, 3457, 3451, 3455, 3458, 3454. Using coding to simplify the data, calculate the (a) mean wavelength, (b) standard deviation, (c) coefficient of variation, (d) estimate of the standard deviation of the mean wavelength, and (e) the standard error in the wavelength.

Chapter 7

Binomial Distribution

In some cases a population consists of only two classes of individuals, for example, alive or dead, even or odd, heads or tails, or simply possessing or not possessing a certain attribute (e.g., a defect). When one trial is made, for example, one ball is drawn at random from a collection containing a proportion p of black balls, the probability of the ball being black is p. The probability per trial is thus fixed. When a random sample of size n is drawn from the population (i.e., by making n trials), the distribution of the two classes of individuals (black balls and others) is, of course, discrete and is of the binomial type.

The name of the distribution arises from a similarity between the distribution of the probabilities of obtaining 0, 1, 2, ... items considered as a success in a sample of size n and the successive terms of the binomial expansion $(q + p)^n$, where p denotes the probability of success in a simple trial, and q (such that $p + q = 1$) the probability of failure. Success means simply encountering a certain class of individual (which in practice may be far from a success, e.g., a defective part), and failure means not encountering this class.

In general terms, it is usual to call the occurrence of some event, as specified, a success; its nonoccurrence—or the occurrence of any other event—would be called a failure.

DERIVATION

Consider n trials in each of which the probability of success is p. Then the probability of failure is $1 - p = q$. To find the probability of r successes, we observe that:

the probability of 1 success in 1 try is p,

the probability of 2 successes in 2 tries is $p \times p$ or p^2,

the probability of 3 successes in 3 tries is $p \times p \times p$ or p^3,

$\vdots$

the probability of r successes in r tries is p^r, and the probability of subsequent $(n - r)$ failures in $(n - r)$ tries is $(1 - p)^{n-r} = q^{n-r}$. From the rule given in Chapter 5, it follows that the probability of r successes followed by $(n - r)$ failures is $p^r(1 - p)^{n-r}$. Here, we have considered only one particular group or combination of r events; i.e., we have started with r successes and finished with $(n - r)$ failures; every other possible ordering of r successes and $(n - r)$ failures will also have the same probability.

The number of possible orderings or the number of selections for r successes and $(n - r)$ failures in n trials is $n!/[r!(n - r)!]$ [see Eq. (5-4)].

Therefore, the probability P_r of an event succeeding r times is

$$P_r = \frac{n!}{r!(n - r)!} p^r(1 - p)^{n-r} \tag{7-1}$$

or

$$P_r = {}_nC_r p^r q^{n-r} \tag{7-1a}$$

It may be observed that this term is similar to the rth term of the binomial expansion $(q + p)^n$ [see Eq. (5-6) or Eq. (5-7)], which can be written

$$(q + p)^n = \sum_{r=0}^{n} {}_nC_r p^r q^{n-r} \tag{7-2}$$

The successive terms of the expansion give the probability P_r of an event succeeding r times in n trials for values of r varying in steps of one from 0 to n.

In many cases we are interested not in the probability of an event succeeding exactly r times but in the probability of its succeeding *at least* r times in n trials. This is given by the *theorem of repeated trials* as

$$P_r + P_{r+1} + \cdots + P_n$$

It should be emphasized that for the binomial distribution to be applicable, the probability of success must be constant from trial to trial and all the trials must be independent events. Thus the conditions (e.g., the method of manufacture) must not change while the samples are being taken and the sampling must be done in a random manner, each selection being independent. However, in industrial work where the binomial distribution is used in lot-by-lot acceptance inspection, these conditions are satisfied from a practical point of view, since the lot size is usually very large compared to the sample size.

Now let us consider several simple examples of binomial distribution.

Example. Two coins are tossed (i.e., we take samples of two from an infinite population of coins). There are three possibilities for each toss or observation:

no heads one head two heads

Two coins can fall in $2^2 = 4$ ways. They give us no heads in only one case (when they both fall tails) so that the probability of no heads is $\frac{1}{4}$. For there to be one head, there are two possible arrangements: a head on the first coin or a head on the second coin; hence the probability of one head is $\frac{2}{4}$. Finally, for there to be two heads both coins have to fall heads, and thus the probability of two heads is $\frac{1}{4}$. We can, therefore, write the probabilities:

no heads	one head	two heads
$\frac{1}{4}$	$\frac{1}{2}$	$\frac{1}{4}$

Denoting the probability of success (heads) in one trial as p and the probability of failure (no heads) as q, we can write the binomial expansion for two tosses:

$$(q + p)^2 = q^2 + 2qp + p^2$$

Since $p = q = \frac{1}{2}$, the expansion gives the probabilities for the three possible cases, respectively:

$$(q + p)^2 = (\tfrac{1}{2})^2 + 2(\tfrac{1}{2})(\tfrac{1}{2}) + (\tfrac{1}{2})^2$$

with a probability of

no heads one head two heads

Example. If four coins are tossed, what are the probabilities of obtaining various numbers of heads?

The number of possible ways in which 4 coins can be tossed is $2^4 = 16$.

For no heads, the number of favorable events is 1 (probability $= \frac{1}{16}$), since to get no heads all four coins have to be tails.

For one head, the number of favorable events is 4 (probability $= \frac{4}{16}$), since

the 1st coin could be tossed as a head,

the 2nd coin could be tossed as a head,

the 3rd coin could be tossed as a head,

the 4th coin could be tossed as a head.

For two heads, the number of favorable events is the number of ways in which one can select 2 items out of 4, i.e.,

$$_4C_2 = \frac{4 \times 3}{2} = 6 \text{ (probability} = \tfrac{6}{16})$$

For three heads, the number of favorable events is $_4C_3 = 4$ (probability $= \frac{4}{16}$).

For four heads, the number of favorable events is $_4C_4 = 1$ (probability $= \frac{1}{16}$).

Compare these values with the binomial expansion of

$$\left(\frac{1}{2}+\frac{1}{2}\right)^4 = \left(\frac{1}{2}\right)^4 + 4\left(\frac{1}{2}\right)^3\frac{1}{2} + \frac{4 \times 3}{2}\left(\frac{1}{2}\right)^2\left(\frac{1}{2}\right)^2 + \frac{4 \times 3 \times 2}{2 \times 3}\left(\frac{1}{2}\right)\left(\frac{1}{2}\right)^3 + \left(\frac{1}{2}\right)^4$$

$$= \frac{1}{16} + \frac{4}{16} + \frac{6}{16} + \frac{4}{16} + \frac{1}{16}$$

with a probability of	no heads	1 head	2 heads	3 heads	4 heads

This distribution is shown in the frequency diagram of Fig. 7-1, and it can be seen that when $p = q$ the distribution of the probabilities is (as expected) symmetrical.

Nonsymmetrical distribution is illustrated by the following case.

Example. Consider a population consisting of equal numbers of balls of three different colors (a type of population favored by statisticians) from which we draw four balls at a time, replacing the balls every time. What is the probability of obtaining in our sample 0, 1, ..., 4 balls of a given color, say black?

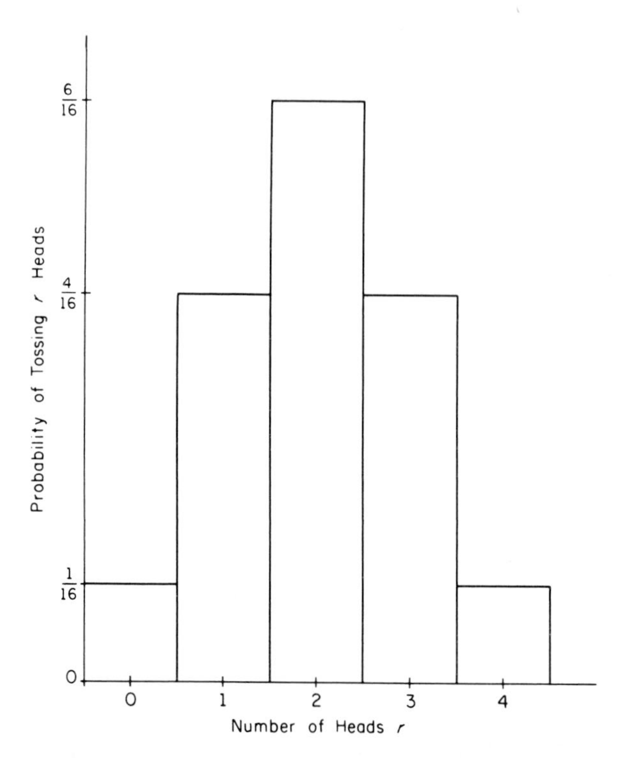

Fig. 7-1. Probability distribution for throwing various numbers of heads in tossing four coins.

For each ball drawn, the probability of success (black ball) is $p = \frac{1}{3}$; the probability of failure is $q = \frac{2}{3}$. Using binomial distribution, we have

$$\left(\frac{2}{3} + \frac{1}{3}\right)^4 = \frac{16}{3^4} + \frac{32}{3^4} + \frac{24}{3^4} + \frac{8}{3^4} + \frac{1}{3^4}$$

	= 0.1975	0.3951	0.2963	0.0988	0.0123
with a probability of:	no black balls	1 black ball	2 black balls	3 black balls	4 black balls

The probability distribution (plotted in Fig. 7-2) is nonsymmetrical.

Example. Seven races are to be held, the same six dogs taking part in each race. What are the probabilities of one particular dog's winning 1, 2, ..., 7 races, assuming that all dogs are "equally good."[1]

1. An assumption of doubtful validity.

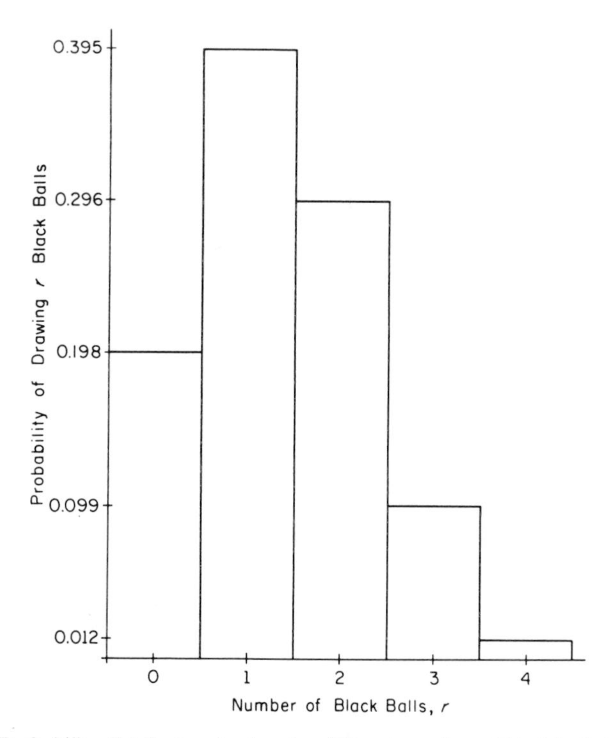

Fig. 7-2. Probability distribution for drawing different numbers of black balls in a sample of 4 balls drawn from a population of balls of 3 colors in equal proportions.

In any one race our dog has a $\frac{1}{6}$ chance of winning (probability of success $= p = \frac{1}{6}$) and a $\frac{5}{6}$ chance of not winning (probability of failure $= q = \frac{5}{6}$). Binomial expansion gives

$$\left(\frac{5}{6}+\frac{1}{9}\right)^7 = \left(\frac{5}{6}\right)^7 + 7\left(\frac{5}{6}\right)^6\left(\frac{1}{6}\right) + \frac{7\times 6}{2}\left(\frac{5}{6}\right)^5\left(\frac{1}{6}\right)^2 + \frac{7\times 6\times 5}{2\times 3}\left(\frac{5}{6}\right)^4\left(\frac{1}{6}\right)^3 + \cdots + \left(\frac{1}{6}\right)^7$$

with a probability of	no wins	1 win	2 wins	3 wins	7 wins

CUMULATIVE TERMS FOR BINOMIAL DISTRIBUTION

From what was said about the theorem of repeated trials (p. 94), it can be seen that it is often advantageous to find *directly* the probability of an event succeeding at least r times in n trials. The relevant data can be presented in tabular form, their derivation being as follows.

The probability P_r of an event succeeding *exactly* r times in n trials is given by Eq. (7-1a),

$$P_r = {}_nC_r p^r q^{n-r}$$

The probability of an event succeeding *at least* r' times in n trials is given by

$$\sum_{r=r'}^{r=n} P_r = \sum_{r=r'}^{r=n} {}_nC_r p^r q^{n-r} \tag{7-3}$$

The values of the summation of Eq. (7-3) are given in Table A-15 for p ranging from 0.05 to 0.50; with n between 2 and 20, and r' between 1 and 20.

For $p > 0.5$, we can utilize the fact that the probability is

$$P_r = 1 - \sum_{r=n-r'+1}^{r=n} {}_nC_r q^r p^{n-r} \tag{7-4}$$

Example. For a probability of success $p = 0.75$ and 4 trials, what is the probability of at least 3 successes? We look up the value in Table A-15 for $n = 4$, $r' = 2$ (from $r = n - r' + 1$, i.e., $3 = 4 - r' + 1$) and $q = 0.25$ (q interchanges with p in the same table) and find 0.2617. Therefore, the answer is $1 - 0.2617 = 0.7383$.

PASCAL'S TRIANGLE

The binomial coefficients can be obtained from Pascal's triangle; this is constructed so that each term is the sum of the two terms immediately above and to either side, as shown in Fig. 7-3. The second term of each line corresponds to the value of n in $(q + p)^n$, and the sum of coefficients in any line is equal to 2^n.

The coefficients of the binomial expansion can, of course, be computed quite rapidly using tables of factorial values. If, however, appropriate factorial values are not available and n is large, Stirling's formula can be used. This gives the factorial:

$$n! = e^{-n} n^n \sqrt{2\pi n} \tag{7-5}$$

The error involved is less than 0.1 percent for n greater than 100 and less than 1 percent for n greater than 10.

1
1 1
1 2 1
1 3 3 1
1 4 6 4 1
1 5 10 10 5 1
1 6 15 20 15 6 1
1 7 21 35 35 21 7 1
1 8 28 56 70 56 28 8 1
1 9 36 84 126 126 84 36 9 1
1 10 45 120 210 252 210 120 45 10 1
1 11 55 165 330 462 462 330 165 55 11 1
1 12 66 220 495 792 924 792 495 220 66 12 1

Fig. 7-3. Pascal's triangle.

Binomial distribution is used extensively in quality testing; a very simple illustration follows.

Example. In the production of wire connections consisting of 4 wires in parallel, it was found that on the average 1 wire in 10 was not tightened. If the strength of each wire is 250 newtons, find the proportion of connections that can withstand 750 newtons.

We assume that a wire which has not been tightened carries no load.

$$\text{probability of a good wire} = p = 0.9$$

$$\text{probability of a defective wire} = q = 0.1$$

Using the binomial distribution, we obtain

$$(0.1 + 0.9)^4 = (0.1)^4 + 4(0.1)^3(0.9) + \frac{4 \times 3(0.1)^2(0.9)^2}{2}$$

$$= 0.0001 \qquad 0.0036 \qquad 0.0486$$

with a probability of: all defective; 3 defective, 1 good; 2 defective, 2 good

$$+ \frac{4 \times 3 \times 2(0.1)(0.9)^3}{2 \times 3} + (0.9)^4$$

$$0.2916 \qquad 0.6561$$

1 defective, 3 good; all good

We can say, therefore, that of all connections made,

$$65.61\% \text{ will have a strength of } 4 \times 250 = 1000 \text{ N}$$

$$65.61 + 29.16 = 94.77\% \text{ will have a strength of } 3 \times 250 = 750 \text{ N or more}$$

$$94.77 + 4.86 = 99.63\% \text{ will have a strength of } 2 \times 250 = 500 \text{ N or more}$$

$$99.63 + 0.36 = 99.99\% \text{ will have a strength of } 1 \times 250 = 250 \text{ N or more}$$

$$0.01\% \text{ will have no strength}$$

We can readily obtain the above results from Table A-15. Thus, for all four connections being good, $r = 4$ and $r' = 1$ (from $r = n - r' + 1$, n being equal to 4). Entering Table A-15 and using $q = 0.1$ (instead of p, since $p = 0.9 > 0.5$) we find the value of 0.3439; the required probability is then calculated by Eq. (7-4) to be $1 - 0.3439 = 0.6561$; or 65.61 percent of all connections made will have a strength of 1000 N. For at least three connections being good, $r = 3$ with $r' = 2$, and the value from the table is 0.0523. The corresponding probability is $(1 - 0.0523)$; or 94.77 percent will have a strength of 750 N or more. The other values can be obtained in a similar manner.

Example. Wireless sets are manufactured with 25 soldered joints each. On the average, 1 joint in 500 is defective. How many sets can be expected to be free from defective joints in a consignment of 10,000 sets?

Let the probability of a defective joint ("success") be $p = 0.002$. Then $q = 0.998$. Using the binomial expansion,

$$(0.998 + 0.002)^{25} = \underbrace{(0.998)^{25}}_{\text{no defective joints}} + \underbrace{25 \times (0.998)^{24} \times 0.002}_{\text{1 joint defective}} + \cdots$$

Thus the proportion of sets with no defective joints is $(0.998)^{25} = 0.95118$ so that in 10,000 sets, 9512 would be expected to be free from defective joints. It is clear that some sets would have more than one defective joint.

MEAN AND STANDARD DEVIATION

Let us now further consider the properties of the binomial distribution. If p is the proportion of successes in the population, then the mean number of successes in n trials is

$$\mu = np \tag{7-6}$$

This is obvious, as the mean number of successes in n trials is equal to the probability of success in one trial times the number of trials.

The standard deviation for a binomial frequency distribution is

$$\sigma = \sqrt{npq} \tag{7-7}$$

Since q is not independent but is equal to $(1 - p)$, we can see that the binomial distribution can be expressed in terms of two parameters, n and p.

Equations (7-6) and (7-7) will now be derived using the definition of expectation for the mean and variance introduced in Chapter 5. Thus, from Eq. (5-19),

$$\begin{aligned} E(r) = \mu = \sum_{r=0}^{n} rP_r &= \sum_{r=0}^{n} r \frac{n!}{r!(n-r)!} p^r (1-p)^{n-r} \\ &= np \sum_{r=1}^{n} \frac{(n-1)!}{(r-1)!(n-r)!} p^{r-1}(1-p)^{n-r} \\ &= np[p+q]^{n-1} \end{aligned}$$

or

$$\mu = np \tag{7-6a}$$

For the variance σ^2 we use Eq. (5-21):

$$\begin{aligned} E(r-\mu)^2 = \sigma^2 &= \sum_{r=0}^{n} (r-\mu)^2 P_r \\ &= \sum_{r=0}^{n} r^2 P_r - 2\mu \sum_{r=0}^{n} rP_r + \mu^2 \sum_{r=0}^{n} P_r \\ &= \sum_{r=0}^{n} r^2 P_r - 2\mu(\mu) + \mu^2 \end{aligned}$$

since

$$\sum_{r=0}^{n} P_r = 1 \qquad \text{and} \qquad \sum_{r=0}^{n} rP_r = \mu = np$$

Thus

$$\begin{aligned} \sigma^2 &= \sum_{r=0}^{n} r^2 P_r - (np)^2 \\ &= \sum_{r=0}^{n} [r(r-1) + r] P_r - (np)^2 \\ &= \sum_{r=2}^{n} r(r-1) \frac{n!}{r!(n-r)!} p^r (1-p)^{n-r} + \sum_{r=0}^{n} rP_r - (np)^2 \\ &= n(n-1)p^2 \sum_{r=2}^{n} \frac{(n-2)!}{(r-2)!(n-r)!} p^{r-2}(1-p)^{n-r} + np - (np)^2 \\ &= n(n-1)p^2[p+q]^{n-2} + np - (np)^2 \end{aligned}$$

or

$$\sigma^2 = n(n-1)p^2 + np - (np)^2 = np(1-p) = npq$$

Hence

$$\sigma = \sqrt{npq} \tag{7-7a}$$

A simple verification of Eqs. (7-6a) and (7-7a) is afforded by the following example.

Example. Consider 32 trials, each consisting of tossing four coins. The theoretical frequencies are given by $32(\frac{1}{2} + \frac{1}{2})^4$, as shown in Table 7-1.

TABLE 7-1

Number of heads x_i	*Theoretical frequency* f_i	$f_i x_i$	$f_i x_i^2$
0	2	0	0
1	8	8	8
2	12	24	48
3	8	24	72
4	2	8	32
Totals	$\sum f_i = 32$	$\sum f_i x_i = 64$	$\sum f_i x_i^2 = 160$

Using the general expressions for mean and standard deviation [Eqs. (3-2) and (4-4)], we have

$$\text{mean} = \mu = \frac{\sum f_i x_i}{\sum f_i} = \frac{64}{32} = 2$$

(as, indeed, expected)

standard deviation $= \sigma$

$$= \sqrt{\frac{\sum f_i x_i^2 - [(\sum f_i x_i)^2 / \sum f_i]}{\sum f_i}} = \sqrt{\frac{160 - [(64)^2/32]}{32}} = 1$$

We can now check these results against Eqs. (7-6) and (7-7): $n = 4$, $p = \frac{1}{2}$, and $q = \frac{1}{2}$. Then

$$\mu = np = 4 \times \tfrac{1}{2} = 2$$

and

$$\sigma = \sqrt{npq} = \sqrt{4 \times \tfrac{1}{2} \times \tfrac{1}{2}} = 1$$

COMPARISON OF EXPERIMENTAL AND BINOMIAL DISTRIBUTIONS

Let us now consider an actual experiment in which the results in Table 7-2 were obtained.

TABLE 7-2

Number of heads x_i	*Observed frequency* f_i	$f_i x_i$	$f_i x_i^2$
0	0	0	0
1	8	8	8
2	15	30	60
3	6	18	54
4	3	12	48
Totals	$\sum f_i = 32$	$\sum f_i x_i = 68$	$\sum f_i x_i^2 = 170$

Then

$$\mu = \frac{68}{32} = 2.12$$

and

$$\sigma = \sqrt{\frac{170 - [(68)^2/32]}{32}} = 0.89$$

A comparison of the observed distribution with the preceding theoretical distribution is shown in Fig. 7-4.

We should note that if the comparison is made in terms of frequencies rather than of probabilities, the expected frequency has a mathematical and not a strictly physical meaning, as fractions of, say, heads are not possible.

A practical problem of this type arises in acceptance tests. For example, let us imagine that a manufacturer delivers a product on the understanding that 90 percent of the items are free from defect. We take a sample of 4 items and find 2 of them defective. Are we justified in rejecting the entire consignment?

To answer this question, we have to know the probability P of obtaining at least 2 defectives in a sample of 4. We have $n = 4$, $p = 0.1$, and $q = 0.9$. Now

$$\begin{aligned} P &= P_2 + P_3 + P_4 \\ &= {}_4C_2(0.1)^2(0.9)^2 + {}_4C_3(0.1)^3(0.9) + {}_4C_4(0.1)^4 \\ &= 0.0523 \end{aligned}$$

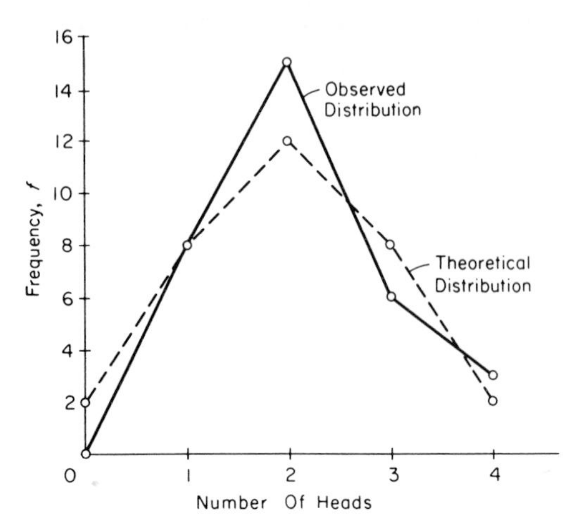

Fig. 7-4. Theoretical and experimental distribution for 32 throws of 4 coins.

Thus there is a probability of only 0.0523 (or approximately 1 in 20) of drawing at least 2 defectives in a sample of 4, and we would be far from unreasonable if we rejected the consignment. This answer also can be directly obtained from Table A-15: with $n = 4$, $r = r' = 2$, $p = 0.1$, we read the required probability as 0.0523.

It should be stressed that an intuitive approach to problems of this type can often be misleading. For example, what answer would the reader expect to the following problem?

Lottery tickets are sold, it being advertised that every fourth ticket carries a prize. If we buy 4 tickets, this being a random sample, what is the probability that we get at least one winning ticket?

The probability of having at least one winning ticket is

$$P = 1 - P_0$$

where P_0 is the probability of having no winning ticket in a sample of 4, and the term 1 represents all the possible cases. We have $p = 0.25$ and $n = 4$. Thus

$$P = 1 - {}_4C_0(0.25)^0(0.75)^4$$

$$= 0.684$$

i.e., there is a 68.4 percent probability of having at least one winning ticket. This may interest those who did not get a winning ticket under similar circumstances and ended up by suspecting the promoters.

The answer to our problem could also be obtained by adding the probabilities of drawing 1, 2, 3, and 4 winning tickets, but the method used here involves less effort. Of course, Table A-15 gives the answer at once with $p = 0.25$, $n = 4$, and $r = r' = 1$.

Another comparison between an experimental distribution and the theoretical binomial distribution may be of interest.

Example. Ten tosses of a suspected die gave the results 1, 1, 1, 6, 1, 1, 3, 1, 1, 4. What is the probability of at least this many aces if the die is true?

The event "at least 7 aces" can materialize in four mutually exclusive ways: 7 aces, 8 aces, 9 aces, and 10 aces.

The probability of throwing at least 7 aces, P, is the sum of the probabilities for 7, 8, 9, and 10 aces, i.e.,

$$P = P_7 + P_8 + P_9 + P_{10}$$

We have $n = 10$, and the probability of throwing an ace in any one throw is $p = \frac{1}{6}$. Therefore,

$$\begin{aligned} P &= {}_{10}C_7\left(\frac{1}{6}\right)^7\left(\frac{5}{6}\right)^3 + {}_{10}C_8\left(\frac{1}{6}\right)^8\left(\frac{5}{6}\right)^2 + {}_{10}C_9\left(\frac{1}{6}\right)^9\left(\frac{5}{6}\right) + {}_{10}C_{10}\left(\frac{1}{6}\right)^{10} \\ &= 0.00027 \end{aligned}$$

This probability is so small that it is most unlikely that the observed 7 aces in 10 throws could be obtained with $p = \frac{1}{6}$. We conclude, therefore, that the hypothesis $p = \frac{1}{6}$ should be rejected, i.e., the die is biased or "loaded."

We can, of course, test the agreement between the two distributions quantitatively in terms of probability; the appropriate methods are dealt with in Chapter 12.

Solved Problems

7-1. If the probability of a weld's being defective is 0.1, what is the probability of obtaining exactly 0, 1, 2, and 3 defective welds in a sample of 5 welds? What is the probability of obtaining less than 2 defective welds? If a structure has 100 welds, how many of them would you expect to be defective? What is the probability of there being at least 90 welds free from defect in a structure containing 100 welds?

SOLUTION

The probability of a defective weld is $p = 0.1$; then $q = 0.9$. Using the binomial expansion, we obtain

$$(0.9 + 0.1)^5 = (0.9)^5 + 5(0.9)^4(0.1) + \frac{5 \times 4}{2!}(0.9)^3(0.1)^2 + \frac{5 \times 4 \times 3}{3!}(0.9)^2(0.1)^3 + \cdots$$

	0.59049	0.32805	0.0729	0.0081
with a probability of	all good	one defective	two defectives	three defectives

probability of obtaining less than two defective welds

$$= 0.59049 + 0.32805$$

$$= 0.91854$$

Of 100 welds we would expect $0.1 \times 100 = 10$ to be defective (on the average). The probability of there being at least 90 welds free from defect is

$$\sum_{r=90}^{100} P_r$$

where $P_r = {}_{100}C_r(0.1)^{100-r}(0.9)^r$

7-2. If the probability of a structure's collapsing after 30 years of service is 0.01 find the probability that out of 10 such structures the following will collapse after 30 years of service.

a. None.
b. Exactly one.
c. Not more than one.
d. More than one.
e. At least one.

SOLUTION

The probability of a sound structure is $p = 0.99$, and the probability of a collapse is $q = 0.01$. Therefore,

$$(0.99 + 0.01)^{10} = (0.99)^{10} + 10(0.99)^9(0.01) + \cdots$$

	$= 0.9036$	0.0913
with a probability of	no collapse	one collapse

Logarithms are used to obtain the above results. For example, put

$$(0.99)^{10} = y$$

Then

$$\log y = 10 \log 0.99 = \bar{1}.9956 \times 10$$

$$= \overline{10} + 9.956$$

$$= \bar{1}.9560$$

Hence $y = 0.9036$. Thus the probabilities are

a. 0.9036
b. 0.0913
c. $0.9036 + 0.0913 = 0.9949$
d. $1 - 0.9949 = 0.0051$
e. $1 - 0.9036 = 0.0964$

Problems

7-1. On checking assembly lines, it was found that 1 of the 10 lines produced defective items. The products of all the lines were mixed by then, and the defectives could not be easily separated out. If a random sample of 10 items is taken, what is the probability of its containing the following number of defective items? (a) no defective items, (b) one defective item, (c) not more than two defective items, (d) more than one defective item?

7-2. If the probability of a child's being male is 0.55, what is the probability of having three daughters in succession?

7-3. A product is supposed to contain 5 percent of defective items. We take a sample of 10 items and find it to contain 2 defectives. Are we justified in suspecting that the consignment is not up to specification?

7-4. A manufacturing process is intended to produce precast units with no more than 2 percent defective. It is checked every day by testing 10 units selected at random from the day's production. If one or more of the 10 fails, the process is halted and carefully examined. If, in fact, its probability of producing a defective unit is 0.01, (a) What is the probability of obtaining one or more defective units? (b) What is the probability of obtaining no defectives in a given test? (c) Find the mean and the standard deviation of the number of defective units in a sample of 10 units.

7-5. The foreman of a casting section in a certain factory finds that on the average 1 in every 5 castings made is defective. If the section makes 8 castings a day, what is the chance that 2 of these will be defective? What is the chance that 5 or more defective castings are made in one day?

7-6. Samples of 6 items are drawn at random from a supply source that contains 9 percent defectives. Draw a histogram showing the probabilities of having 0, 1, . . . , 6 defectives in the sample.

7-7. In a game of Russian roulette one chamber of a six-chamber gun is loaded. The cylinder is spun around, and the trigger is pulled. Three men are playing the "roulette," taking 3 turns each at pulling the trigger. What is the probability that (a) a particular man will survive the game? (b) one of the three men will survive the game? (c) all the men will survive?

7-8. From seven different concrete mixes a number of compression test specimens of two sizes were made. The results are given in Table 7-3. Test the hypothesis that there is no difference between the strength of specimens of 10-cm and 20-cm size.

[HINT: Calculate the probability of obtaining the observed number of differences of the same sign if positive and negative differences occur equally often in the parent population.]

TABLE 7-3

	10-cm specimens		*20-cm specimens*	
Mix	*Sample size*	*Mean strength* (MN/m^2)	*Sample size*	*Mean strength* (MN/m^2)
A	6	13.8	6	11.3
B	27	23.2	21	21.4
C	15	28.3	12	27.0
D	6	36.0	9	31.8
E	12	44.3	7	42.6
F	18	58.0	6	50.7
G	11	72.5	9	63.4

Source: A. M. Neville, "Some Aspects of the Strength of Concrete," Part II, *Civil Engineering and Public Works Review*, Vol. 54, no. 640, Nov. 1959, pp. 1308–1310. The original data was in kg/cm^2 units.

7-9. From previous experience it is known that the probability of a specific bridge failing under load is one in five hundred. Strengthening of the bridge would cost $15,000. If losses due to bridge failure are estimated at $10,000,000 for lawsuits, etc., should the present bridge be strengthened?

7-10. A car manufacturing company has detected that 5 percent of its new model cars have an idling problem. To solve this problem a new nozzle was introduced in a pilot lot of a hundred cars.

a. How many cars would one expect to have the idling problem if the new nozzle has no effect?
b. Would you conclude that the new nozzle has solved the problem if it was found that only two cars in the pilot lot are now experiencing the idling problem? [HINT: Find the probability of getting two or less defectives in a sample size of 100, assuming that the new nozzle has no effect.]

Chapter 8

Poisson Distribution

The Poisson distribution represents the probability of an isolated event occurring a specified number of times in a given interval of time (or space) when the rate of occurrence in a continuum of time (or space) is fixed. The occurrence of events must be affected by chance alone, and the Poisson distribution, therefore, is such that information about the position of one event is of no help in predicting the position of any other specific event; furthermore, data on one interval of time (or space) are of no help in predicting how many events will occur in any other interval. A characteristic feature of the Poisson distribution is the fact that only the occurrence of an event can be counted; its nonoccurrence cannot because it has no physical meaning. Thus the total number of events n cannot be measured, and, in consequence, the binomial distribution is not precisely applicable.

Some of the phenomena that follow the Poisson distribution are flaws in castings, the number of vehicles on a highway, the number of telephone calls, clicks of a Geiger counter, or the celebrated case of cavalrymen killed by a horse-kick (discussed later in this chapter).

TERMS OF THE POISSON DISTRIBUTION

The Poisson distribution is made up of a series of terms:

$$e^{-\mu}, \quad e^{-\mu}\mu, \quad e^{-\mu}\frac{\mu^2}{2!}, \quad e^{-\mu}\frac{\mu^3}{3!}, \quad e^{-\mu}\frac{\mu^4}{4!}, \ldots$$

representing, respectively, the probability of the occurrence of 0, 1, 2, 3, 4, etc., events, where e is the base of natural logarithms and μ is the mean frequency of occurrence. The sum of all terms of the series is unity, as must be the case with a sum of all probabilities.

If r is the number of occurrences whose probability we require, we can write the general term of the series

$$P_r = \frac{e^{-\mu}\mu^r}{r!} \tag{8-1}$$

The values of $e^{-\mu}$ for values of μ between 0.01 and 5 are given in Table A-2, but it is clear that such a computation is tiresome, and the use of tabulated or plotted values such as shown in Fig. 8-1 is preferable.

The Poisson distribution is asymmetrical (skewed), and the relation between the mode and the mean μ is such that the mode is

at 0 occurrence when $\mu < 1$,

at 1 occurrence when $1 \leq \mu < 2$,

at 2 occurrences when $2 \leq \mu < 3$, etc.

The terms of the Poisson distribution, Eq. (8-1), are derived from a Poisson process in Appendix B. A Poisson process is a random physical mechanism in which events occur randomly on a time scale (or distance scale). For example, the occurrence of accidents at a specific road junction follows a Poisson process. It should be remembered that we cannot predict exactly how many accidents will take place in a particular time interval, but we can predict the *pattern* of accidents in a large number of such time intervals.

Example. It is assumed that cosmic-ray counts from Geiger counters are completely random and follow a Poisson distribution. Check whether this is true for the following data on counts recorded in 50 consecutive periods of 60 s.

1, 0, 2, 4, 1, 1, 2, 1, 3, 0, 0, 2, 1, 0, 2, 3, 3,

0, 4, 1, 2, 0, 2, 1, 1, 4, 2, 1, 1, 3, 1, 1, 4, 2,

0, 0, 3, 2, 1, 2, 3, 1, 0, 2, 3, 1, 0, 1, 0, 3

The data are arranged in tabular form in Table 8-1.

$$\mu = np = \frac{\sum fr}{\sum f} = \frac{0 \times 11 + 1 \times 16 + 2 \times 11 + 3 \times 8 + 4 \times 4}{50} = \frac{78}{50}$$

$$= 1.56 \qquad \text{average number of counts per 60 s interval}$$

Substituting $\mu = 1.56$ in Eq. (8-1), we have

$$P_r = \frac{e^{-1.56}(1.56)^r}{r!} \qquad \text{for } r = 0, 1, 2, 3, 4$$

The values of P_r are calculated and the expected number of 60-s intervals is obtained from $P_r \times \sum f$. The results are shown in Table 8-1. Comparison

TABLE 8-1

Count in 60 s interval (r)	0	1	2	3	4	> 4	Total
Observed number of 60 s intervals with r counts, f	11	16	11	8	4	0	50
Probability P_r *by Poisson distribution [Eq. (8-1)]*	0.23	0.33	0.26	0.11	0.05	0.02	1.0
Expected numbers of intervals $P_r \times \sum f$	11.5	16.5	13.0	5.5	2.5	1.0	50.0

between the observed and expected values indicates good agreement. The question of how good this agreement is can be answered by the rigorous χ^2 test, considered in Chapter 12. This conclusion is of interest because it tells us that our observed data are the result of chance happenings. A non-Poisson distribution of cosmic-ray counts would suggest that randomness or chance of the counts had changed during the test or that some malfunction of the counter has occurred; this would then initiate action to correct the situation.

To illustrate the application of the Poisson distribution to the problem of traffic accidents, let us consider the following situation. Accident histories of a number of drivers, selected at random, were collected for a 5-year period. The average number of accidents for this period was calculated to be $\mu = 0.25$ accidents. If we wish to know whether a driver having 3 accidents in a 5-year period is an accident-prone driver, we calculate P_r for $r = 3$ and $\mu = 0.25$.

$$P_r = \frac{e^{-\mu}\mu^r}{r!} = \frac{e^{-0.25}(0.25)^3}{3!} = 0.002$$

This means that the chances are 500 to 1 against an *average* driver's having 3 accidents. Therefore, we conclude that a driver who has this many accidents is a bad risk.

DERIVATION FROM BINOMIAL DISTRIBUTION

The Poisson distribution can also be deduced from the binomial distribution, provided that n is large ($\rightarrow \infty$), p is very small ($\rightarrow 0$), and np is finite. It was shown earlier [Eq. (7-1)] that in n trials the probability of an event succeeding r times is

$$P_r = \frac{n!}{r!(n-r)!} p^r q^{n-r} \tag{8-2}$$

When n is large compared with r,

$$\frac{n!}{(n-r)!} = n(n-1)(n-2) \cdots (n-r+1)$$
$$\simeq n^r$$

Therefore, the probability of r successes becomes

$$P_r = \frac{n^r}{r!} p^r q^{n-r} \tag{8-3}$$

Now if p is very small and r is not large,

$$q^r = (1-p)^r \simeq 1$$

and

$$q^{n-r} \simeq q^n = (1-p)^n$$

Hence

$$P_r = \frac{(np)^r}{r!}(1-p)^n$$

$$= \frac{(np)^r}{r!}\left[1 - np + \frac{n(n-1)(-p)^2}{2!} + \frac{n(n-1)(n-2)(-p)^3}{3!} + \cdots\right]$$

$$\simeq \frac{(np)^r}{r!}\left[1 - np + \frac{(np)^2}{2!} - \frac{(np)^3}{3!} + \cdots\right]$$

Thus

$$P_r = \frac{(np)^r}{r!} e^{-np} \tag{8-4}$$

This, then, is the probability of r successes in n trials.

MEAN AND STANDARD DEVIATION

The mean number of occurrences of an event per unit of time (or space) is

$$\mu = np \tag{8-5}$$

and the standard deviation of the numbers of events is

$$\sigma = \sqrt{np} \tag{8-6}$$

Thus the mean and variance are equal to one another:

$$\mu = \sigma^2 = np \tag{8-7}$$

Equations (8-5) and (8-7) will now be derived using Eqs. (5-19) and (5-21). From Eq. (5-19),

$$\text{mean} = E(r) = \sum_{r=0}^{\infty} rP_r = \sum_{r=0}^{\infty} \frac{re^{-\mu}\mu^r}{r!}$$

$$= 0 + \mu e^{-\mu} + 2\frac{\mu^2 e^{-\mu}}{2!} + 3\frac{\mu^3 e^{-\mu}}{3!} + \cdots$$

$$= \mu e^{-\mu}\left[1 + \mu + \frac{\mu^2}{2!} + \frac{\mu^3}{3!} + \cdots\right]$$

$$= \mu e^{-\mu} e^{\mu} = \mu. \qquad [8\text{-}5]$$

The variance σ^2 is given by Eq. (5-21):

$$\sigma^2 = E(r-\mu)^2 = \sum_{r=0}^{\infty}(r-\mu)^2 P_r$$

$$\sigma^2 = \sum_{r=0}^{\infty} r^2 P_r - 2\mu \sum_{r=0}^{\infty} r P_r + \mu^2 \sum_{r=0}^{\infty} P_r$$

Now

$$-2\mu \sum_{r=0}^{\infty} r P_r = -2\mu(\mu) = -2\mu^2,$$

$$\mu^2 \sum_{r=0}^{\infty} P_r = \mu^2(1) = \mu^2,$$

and

$$\begin{aligned}
\sum_{r=0}^{\infty} r^2 P_r &= \sum_{r=0}^{\infty} [r(r-1)+r]P_r \\
&= \sum_{r=0}^{\infty} r(r-1)P_r + \mu \\
&= \left[0 + 0 + 2\frac{\mu^2 e^{-\mu}}{2!} + 6\frac{\mu^3 e^{-\mu}}{3!} + \cdots\right] + \mu \\
&= \mu^2 e^{-\mu}\left[1 + \mu + \frac{\mu^2}{2!} + \cdots\right] + \mu \\
&= \mu^2 e^{-\mu}(e^{\mu}) + \mu = \mu^2 + \mu
\end{aligned}$$

Thus, collecting the terms, we have

$$\sigma^2 = \mu^2 + \mu - 2\mu^2 + \mu^2$$

or

$$\sigma^2 = \mu \qquad [8\text{-}7]$$

It may be observed that Eq. (8-5) is identical with Eq. (7-6), derived for the binomial distribution. Likewise, Eqs. (8-6) and (7-7) are identical, provided that we remember that the Poisson distribution was derived for $q \simeq 1$.

It is important to note that the Poisson distribution contains only one parameter, np, the mean occurrence of an event, and we do not know the value of n. On the other hand, in the binomial distribution we know the number of times an event occurs and the number of times an event does not occur.

POISSON DISTRIBUTION AS AN APPROXIMATION TO BINOMIAL DISTRIBUTION

The Poisson distribution can thus be used as an approximation to the binomial distribution when the sample size n is large and the probability of success p is small (the same applies when q is small, p and q being of course interchangeable), i.e., when the binomial distribution is highly skewed. As a guide, we can say that a good approximation is obtained when $n \geq 20$ and $p \leq 0.05$, and the approximation improves with a decrease in p.

Example. In making glass, undissolved particles called "stones" sometimes occur. Let us assume that there is an average of 1 stone per kilogram of glass made. If there are 100,000 "particles" in 1 kg of glass, then the probability of a stone is

$$p = 10^{-5}$$

If we are making glass lenses of $\frac{1}{10}$ kg mass each, then there will be on the average 1 stone in every 10 lenses. Thus $np = 0.1$.

What is the proportion of lenses free from stones? The probability P_r of finding r stones in a lens is given by Eq. (8-4):

$$P_r = \frac{(np)^r}{r!} e^{-np}$$

and is tabulated as follows:

Number of stones in a lens, r	0	1	2	3	4
Probability P_r	0.9048	0.09048	0.004524	—	—

Therefore,

90.48% of all lenses made will be free from stones
9.05% of all lenses made will contain 1 stone
0.45% of all lenses made will contain 2 stones

and so on.

If each lens had a mass of 1 kg, then we would have $np = 1$, and hence the following values of probability of obtaining a lens containing r stones:

Number of stones in a lens, r	0	1	2	3	4	5
Probability P_r	0.367	0.367	0.183	0.061	0.015	—

Therefore,

36.7 percent of all lenses made will be free from stones
36.7 percent of all lenses made will contain 1 stone
18.3 percent of all lenses made will contain 2 stones
6.1 percent of all lenses made will contain 3 stones
1.5 percent of all lenses made will contain 4 stones

and so on.

The Poisson distribution (instead of binomial expansion) is particularly useful when n is extremely large and p very small, as it then becomes virtually impossible to compute the binomial terms. For example, assume that we have n radioactive nuclei such that the probability of one of these decaying in a time interval t is p. We want to calculate the probability of r of these undergoing decay in time t.

In terms of binomial distribution the required probability is [from Eq. (7-1a)]:

$$P_r = {}_nC_r p^r (1-p)^{n-r}$$

If $n = 10^{23}$ and $p = 10^{-22}$, the probability becomes

$$P_r = \frac{(10^{23})!}{(10^{23}-r)!\,r!} p^r (1-p)^{10^{23}-r}$$

which is difficult to evaluate. We prefer, therefore, to work in terms of $np = 10$ and use the Poisson approximation [Eq. (8-4)]:

$$P_r = \frac{10^r}{r!} e^{-10}$$

SIGNIFICANCE OF POISSON DISTRIBUTION

In many cases we may suspect that a set of occurrences is completely random in nature, and we may want to determine whether this is so. We proceed to compare the experimental distribution with an assumed Poisson distribution; if the agreement is good, we conclude that the distribution of occurrences is influenced by chance alone. If the agreement is not good, we suspect that some definite influences may exist. We then have to study the data further, e.g., by examining the periods of tests and seeking a correlation between these and some other, possibly intermittent, influence.

The goodness of fit of the observed distribution to the assumed Poisson distribution may be tested by the χ^2 test, discussed in Chapter 12. At this stage we shall only make a simple comparison.

Example. Records of deaths due to horse-kicks were kept in 10 army corps over a period of 20 years:[1]

Number of deaths per corps per annum r	0	1	2	3	4	5 or more
Number of corps per annum with r deaths f	109	65	22	3	1	0

Compare these values with a Poisson distribution derived from the observed mean frequency of death.

We take an army corps per annum as a unit. We have thus $20 \times 10 = 200$ observations.

TABLE 8-2

Number of deaths per corps per annum r	0	1	2	3	4	5
Probability of an army corps with r deaths P_r	$e^{-0.61}$ $= 0.545$	$0.61e^{-0.61}$ $= 0.331$	$\frac{(0.61)^2e^{-0.61}}{2}$ $= 0.101$	$\frac{(0.61)^3e^{-0.61}}{2 \times 3}$ $= 0.020$	$\frac{(0.61)^4e^{-0.61}}{2 \times 3 \times 4}$ $= 0.003$	...
Calculated number of army corps per annum with r deaths $= 200P_r$	109	66	20	4	$0.6 \simeq 1$	...
Observed number of army corps per annum with r deaths	109	65	22	3	1	0

1. Obtained by Bortkewitch and quoted in R. A. Fisher, *Statistical Methods for Research Workers* (New York: Hafner Publishing Co., 1958), p. 55.

The mean number of deaths per corps per annum is

$$np = \frac{\sum fr}{\sum f} = \frac{0 \times 109 + 1 \times 65 + 2 \times 22 + 3 \times 3 + 4 \times 1 + 5 \times 0}{200}$$

$$= \frac{122}{200} = 0.61$$

Substituting in Eq. (8-4), we obtain the data in Table 8-2.

As shown by the two last lines of Table 8-2, the agreement is extremely close, and we conclude that the distribution of deaths is due to chance alone and not to such factors as the locality where different corps are stationed or the personality of the general commanding.

POISSON PROBABILITY PAPER

A graph paper ruled so that the abscissae represent, to a logarithmic scale, the expected average number of events and the ordinates give the probability of occurrence to a Poisson probability scale is called a Poisson probability

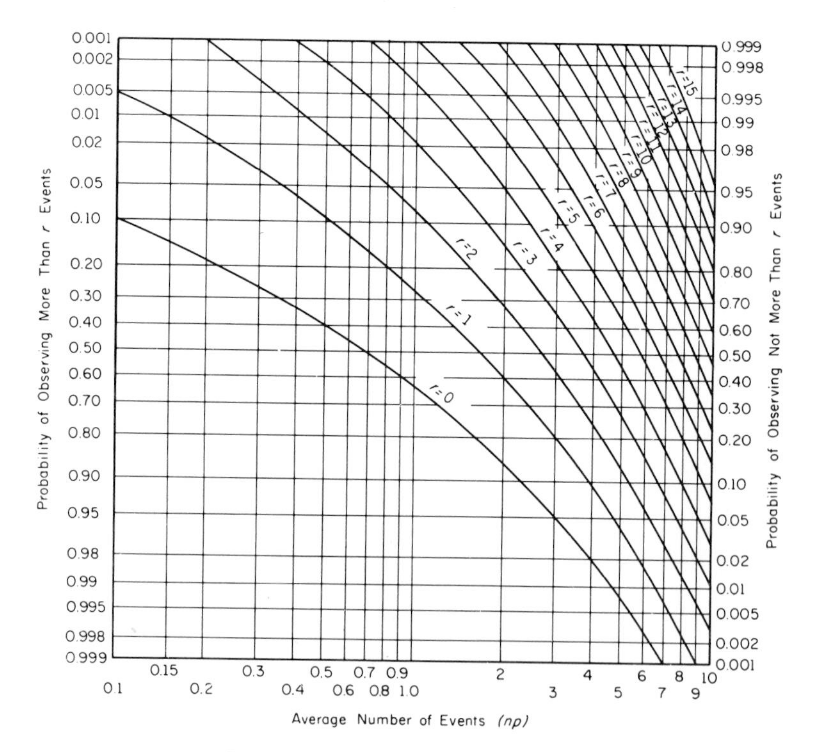

Fig. 8-1. Cumulative Poisson probability.

paper. Lines can be drawn showing the probability of an event occurring not more than r times for values of $r = 0, 1, 2, 3, \ldots 15$, as shown in Fig. 8-1.

If we require the probability of an event's occurring exactly a certain number of times, we have to use differences. For example, if the average number of events is 1, and we require the probability of an event occurring 3 times, we find from Fig. 8-1 that the probability of its occurring not more than 3 times is 0.981. The probability of its occurring not more than 2 times is 0.920. Hence the probability of its occurring 3 times is $0.981 - 0.920 = 0.061$.

Poisson probability paper can also be used to check whether an actual distribution agrees closely with the Poisson distribution, and, if so, what is the expected number of occurrences. From the experimental data we calculate the frequency and (by dividing by the total frequency) the probability of an event's occurring not more than 1, 2, 3, etc., times. For a Poisson distribution the points corresponding to a calculated probability and the average number of events would lie on a vertical line; the proximity of the actual points to such a line indicates the agreement with a Poisson distribution, and the abscissa of the line gives the expected (average) number of events.

The paper is generally valuable when we can use the Poisson distribution as an approximation to the binomial distribution, i.e., when the latter is highly skewed.

Solved Problems

8-1. The number of cars passing over a toll bridge during the time interval 10 to 11 A.M. is 1200. The cars pass individually and collectively at random. Write an expression for the probability that not more than 4 cars will pass during the 1-min interval 10:45 to 10:46. Derive an expression for the probability that 5 or more cars will pass during the same interval.

SOLUTION

number of cars in 60 min = 1200

$$\text{mean number of cars in 1-min interval} = \frac{1200}{60} = 20 = np$$

The probability of not more than 4 cars passing in that given interval

= sum of probabilities of 0 cars to and including 4 cars

$$= \sum_{r=0}^{4} \frac{(np)^r e^{-np}}{r!} = \left[\frac{(20)^0}{0!} + \frac{(20)^1}{1!} + \frac{(20)^2}{2!} + \frac{(20)^3}{3!} + \frac{(20)^4}{4!}\right] e^{-20}$$

Hence the probability of 5 or more cars

$$= 1 - \sum_{r=0}^{4} \frac{(20)^r e^{-20}}{r!}$$

8-2. If the probability of a concrete beam failing in compression is 0.05, use the Poisson approximation to obtain the probability that from a sample of 50 beams (a) At least three will fail in compression. (b) No beam will fail in compression.

SOLUTION

Here, $p = 0.05$. Mean number of failures in compression $= np = 0.05 \times 50 = 2.5$.

a. The probability of at least 3 beams failing

$$= 1 - \sum_{r=0}^{2} \frac{(np)^r e^{-np}}{r!}$$

$$= 1 - \left[\frac{(2.5)^0 e^{-2.5}}{0!} + \frac{(2.5)^1 e^{-2.5}}{1!} + \frac{(2.5)^2 e^{-2.5}}{2!}\right]$$

$$= 1 - e^{-2.5}[1 + 2.5 + 3.125]$$

$$= 1 - 0.082 \times 6.625 = 0.4575$$

b. The probability of no beam failing in compression

$$= \frac{(2.5)^0 e^{-2.5}}{0!} = 0.082$$

Alternatively, we can use Poisson probability paper (see Fig. 8-1). For $np = 2.5$, the probability of at least 3 beams failing = 1 − the probability of not more than 2 beams failing. For $r = 2$, probability = 0.54. Therefore, the required probability of at least 3 beams failing is $1 - 0.54 = 0.46$. This can be found directly from the vertical scale on the left in Fig. 8-1.

For $r = 2$, probability = 0.46. Also, from Fig. 8-1 the probability of no failure, for $np = 2.5$ and $r = 0$, is 0.08.

8-3. If the probability of a Bailey bridge collapsing after 10 years of service without maintenance is 0.02, find the probability that out of 30 such bridges (having 10 years of service) the following will collapse. (a) none, (b) exactly two, (c) not more than one, (d) more than two, (e) at least one.

SOLUTION

Since the probability of success (bridge collapsing) $p < 0.05$ and the sample size $n > 20$, we can conveniently use the Poisson distribution as a good approximation to the binomial distribution. We have $p = 0.02$ and $n = 30$. Therefore,

$$np = 0.6$$

Using the cumulative Poisson probability graph of Fig. 8-1 with $np = 0.6$ and the appropriate r, we obtain:

a. For $r = 0$, probability = 0.54 or 54 percent.
b. For $r = 2$, probability $= 0.975 - 0.875 = 0.10$ or 10 percent.
c. For $r \ngtr 1$, probability = 0.875 or 87.5 percent.
d. For $r > 2$, probability = 0.025 or 2.5 percent.
e. For $r \geq 1$, read off for $r > 0$ (i.e., left-hand ordinate), probability = 0.44 or 44 percent.

Problems

8-1. The numbers of road accidents per day reported in a given city on 100 consecutive days are as follows:

Number of accidents	0	1	2	3	4	5	6
Number of days	19	26	26	15	9	4	1

a. Check whether the distribution of accidents can be considered to be a Poisson distribution.
b. Compare the standard deviation of the given data with that of the Poisson distribution.

8-2. The probability for people of a certain age of dying within a year of their birthday is 0.0038. If there are 1000 people of this age in a certain town, what is the probability of 10 of them dying during the year?

8-3. Cosmic ray counts are believed to be completely random, i.e., follow a Poisson distribution. Is this true of the following data?

Number of counts in 1 min, r	0	1	2	3	4	5	6	7
Number of minutes having r counts	40	70	41	20	13	0	1	0

8-4. If a product is supposed to contain 2 percent of defective items, would we be justified in rejecting a consignment if, after testing 50 items, we find it to contain 5 defectives?

8-5. If the proportion of defective bearings being manufactured is $\frac{1}{25}$, approximate by means of a Poisson distribution the probability that a random sample of 75 bearings will contain 3 or fewer defectives.

8-6. The number of failures of telephones connected to a private exchange is shown in Table 8-3.

TABLE 8-3

Number of failures reported in a day, r	*Number of days with r failures*
0	101
1	60
2	31
3	8
4	0
5	1
6	0

Test the assumption that the failures occur randomly.

8-7. A study of four block faces containing 52 one-hour parking spaces was carried out and the results are given in Table 8-4.

TABLE 8-4

Number of vacant 1-hour parking spaces per observation period	0	1	2	3	4	5	≥ 6
Observed frequency	30	45	20	15	7	3	0

Assuming that the data follow a Poisson distribution, determine the

a. Mean number of vacant parking spaces,
b. Standard deviation, and
c. Probability of finding one or more vacant one-hour parking spaces.

8-8. A large batch of piston rings manufactured by a certain machine is examined by taking samples of 6 rings. It is found that the numbers of samples containing 0, 1, 2, 3, 4, 5, defective rings are 60, 36, 15, 7, 2, 0, respectively. Does this data follow a Poisson distribution?

Chapter 9

Normal Distribution

Chapters 7 and 8 dealt with distributions of occurrences of distinct events, i.e., with discrete variables. We shall now return to a consideration of quantities that vary continuously, and, specifically, to the properties of populations (and of samples drawn from such populations) whose individual members vary due to what is commonly referred to as errors.

ERRORS

In Chapter 2 we introduced the classification of errors into two broad types: *systematic* errors and *random* errors. The former arise from causes that act consistently under the given circumstances; for example, a rule calibrated at one temperature will read systematically incorrectly at another, and the same applies in the simpler case of a false zero through the end of the rule having been cut off. Such an error can and should be avoided by suitable experimental techniques.

But even when all systematic errors have been eliminated, there still remain accidental or random errors, which consist of a large number of very small effects, such as imprecision in an estimate of a fraction of a division on a scale, or a small, natural (random) variation in temperature from the standard at which the equipment has been calibrated (but not a seasonal variation that would introduce a systematic error and should be allowed for). Some of these effects are positive, others negative—i.e., they affect the value of the observation being made in a plus or minus direction with an equal probability. Thus the probability p of the occurrence of a positive error is $\frac{1}{2}$, and the probability q of its nonoccurrence (i.e., occurrence of a negative error) is also $\frac{1}{2}$. This fact is of fundamental importance and will be used in deriving the equation to the normal curve from the binomial distribution.

It is convenient to assume that all the small contributory errors are of equal absolute magnitude $|E|$, and in any particular case there are $2n$ of them. Thus the total error will range from $-2nE$ (when all the contributory errors are negative) to $+2nE$ (when they are all positive). In any intermediate case, there will be a surplus (or a deficiency) of positive errors equal in number to $2r$, so that the resultant positive error is $2rE = X$. It is this error that causes variation between observations even under most carefully controlled conditions.

GAUSS FUNCTION

The distribution of these errors can be derived from mathematical considerations, and is given by the so-called Gauss function:

$$y = Ce^{-h^2X^2} \qquad (9\text{-}1)$$

where X is the error (i.e., deviation from the mean or "true" value);

y the probability of occurrence of this error (or, strictly speaking, of an error in the range X to $X + \Delta X$);

e the base of natural logarithms;

C the constant that determines, as will be shown later, the maximum height of the curve; and

h the constant that determines the spread of the curve, i.e., expresses the precision of the measurement, and is known as the *precision constant*.

The Gauss function, given by Eq. (9-1), can also be viewed from another standpoint, namely, as an empirical (approximate) formula for the distribution of many physical quantities that have a continuously variable magnitude. These twin *raisons d'être* of the Gauss function make it particularly important in statistical work. We may note in passing that the function was derived not only by Gauss, but also by Laplace and de Moivre.

The distribution described by the Gauss function [Eq. (9-1)] is commonly known as the *normal* distribution, but the name should not be construed to mean that the distribution is any more normal than other distributions.

DERIVATION OF THE FUNCTION

A general mathematical derivation of the Gauss function is not considered appropriate in this book, but a derivation from the binomial distribution should be of engineering interest.

Let us consider the binomial distribution $(q + p)^n$ with $p = q = \frac{1}{2}$. The mean is then $\mu = np = \frac{1}{2}n$. The frequency polygon for this distribution with $n = 8$ is shown in Fig. 9-1. As n increases indefinitely, the frequency polygon will approach a smooth curve, symmetrical about a vertical line through the mean, and we can consider the normal curve to be the limit of the binomial expansion $(\frac{1}{2} + \frac{1}{2})^n$ as $n \to \infty$. The normal curve is plotted in Fig. 9-1 to

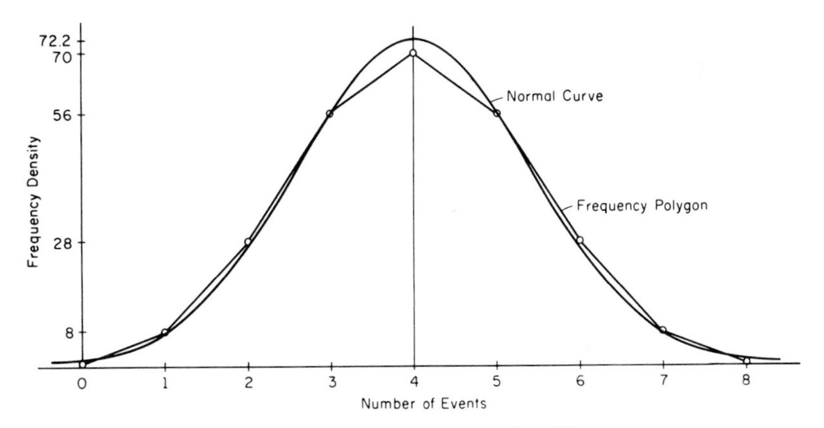

Fig. 9-1. Frequency polygon for binomial distribution $(\frac{1}{2} + \frac{1}{2})^8$ and the normal distribution curve, drawn to the same scale.

the same scale as the frequency polygon; i.e., the area under each is the same.

Let us consider the expansion

$$\left(\frac{1}{2} + \frac{1}{2}\right)^{2n} = 2^{-2n}(1 + {}_{2n}C_1 + {}_{2n}C_2 + \cdots + {}_{2n}C_{n+r} + \cdots + 1) \quad (9\text{-}2)$$

The successive terms in brackets are plotted in Fig. 9-2 as ordinates at intervals Δx, beginning with the first term at the origin $x = 0$. The maximum term in Eq. (9-2) is $2^{-2n} \times {}_{2n}C_n$, and this is the ordinate at the mean μ. It is convenient to transfer the origin to the mean, so that

$$\left.\begin{aligned} X &= x - \mu \\ \Delta X &= \Delta x \end{aligned}\right\} \quad (9\text{-}3)$$

The y coordinate remains unaltered.

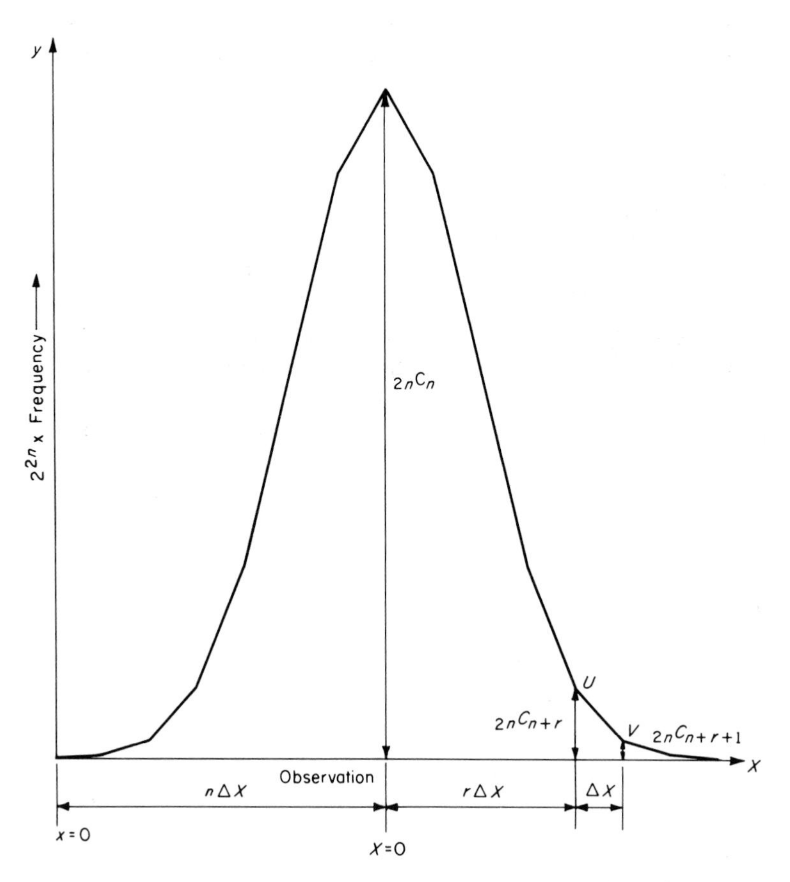

Fig. 9-2. Frequency polygon for binomial distribution $(\frac{1}{2} + \frac{1}{2})^{2n}$.

Consider points U and V in Fig. 9-2. Their ordinates are:

$$y_U = {}_{2n}C_{n+r} \qquad \text{and} \qquad y_V = y_U + \Delta y = {}_{2n}C_{n+r+1}$$

But

$$_{2n}C_{n+r+1} = \frac{_{2n}C_{n+r}(n-r)}{n+r+1}$$

(see footnote[1]).

1. Since $_mC_{k+1} = \dfrac{m!}{(m-k-1)!\,(k+1)!} = \dfrac{m!}{(m-k)!\,k!} \times \dfrac{(m-k)}{(k+1)} = {}_mC_k \times \dfrac{(m-k)}{(k+1)}$.

Therefore,

$$\begin{aligned}\Delta y &= y_V - y_U \\ &= {}_{2n}C_{n+r+1} - {}_{2n}C_{n+r} \\ &= {}_{2n}C_{n+r}\left(\frac{n-r}{n+r+1} - 1\right)\end{aligned}$$

or

$$\begin{aligned}\Delta y &= {}_{2n}C_{n+r} \times \frac{n-r-n-r-1}{n+r+1} \\ &= {}_{2n}C_{n+r} \times \left(\frac{-2r-1}{n+r+1}\right) \\ &= y \times \left(\frac{-2r-1}{n+r+1}\right)\end{aligned}$$

Hence

$$\frac{\Delta y}{\Delta X} = \frac{y}{\Delta X}\left(\frac{-2r-1}{n+r+1}\right) \tag{9-4}$$

The abscissa of point U is

$$X = r\,\Delta X$$

Hence

$$r = \frac{X}{\Delta X}$$

Substituting in Eq. (9-4),

$$\frac{\Delta y}{\Delta X} = \frac{y}{\Delta X}\left(\frac{-(2X/\Delta X) - 1}{n + (X/\Delta X) + 1}\right)$$

or

$$\frac{\Delta y}{\Delta X} = \frac{y}{\Delta X}\left(\frac{-2(X + \Delta X)}{n\,\Delta X + X + \Delta X}\right) \tag{9-5}$$

It is shown in Appendix C that in the expansion $(\frac{1}{2} + \frac{1}{2})^{2n}$ the mean of all the terms is

$$\mu = n\,\Delta x = n\,\Delta X \tag{9-6}$$

and the variance is

$$\sigma^2 = \frac{n}{2}(\Delta x)^2 = \frac{n}{2}(\Delta X)^2 \tag{9-7}$$

Substituting in Eq. (9-5), we obtain

$$\frac{\Delta y}{\Delta X} = \frac{-y(2X + \Delta X)}{2\sigma^2 + X\,\Delta X + (\Delta X)^2}$$

As $n \to \infty$,

$$\Delta X \to 0 \qquad \text{and} \qquad \frac{\Delta y}{\Delta X} \to \frac{dy}{dX}$$

Therefore,

$$\frac{dy}{dX} = -y\frac{X}{\sigma^2}$$

or

$$\frac{dy}{y} = -\frac{X\,dX}{\sigma^2}$$

Integrating,

$$y = Ce^{-(X^2/2\sigma^2)} \tag{9-8}$$

Putting

$$h = \frac{1}{\sqrt{2}\,\sigma} \tag{9-9}$$

(h being the precision constant of the Gauss function), we can write Eq. (9-8) as

$$y = Ce^{-h^2X^2}$$

which is Eq. (9-1).

THE NORMAL CURVE

The equation to the normal distribution curve can be written in several different forms. If we require the normal *frequency distribution* curve, the equation must satisfy the condition that the area under the curve is equal to the total number of observations N. Thus

$$\int_{-\infty}^{+\infty} y\,dX = N \tag{9-10}$$

The integration extends between $-\infty$ and $+\infty$ because, when we postulated in our derivation that $n \to \infty$, we extended the curve to infinity in either direction from the mean μ.

Such a curve can be directly compared with a histogram because the area under the histogram is also equal to the total number of observations.

Substituting in Eq. (9-10) from Eq. (9-8), we have

$$C \int_{-\infty}^{+\infty} e^{-X^2/2\sigma^2}\, dX = N \tag{9-11}$$

It is shown in Appendix D that

$$\int_{-\infty}^{+\infty} e^{-X^2/2\sigma^2}\, dX = \sigma\sqrt{2\pi} \tag{9-12}$$

Hence the maximum height of the curve is

$$C = \frac{N}{\sigma\sqrt{2\pi}} \tag{9-13}$$

and the equation to the normal frequency distribution curve can be written:

$$y = \frac{N}{\sigma\sqrt{2\pi}} e^{-X^2/2\sigma^2} \tag{9-14}$$

As previously defined, σ is the standard deviation of the population.

If we operate in terms of actual observations x (rather than deviations from the mean X), we can substitute from Eq. (9-14):

$$y = \frac{N}{\sigma\sqrt{2\pi}} e^{-(x-\mu)^2/2\sigma^2} \tag{9-15}$$

For many purposes it is preferable to deal with probability rather than frequency distribution. Since the sum of all probabilities is unity, the area under the normal *probability distribution* curve must be equal to unity. Such a curve is said to be *normalized*, and is given by

$$p(x) = \frac{1}{\sigma\sqrt{2\pi}} e^{-(x-\mu)^2/2\sigma^2} \tag{9-16}$$

where $p(x)$ is the *probability density* for the deviation $(x - \mu)$ and represents the rate of change of probability with x.

Since the mean μ is constant for any one distribution, the effect of μ is to move the position of the normal curve along the x-axis but not to change the shape of the curve, which depends on the value of standard deviation σ only. Thus σ determines the horizontal spread, and it is for many purposes convenient to use σ as a unit of deviation from the mean. We put

$$z = \frac{x - \mu}{\sigma} = \frac{X}{\sigma} \tag{9-17}$$

Hence

$$dz = \frac{1}{\sigma} dX = \frac{1}{\sigma} dx$$

Now, from our definitions of probability density and probability distribution of a continuous variable (see Fig. 5-7), changing the variable from x to z does not alter the probability sought; i.e., the probability that a value x_1 will occur in the interval between x and $x + \Delta x$ is equal to the probability that z_1 will occur in the interval between z and $z + \Delta z$. Thus we can deduce by means of calculus that

$$p(x)\, dx = f(z)\, dz$$

or

$$f(z) = p(x) \frac{dx}{dz}$$

where $p(x)$ and $f(z)$ are the probability densities in terms of the variables x and z, respectively. Thus from Eqs. (9-16) and (9-17) we have

$$f(z) = p(x) \frac{dx}{dz} = \frac{1}{\sigma\sqrt{2\pi}} e^{-z^2/2}\, \sigma$$

Hence

$$f(z) = \frac{1}{\sqrt{2\pi}} e^{-z^2/2} \tag{9-18}$$

It should be noted that the variable z has a mean of zero with a standard deviation of unity.

This equation for a normal probability distribution is said to be in a standard form,[2] and it is for this form that most statistical tables have been prepared. Table A-3 gives the values of $f(z)$. (To standardize a variate, we transform it, using Eq. (9-17); i.e., we subtract the mean from all values and divide the results by the standard deviation.)

From the above argument it is clear that μ and σ are the parameters of the normal distribution.

PHYSICAL SIGNIFICANCE OF STANDARD DEVIATION

Consider Eq. (9-16), using $X = x - \mu$. In order to find the point at which the curvature of this curve changes sign, we equate the second differential coefficient of the equation to zero. Thus

$$\frac{d^2p(X)}{dX^2} = \frac{-1}{\sigma^3\sqrt{2\pi}} \left(e^{-X^2/2\sigma^2} - \frac{X^2}{\sigma^2} e^{-X^2/2\sigma^2} \right) = 0$$

2. It applies to x when $\mu = 0$ and $\sigma = 1$.

whence

$$1 - \frac{X^2}{\sigma^2} = 0$$

or

$$X = \pm\sigma$$

In other words, the point of inflection occurs on either side of the mean at a distance equal to the standard deviation, i.e., $z = 1$. This establishes a physical significance of the standard deviation and should be remembered when sketching the normal curve.

AREA UNDER THE NORMAL CURVE

The area under the *normal probability* curve between the mean ($z = 0$) and $z = z$ is

$$F(z) = \int_0^z f(z)\,dz$$

Thus from Eq. (9-18),

$$F(z) = \frac{1}{\sqrt{2\pi}} \int_0^z e^{-z^2/2}\,dz \tag{9-19}$$

Figure 9-3 shows the normal probability curve, the abscissas being marked in terms of both X and z. The areas under the curve $F(z)$ corresponding to deviations in steps of one standard deviation are written on the figure. It may be noted that Eq. (9-19) cannot be integrated directly and $F(z)$ must be obtained by numerical methods (see Appendix D). The full range of

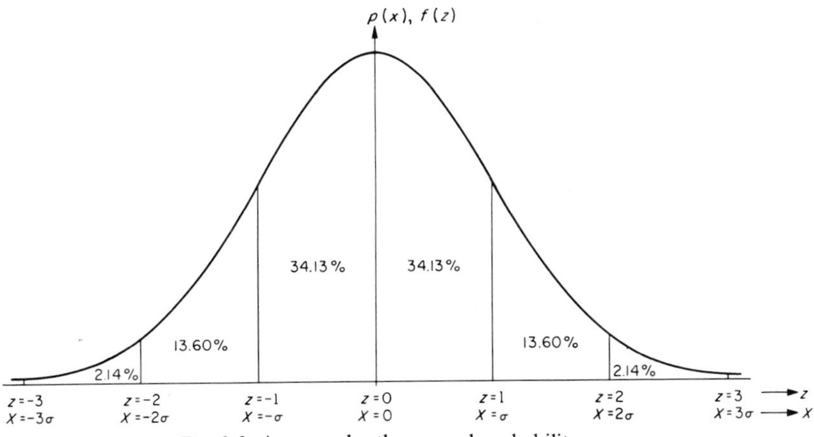

Fig. 9-3. Areas under the normal probability curve.

TABLE 9-1

Areas Under the Normal Curve. [See Eq. (9-19)]

z	$F(z)$	$2 \times F(z)$
0	0.0000	0.0000
0.25	0.0987	0.1974
0.50	0.1915	0.3830
0.67	0.2486	0.4972
0.6745	0.2500	0.5000
0.68	0.2518	0.5035
1.00	0.3413	0.6826
1.96	0.4750	0.9500
2.00	0.4772	0.9544
2.50	0.4938	0.9876
2.576	0.4950	0.9900
3.00	0.4987	0.9974

values of $F(z)$ is given in Table A-4. A selection of values of $F(z)$ is given in Table 9-1, together with areas for the range $(\mu - z\sigma, \mu + z\sigma)$, i.e., corresponding to a deviation from the mean not exceeding $\pm z\sigma$. The latter areas are equal to $2 \times F(z)$ and give the probability of a given deviation in *either* direction not being exceeded, while $F(z)$ gives values for a positive *or* a negative deviation not being exceeded. The distinction is obvious, but it is important to remember it at all times.

The probability of a deviation's lying between two values of z is given by the area under the normal curve between these limits. Since Table A-4 (and, indeed, all statistical tables) gives the values of $F(z)$ with one limit of $z = 0$ only,[3] we have to use differences. For example, the probability of a deviation z such that $1 \leq z \leq 2$ is obtained by subtraction of $F(z)$ for $z = 1$ from $F(z)$ for $z = 2$, namely, $0.4772 - 0.3413 = 0.1359$. Thus the probability of an observation's having a deviation from the mean not smaller than σ and not greater than 2σ is 0.1359.

Example. Find the probabilities of a set of observations, believed to be normally distributed, having values that fall outside the range specified in column 1 of Table 9-2. Hence find the number per 1000 outside the specified range.

To obtain column 2, we find $F(z)$ from Table A-4.

For example, in order to find the probability for the range $-\frac{1}{4}\sigma$ to $+\frac{1}{4}\sigma$, we observe that the area under the normal curve between $z = 0$ and $z = \frac{1}{4}$ is 0.0987. Since the curve is symmetrical, the probability for the range $-\frac{1}{4}\sigma$ to

3. Or $z = -\infty$ but not finite values.

TABLE 9-2

(1) Range	(2) Probability of falling inside the range	(3) Number of observations per 1000 outside the specified range
$-\frac{1}{4}\sigma$, $+\frac{1}{4}\sigma$	0.1974	803
$-1\frac{1}{2}\sigma$, $+\frac{1}{2}\sigma$	0.6247	375
$-\frac{1}{2}\sigma$, $+1\frac{1}{2}\sigma$	0.6247	375
$-\frac{1}{2}\sigma$, $+3\sigma$	0.6902	310
$+2\sigma$, $+4\sigma$	0.0228	977

$+\frac{1}{4}\sigma$ is $2 \times 0.0987 = 0.1974$. To obtain column 3, we multiply the probability of falling inside the range by 1000 and subtract the result from 1000.

To find the probability for the range $-1\frac{1}{2}\sigma$ to $+\frac{1}{2}\sigma$, we have to find $F(z)$ for 0 to $\frac{1}{2}\sigma$ (0.1915), and $F(z)$ for 0 to $1\frac{1}{2}\sigma$ (0.4332) and add the two results (obtaining 0.6247). We utilize the fact that $F(z)$ for 0 to $-1\frac{1}{2}\sigma$ is the same as $F(z)$ for 0 to $+1\frac{1}{2}\sigma$. For the same reason we can write the probability for the range $-\frac{1}{2}\sigma$ to $1\frac{1}{2}\sigma$ without further calculations.

For the probability for the range $+2\sigma$ to $+4\sigma$, we have to subtract $F(z)$ for 0 to 2σ (0.4772) from $F(z)$ for 0 to 4σ (0.49997); the answer is 0.02277.

Solved Problems

9-1. The finished inside diameter of a piston ring is normally distributed with a mean of 4.50 cm and a standard deviation of 0.005 cm. What is the probability of obtaining a diameter exceeding 4.51 cm?

SOLUTION

Given $\mu = 4.50$ cm and $\sigma = 0.005$ cm. The deviation from the mean is

$$x - \mu = 4.51 - 4.50 = 0.01$$

Hence

$$z = \frac{x - \mu}{\sigma} = \frac{0.01}{0.005} = 2.0$$

From Table A-4 for $z = 2.0$, $F(z) = 0.4772$. Therefore, the required probability

$$= 0.5000 - 0.4772$$

$$= 0.0228$$

$$= 2.28 \text{ percent (a chance of 1 in 44)}$$

9-2. The resistance of a foil strain gauge is normally distributed with a mean of 120.0 ohms and a standard deviation of 0.4 ohm. The specification limits are 120.0 ± 0.5 ohms. What percentage of gauges will be defective?

SOLUTION

Given $\mu = 120.0$ ohms and $\sigma = 0.4$ ohm. The allowable deviation from the mean is

$$x - \mu = \pm 0.5 \text{ ohm}$$

Hence

$$z = \frac{x - \mu}{\sigma} = \frac{0.5}{0.4} = 1.25$$

From Table A-4, for $z = 1.25$, $F(z) = 0.3944$. Therefore, the probability of a defective gauge

$$= 1 - 2 \times 0.3944$$
$$= 1 - 0.7888$$
$$= 0.2112$$
$$= 21.12 \text{ percent}$$

9-3. The measurement of the inside diameter of a cast-iron pipe is normally distributed with a mean of 5.01 cm and standard deviation of 0.03 cm. The specification limits are 5.00 ± 0.05 cm. What percentage of pipes is not acceptable?

SOLUTION

Refer to Fig. 9-4. For the upper range,

$$z = \frac{5.05 - 5.01}{0.03} = 1.33$$

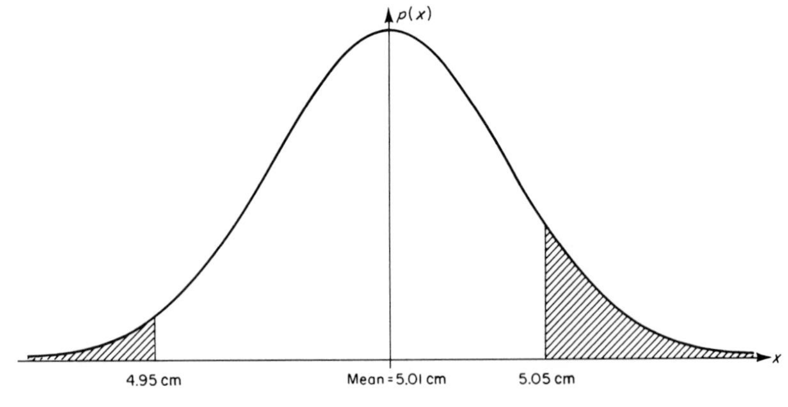

Fig. 9-4

From Table A-4, for $z = 1.33$, $F(z_1) = 0.4082$. For the lower range,

$$z = \frac{5.01 - 4.95}{0.03} = \frac{0.06}{0.03} = 2.0$$

The corresponding $F(z_2) = 0.4772$. The probability of falling within the specified limits

$$= F(z_1) + F(z_2)$$
$$= 0.4082 + 0.4772$$
$$= 0.8852$$

Probability of falling outside the limits

$$= \text{percentage of unacceptable pipes}$$
$$= 1 - 0.8852 = 0.1148$$
$$= 11.48 \text{ percent}$$

9-4. A teacher of a large class assigns grades using a system commonly known as "grading on the curve" in the following manner:

a. If the mark $> \bar{x} + 1.6\sigma$ the grade is A.
b. If $\bar{x} + 0.4\sigma \leq$ mark $\leq \bar{x} + 1.6\sigma$ the grade is B.
c. If $\bar{x} - 0.4\sigma \leq$ mark $\leq \bar{x} + 0.4\sigma$ the grade is C.
d. If $\bar{x} - 1.6\sigma \leq$ mark $\leq \bar{x} - 0.4\sigma$ the grade is D.
e. If mark $< \bar{x} - 1.6\sigma$ the grade is F.

Assuming that the marks are normally distributed with mean $\bar{x}$ and standard deviation σ (valid, since class size is large), what is the percentage of each grade given by the teacher?

SOLUTION

Referring to Fig. 9-5 and using Table A-4, we find that

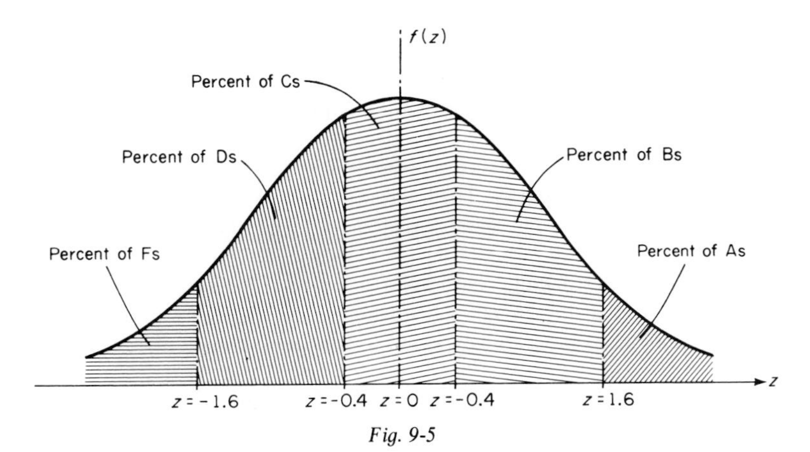

Fig. 9-5

(a) Percent of As: For $z = 1.6$, $F(z) = 0.4452$.

$$\text{area required} = 0.5 - 0.4452$$
$$= 0.0548$$

Therefore, the percent of As is approximately 5.5 percent.
(b) Percent of Bs: For $z = 1.6$, $F(z) = 0.4452$; for $z = 0.4$, $F(z) = 0.1554$.

$$\text{required area} = 0.4452 - 0.1554 = 0.2898$$

Therefore, the percent of Bs is approximately 29 percent.
(c) Percent of Cs: For $z = 0.4$, $F(z) = 0.1554$. Therefore, the area between limits $z = 0.4$ to $z = -0.4$

$$= 2 \times F(0.4) = 0.3108$$

Therefore, the percent of Cs is 31 percent.
(d) Percent of Ds: For $z = -1.6$, $F(z) = 0.4452$; for $z = -0.4$, $F(z) = 0.1554$.

$$\text{required area} = 0.4452 - 0.1554 = 0.2898$$

Therefore, the percent of Ds is approximately 29 percent.
(e) Percent of Fs: For $z = -1.6$, $F(z) = 0.4452$.

$$\text{required area} = \tfrac{1}{2} - F(z) = 0.0548$$

Therefore, the percent of Fs is approximately 5.5 percent.

Problems

9-1. If μ is the mean and σ is the standard deviation of the diameters of ball bearings, which follow a normal distribution, what percentage of ball bearings would have the following diameters? (a) Within the range $(\mu \pm 2\sigma)$. (b) Outside the range $(\mu \pm 0.9\sigma)$. (c) Greater than $(\mu - 1.5\sigma)$.

9-2. The strength of a plastic produced by a certain method is known to be normally distributed. If 10 percent of the results exceed 8000 N and 70 percent exceed 6000 N, what are the mean and the standard deviation?

9-3. If x is normally distributed with a mean of 100 and a standard deviation of 18, find the probability of a random observation falling between (a) 115 and 140. (b) 90 and 120.

9-4. If the I.Q. scores of recruits are normally distributed with a mean of 100 and a standard deviation of 13, find (a) The fraction who have an I.Q. greater than 133. (b) The fraction who have an I.Q. greater than 90. (c) The I.Q. exceeded by the upper quartile of recruits.

9-5. The navy uses stockings that have a mean life of 50 days with a variance of 64 $(\text{days})^2$. Assuming that the life of such stockings is normally distributed, of 150,000 pairs issued how many would be expected to need replacement after 42 days? After 63 days?

9-6. Electric bulbs bought for lighting an outdoor rink have a mean life of 3000 hours with a coefficient of variation of 11.3 percent. If it is more economical to replace all the bulbs when 20 percent of them have burned out than to change them as needed, after how many hours should the bulbs be changed?

Assuming that the lamps have not been changed, find the period after which an additional 20 percent of the lamps will have burned out.

9-7. A standard 5000-ohm resistor is measured a large number of times using an ohmmeter. It is found that one-half of all readings lie outside the range 4750 to 5250 ohms. Estimate the precision index h of the ohmmeter.

9-8. An automatic parachute has been equipped with a new altitude detector which has a precision index h of 0.004 m^{-1}. It is known that parachutes opening less than 30 m from the ground will smash and damage their payload. Of the 200 dropped parachutes with detectors set to open them at 300 m, how many will damage their payload?

Chapter 10

Use of Normal Distribution

Two comments on normal distribution may now be in order. In a mathematical derivation of the normal distribution curve, x (or X) is assumed to be a continuous variable. For this reason the probability of x having *exactly* a particular value is zero, and we consider only the probability of x, when chosen at random, falling between x and $x + \Delta x$; this is given by the area $p(x)\,\Delta x$. In the limit this area becomes $p(x)\,dx$. As mentioned before, $p(x)$ represents the probability density. See also Chapter 5.

LIMITS OF PRACTICAL DISTRIBUTIONS

The second comment concerns limits of the normal distribution curve. Equation (9-11) shows these as $-\infty$ and $+\infty$, and yet in many cases the observed values cannot extend so far from the mean; furthermore, negative values of x may sometimes have no physical meaning. To explain this anomaly, we should remember that we use the normal distribution as an approximation to physical observations, the mathematical equation being valid only between finite limits. The approximation is, however, very good as 99.9936 percent of the area under the normal curve lies within a range of four standard deviations from the mean; for three standard deviations the area contained is 99.740 percent (see Table A-4). We are, therefore, justified in the majority of cases in ignoring the *tails* of the distribution beyond $z = 4$, or some other appropriate value.

SOME IMPORTANT PROBABILITIES

Several values of area under the normal probability curve are of especial interest and are shown diagrammatically in Fig. 10-1. Figure 10-1a illustrates the fact that one-half of the area under the curve lies within $\pm 0.6745\sigma$ from the mean. Thus a single observation has an equal chance of falling

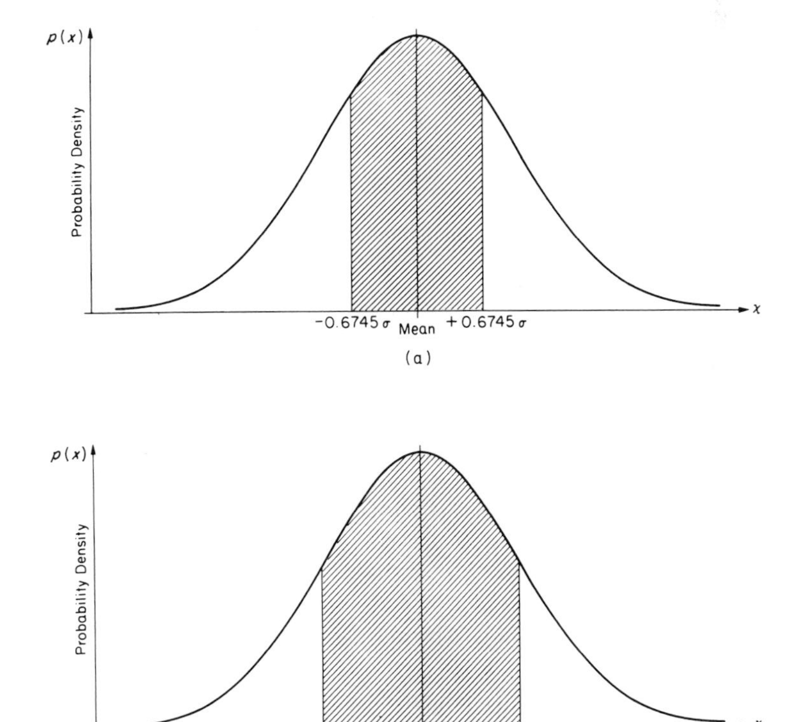

Fig. 10-1. Range of deviations from the mean for specified probabilities of an observation falling within the range: (a) 50 percent; (b) 68.26 percent;

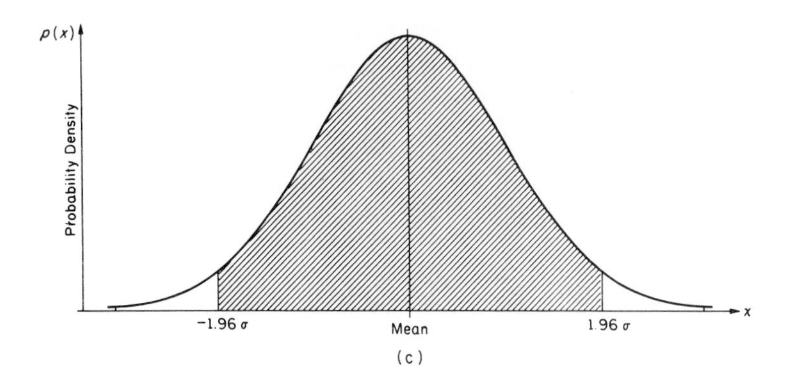

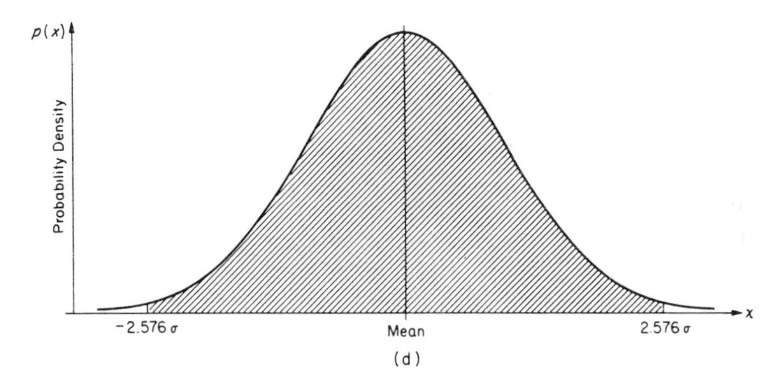

Fig. 10-1. (continued) (c) 95 percent; and (d) 99 percent.

within or without this range. For this reason such a deviation has in the past been called the *probable error*, but the term is not often used nowadays.[1]

Figure 10-1b shows that within a range of $\pm\sigma$ from the mean are contained 68.26 percent of all observations. A deviation of $\pm\sigma$ is referred to as a *standard error* and is often quoted as a measure of precision. Thus if a mean value of a quantity is given, e.g., as 27.05 ± 0.042, we interpret this to mean that the "true" value of the mean lies between $(27.05 - 0.042)$ and $(27.05 + 0.042)$, and there is a 68.26 percent probability of our being correct. In other words, if we present results in this form on a number of occasions, we shall be correct in our statements in 68.26 percent of all cases.

Figures 10-1c and d are of particular interest as they show the values of a deviation, positive or negative, exceeded by chance in 5 percent and 1 percent of all cases, respectively, these percentages being extensively used in

1. The probable error of the mean $\bar{x}$ is a value E such that there is a 50 percent probability that $\bar{x}$ does not differ from μ by more than E.

statistical treatments of data. If we are interested in a deviation of a specified sign only, say positive, the probability would be given by the area under the normal curve from the specified deviation to $+\infty$. Thus the magnitude of a positive deviation that would be exceeded in 1 percent of all cases is given by

$$0.5 + F(z) = 0.99$$

Hence

$$F(z) = 0.49$$

From Table A-4,

$$z = 2.33$$

i.e., a deviation of $+2.33\sigma$ is exceeded by chance in 1 percent of all cases.

FITTING A NORMAL CURVE

In experimental work we frequently obtain a set of observations that we consider as members of a population, but we may have no assurance that the data follow a normal distribution, or any other standard distribution. We may, however, suspect on the basis of past experience that certain observations conform to a given distribution, and it is one of the more important applications of statistics to the problems of measurement to investigate whether the data in hand fit an assumed distribution.

We shall now consider fitting a normal curve by following a numerical example. Before proceeding, however, we should recall two assumptions made in our derivation of the normal frequency curve:

a. The mean and the standard deviation of the normal frequency distribution are equal, respectively, to the mean and standard deviation of the actual observations.
b. The area under the normal frequency distribution curve is equal to the total number of observations, i.e., to the area under the histogram.

Example. Let us fit the normal curve to the theoretical frequency distribution of heads when 8 coins are tossed.

In Fig. 9-1 the binomial expansion of $(\frac{1}{2} + \frac{1}{2})^8$ was shown to have the following values:

Number of successes, r	0	1	2	3	4	5	6	7	8
Frequency, $p(r) \times 2^8$	1	8	28	56	70	56	28	8	1

Thus the total number of observations is $N = \sum$ frequency $= 256$. It should be noted that N is equal to the area under the histogram $= w \sum$ frequency, where w is the class width. In this case w is unity and hence $N = \sum$ frequency.

$$\text{mean} = \mu = np = \tfrac{1}{2} \times 8 = 4$$

and

$$\sigma = \sqrt{npq} = \sqrt{8 \times \tfrac{1}{2} \times \tfrac{1}{2}} = \sqrt{2}$$

Substituting in Eq. (9-15) for the normal frequency distribution,

$$y = \frac{N}{\sigma\sqrt{2\pi}} e^{-(x-\mu)^2/2\sigma^2} \tag{10-1}$$

i.e.,

$$y = \frac{128}{\sqrt{\pi}} e^{-(x-4)^2/4} \tag{10-2}$$

We can now plot y for various values of x and draw a smooth curve through the points. This has been done in Fig. 10-2, which shows also the

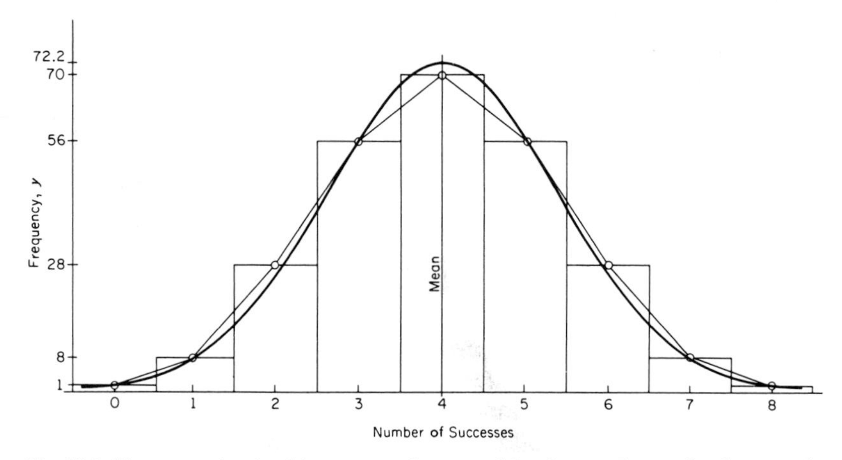

Fig. 10-2. Frequency density: histogram, polygon, and fitted normal curve for the example.

frequency polygon and histogram for the binomial distribution. Since we are dealing with an ideal distribution, the agreement between the curves is, of course, good, the discrepancy being due solely to the discontinuous character of the binomial distribution.

In practice, instead of calculating the values from Eq. (10-2), we use tabulated values of the ordinate of the normal curve. Table A-3 gives the ordinates for the normal curve in the standard form [Eq. (9-18)], i.e.,

$$f(z) = \frac{1}{\sqrt{2\pi}} e^{-z^2/2}$$

In our case the tabulated ordinates would have to be multiplied by

$$\frac{N}{\sigma} = \frac{256}{\sqrt{2}}$$

Let us now consider an example based on experimental results, as distinct from a theoretical distribution.

Example. Two hundred and fifty-five lengths of wire were roughly cut to a length of 100.060 m each. When measured, they were found to range between 100.005 m and 100.125 m. The measured lengths in excess of 100 m were recorded for intervals of 0.5 to 1.5 cm, 1.5 to 2.5 cm, ..., 11.5 to 12.5 cm, with the following observed frequency distribution:

Class midpoint (cm), x_i	1	2	3	4	5	6	7	8	9	10	11	12
Freqency, f_i	2	10	19	25	40	44	41	28	25	15	5	1

In order to fit the normal curve to this distribution, we have to find the mean and the standard deviation of the observed data. It is convenient to assume a fictitious mean $\bar{X}_0 = 6$ cm. We can then use a new variable $X'_i = x_i - \bar{X}_0$ and tabulate the results as follows:

TABLE 10-1

x_i	f_i	*Deviation from fictitious mean* X'_i	$f_i X'_i$	$f_i X'^2_i$
1	2	−5	−10	50
2	10	−4	−40	160
3	19	−3	−57	171
4	25	−2	−50	100
5	40	−1	−40	40
6	44	0	0	0
7	41	1	41	41
8	28	2	56	112
9	25	3	75	225
10	15	4	60	240
11	5	5	25	125
12	1	6	6	36
Totals	$\sum f_i = 255$		$\sum f_i X'_i = -197 + 263 = 66$	$\sum f_i X'^2_i = 1300$

Hence

$$\text{mean} = \bar{X} = \bar{X}_0 + \frac{\sum f_i X_i'}{\sum f_i} = 6 + \frac{66}{255} = 6.26 \text{ cm}$$

and

$$\sigma = \sqrt{\frac{\sum f_i X_i'^2 - [(\sum f_i X_i')^2/n]}{n-1}} = \sqrt{\frac{1300 - (66^2/255)}{254}} = 2.25 \text{ cm}$$

Substituting in Eq. (9-14), we obtain

$$\begin{aligned} y &= \frac{255}{2.25\sqrt{2\pi}} e^{-X^2/2(2.25)^2} \\ &= 45.21 e^{-0.0986 X^2} \end{aligned}$$

where

$$X = x - \bar{X}$$

It should be noted that $N = n = 255$, since the class width of the histogram is unity.

For the purpose of plotting the curve, we multiply the values of the ordinate of the normal curve in the standard form $f(z)$ (Table A-3) by $N/\sigma = 255/2.25$ and obtain:

$z = X/\sigma =$	0	±0.5	±1.0	±1.5	±2.0	±2.5	±3.0
$f(z) =$	0.3989	0.3521	0.2420	0.1295	0.0540	0.0175	0.0044
$y = f(z)(N/\sigma) =$	45.21	39.90	27.43	14.68	6.12	1.98	0.50
$X =$	0	1.12	2.25	3.37	4.50	5.62	6.75
$x =$	6.26	5.14	4.01	2.88	1.76	0.63	−0.49
		7.38	8.51	9.63	10.76	11.88	13.01

Figure 10-3 shows the plot of $y = (N/\sigma) f(z)$ against x.

It is important to note that the ordinates of the normal curve have been calculated for deviations of X from the mean, while the histogram is plotted for the arbitrary intervals of 1 cm, 2 cm, 3 cm, and so on. The height of the histogram blocks is obtained by dividing each class frequency by the class width; the histogram will then be plotted to the same scale as the normal

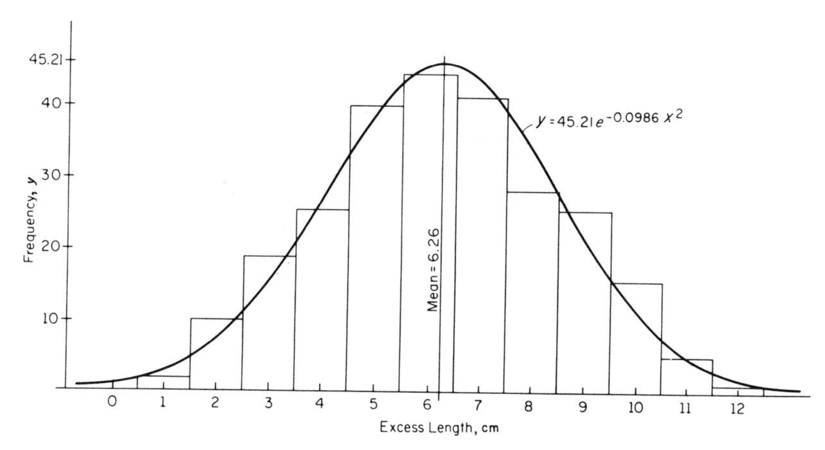

Fig. 10-3. Histogram and fitted normal curve for the example; the lengths plotted are those in excess of 100 m.

curve. Since in our case the class width is unity, the heights are numerically equal to frequencies. Finally, we must not forget that the lengths plotted are those in excess of 100 m.

CONTINUITY CORRECTION

This correction is applied when we approximate a discrete distribution by a continuous one. Figure 10-4 shows a discrete probability distribution of the variable r together with the continuous normal probability distribution. The probability that r will fall in the range $r_1 \leq r \leq r_2$ is the total area of the four rectangles marked A, B, C, and D. The approximation to this area afforded by the normal distribution, *without correction*, is the shaded portion. However, it can be readily observed that an improvement to this approximation can be made if the integration of the area under the normal probability curve is carried out from $r_1 - \frac{1}{2}$ to $r_2 + \frac{1}{2}$ instead of from r_1 to r_2, i.e., from the lower boundary of rectangle A to the upper boundary of rectangle D. Thus it can be seen that in the case of a discrete distribution, where the values of r differ by unity, the continuity correction consists of adding $\frac{1}{2}$ to the upper middle point (to form the upper limit of integration), and subtracting $\frac{1}{2}$ from the lower middle point (to form the lower limit of integration). If the discrete values of the statistic r differ by γ, then the corresponding continuity corrections will be $\pm\gamma/2$ to the integration limits.

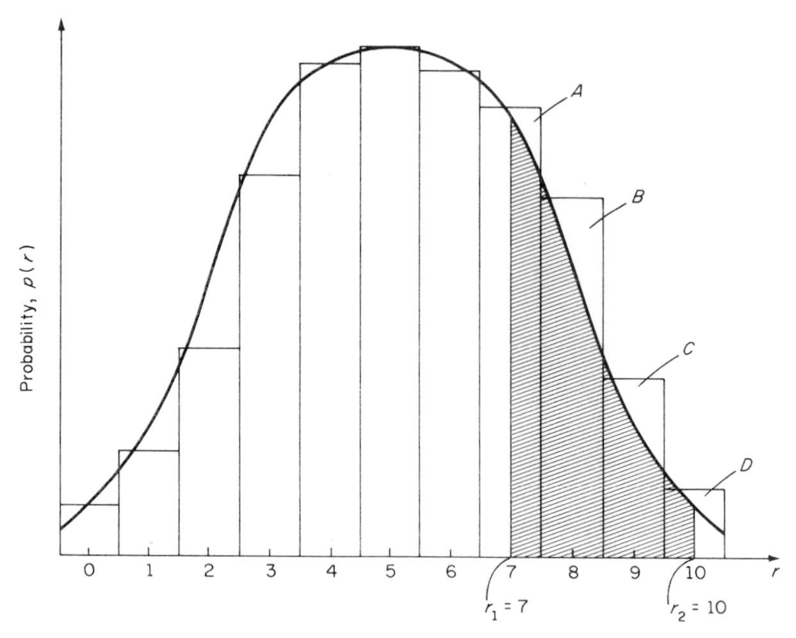

Fig. 10-4. Normal approximation to discrete distribution.

Example. Find the probability of obtaining between 4 and 7 heads in 10 throws of a coin, using the binomial expansion, and approximating by the normal distribution.

Using the binomial distribution, we write by virtue of the theorem of repeated trials, and from Eq. (7-1a),

$$P = \sum_{i=4}^{7} {}_{10}C_i \left(\frac{1}{2}\right)^{10}$$

$$= \frac{210 + 252 + 210 + 120}{1024}$$

$$= 0.7734$$

i.e., the probability of obtaining the required number of heads is 77.34 percent.

In the approximation based on the normal distribution, we obtain a more accurate result if we recognize that the extreme values (4 and 7 heads) represent class midpoints; the actual class intervals extend to the (rather theoretical) values of $3\frac{1}{2}$ and $7\frac{1}{2}$, respectively (Fig. 10-5). Thus the area under

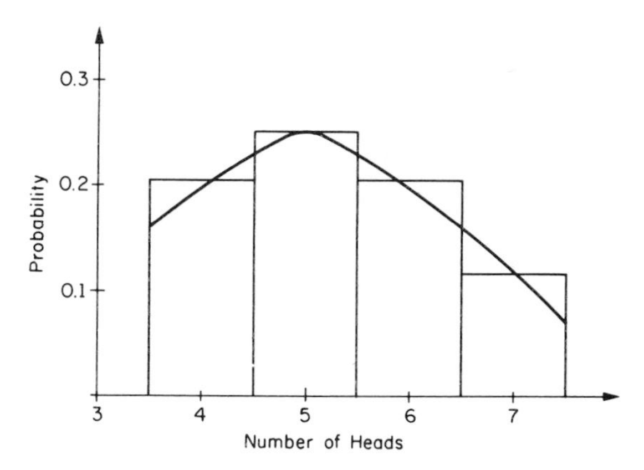

Fig. 10-5. Probability distribution for the example.

the histogram, which is to be equaled by the fitted normal curve, should be taken to these extremes. We have

$$p = q = \frac{1}{2}$$

$$\mu = np = 5$$

$$\sigma = \sqrt{npq} = 1.58$$

$$X_1 = 3.5 - 5 = -1.5$$

$$X_2 = 7.5 - 5 = +2.5$$

Therefore,

$$z_1 = \frac{-1.5}{1.58} = -0.95$$

and

$$z_2 = \frac{2.5}{1.58} = 1.58$$

The areas under the normal curve between these values of z and the mean are found from Table A-4 to be $0.3289 + 0.4430 = 0.7719$. The approximation is thus very close.

PROBABILITY PAPER

We may recall the cumulative frequency curve, mentioned in Chapter 2. When a variate follows a normal distribution, the cumulative frequency plots as an *S*-shaped curve. A curve of such shape is not convenient to use, and it is preferable to rectify it. This is achieved by the use of probability paper.

The construction of this paper is based on the fact that the cumulative frequency represents the area under the normal probability curve between $-\infty$ and the value of the variate up to which the cumulative frequency is required. Thus we make the increments in the ordinates (labeled as the cumulative area under the normal curve) equal for equal increments in the abscissas (representing the variate). On commercial probability paper the ordinates are marked off for probabilities of 0.01, 0.02, ..., 0.1, 0.2, ..., 1.0, 1.2, ..., 2, 3, ..., 20, 22, ..., 50 and on to 100 percent. Since the normal distribution is symmetrical, the spacing is symmetrical about the ordinate of 50 percent. The abscissas represent the variate to a linear scale. The normal probability paper is shown in Fig. 10-6.

We may note that the areas used in constructing the probability paper are the tail areas, while the area $F(z)$ given in Table A-4 is reckoned from the mean, i.e., from the axis of symmetry of the normal curve; this is shown diagrammatically in Fig. 10-7. The relation between the two is simply:

$$\text{tail area} = 0.5 - F(z)$$

When normal probability paper is used, a normal cumulative frequency plots as a straight line passing through the centroidal point (mean value of variate, 50 percent probability), and the paper is useful, therefore, in visual testing of normality of a set of observations.

As an example, Fig. 10-6 shows the cumulative frequency distribution for the data of Table 2-3, with the "best" straight line drawn by eye. Since we are interested in strength lower than a given value, the abscissas represent upper class boundaries. The mean is found as the abscissa of the point corresponding to a 50 percent cumulative frequency, in our case 6.9 MN/m^2, which agrees closely with the value obtained by calculation following Eq. (3-3).

Since the area under a normal curve between the mean and a deviation equal to the standard deviation is equal to 34.13 percent of the total area under the curve (see Fig. 9-3), we can read off an estimate of the standard deviation from our graph by finding the difference in abscissas between the mean and a point whose ordinate is $50 - 34.13 \simeq 16$ percent. This is illustrated in Fig. 10-6. In our case, the standard deviation is estimated to be 1.2 MN/m^2, compared with 1.39 MN/m^2 obtained by calculation following Eq. (4-8).

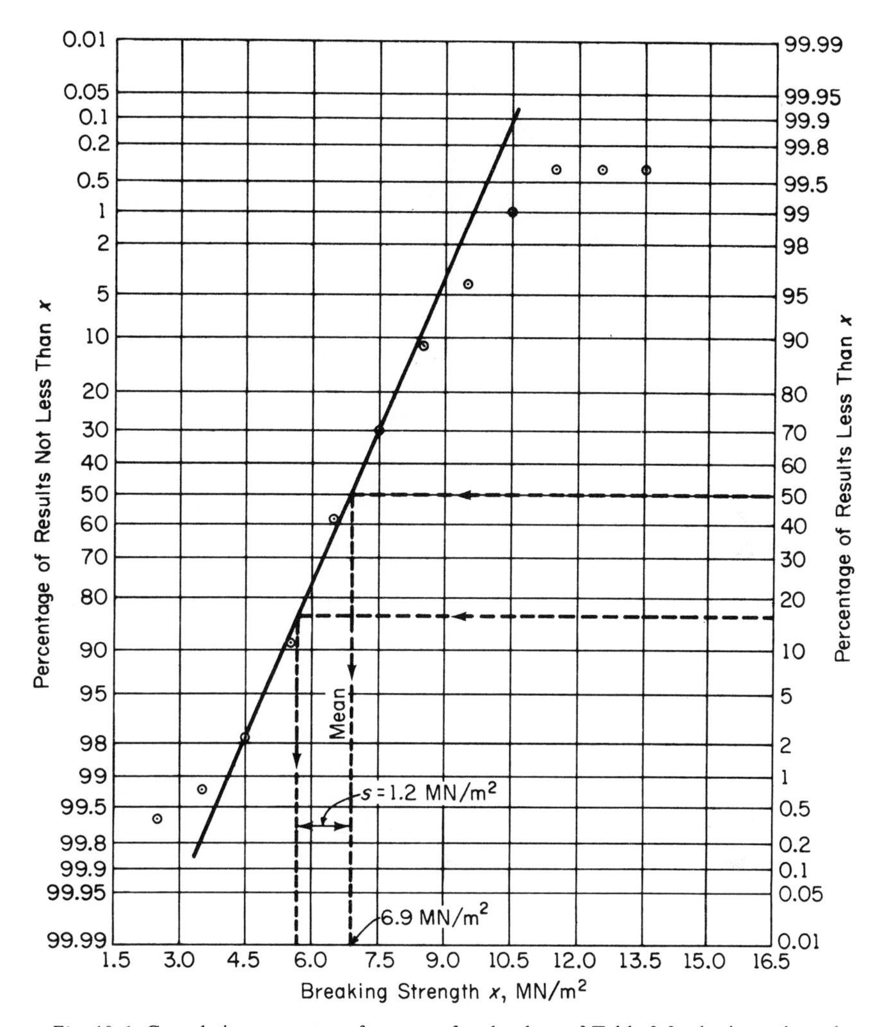

Fig. 10-6. Cumulative percentage frequency for the data of Table 2-3; abscissas show the upper class boundaries.

The agreement between the values read off Fig. 10-6 and the calculated values is good, and indeed the diagram shows that the actual distribution departs only very slightly from normal. We should note that in drawing the "best" line we pay more attention to points near the center of the distribution than to those near the extremes. Tails are bound to show scatter, since the actual number of observations in that region is usually very small; for example, the normal distribution may require 0.4 percent of observations to

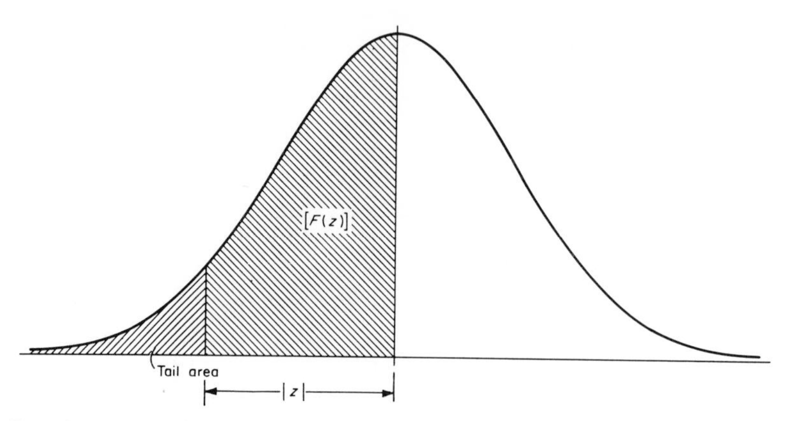

Fig. 10-7. Diagrammatic representation of tail area used in construction of probability paper and $F(z)$ of Table A-4.

have a strength less than 3.50 MN/m^2. This would correspond to $0.4 \times 10^{-2} \times 270 = 1.08$ observations. The actual number must be an integer—if 1, it is too low; if 2, it is too high.

To offer general guidance, we can say that a line which, even though straight, does not pass through the centroidal point (i.e., a point whose coordinates are the true sample mean and the 50 percent probability) signifies a skewed distribution.

If the points are so distributed that those corresponding to probabilities somewhat below 50 percent "less than" lie below a straight line through the centroidal point, and those corresponding to probabilities above 50 percent "less than" lie above the straight line, the distribution is more peaked than normal. If the deviations are in the opposite direction, the distribution is less peaked than normal. However, to establish the significance for departures from normality not less than about 500 observations are needed.

MEAN DEVIATION IN NORMAL DISTRIBUTION

The use of mean deviation d_m was discouraged in Chapter 4, but in the case of the normal distribution, d_m may be of use when only a very approximate value of standard deviation σ is required. It can be shown that when the variate is normally distributed,

$$\sigma = \sqrt{\frac{\pi}{2}}\, d_m$$

$$\sigma \simeq 1.25\, d_m$$

The mean deviation d_m has the advantage, of course, of being obtained very rapidly from Eq. (4-9).

CENTRAL LIMIT THEOREM

The importance of the normal distribution lies not only in the fact that numerous variables are actually nearly normally distributed, but also in the large body of statistical methods and tables derived for the normal distribution and often applicable approximately even to distributions departing from normal. In particular, numerous statistical techniques concerned with sampling involve the use of normal distribution.

We mentioned in Chapter 6 that sample means follow approximately normal distribution even if the underlying distribution is not normal, and we shall now present this more formally as the central limit theorem:[2] If we have $x_1, x_2, \ldots, x_n$ identically distributed independent random variables, each with mean μ and finite variance σ^2, then the variable y, given by

$$y = \sum_{i=1}^{n} x_i$$

will have a distribution $z = (y - n\mu)/\sqrt{n}\,\sigma$ which approaches a normal distribution with mean 0 and variance 1 as n becomes indefinitely large.[2] We can rearrange the expression for z by writing

$$z = \frac{y - n\mu}{\sqrt{n}\,\sigma} = \frac{y/n - \mu}{\sigma/\sqrt{n}} = \frac{\sum x_i/n - \mu}{\sigma/\sqrt{n}} = \frac{\bar{x} - \mu}{\sigma/\sqrt{n}}$$

which leads us to the conclusion that the mean of n identically distributed independent random variables will be approximately normally distributed, regardless of the underlying distribution of the individual variables. Thus we can state that if we draw samples of size n from a population with a mean μ and a finite variance σ^2, with an increase in n the distribution of sample means approaches a normal distribution with a mean μ and variance σ^2/n. We can see that virtually the only limitation on the underlying distribution is that the variance be finite, and this is satisfied in nearly all engineering and scientific problems.

For example, in Chapter 14 (p. 239) we deal with errors of measurements and we assume that such errors are normally distributed. This is quite valid by the central limit theorem, since these errors are usually composed of the sum of many small independent components. (See also p. 123.)

How good the approximation is for a given sample size depends on the shape of the underlying distribution. We can look at the problem another way and say that the further the underlying distribution is removed from normal, the larger the samples need to be for their mean to be nearly normally distributed. However, even if the underlying distribution is rectan-

2. Proof of this theorem can be found in H. Cramer, *Mathematical Methods of Statistics*, (Princeton, N.J.: Princeton University Press, 1946).

gular or triangular, the means of samples of four items or more are approximately normally distributed. The approximation is least accurate near the tails, and care is necessary when that part of the distribution is of importance.

Thus we can see how the use of normal distribution is extended by the central limit theorem. Furthermore, many statistical tests that have been derived for normal distribution (such as tests of significance and analysis of variance, dealt with in Chapters 13 and 18) remain valid when applied to sample means of distributions which depart from normality.

TRANSFORMATION TO A NORMALLY DISTRIBUTED VARIATE

If, however, a distribution is known not to be normal and its shape is known, it would be foolish to disregard this knowledge. But even then it may be possible to transform the data to a form that is normally distributed. For example, in fatigue tests the variation in the number of cycles which metal members survive before failing is such that logarithms of these life values are normally distributed.

The log transformation is applicable in many other cases, especially when the range of observations covers several orders of magnitude. The measurement of acidity by pH is an example of such a transformation in ordinary scientific work.

Transformation of the observed variable x other than by logarithms may be required to achieve a normal distribution relation; the more common transforms are $1/x$, $\sqrt{x}$, $\sqrt[3]{x}$, and so on.

The need for transformation may arise from the method of measurement used. For example, assume that spherical components are manufactured and their size (diameter) is known to be normally distributed. If we measure the components by weight, we shall find that the weights are not normally distributed, since they are proportional to a third power of the linear dimension. Therefore, changing the variable to the cube root of the weight will produce normal distribution. The same would apply in particle grading, where the size of particles is determined by a linear test but is measured by weight.

If we have no prior information as to which transformation should be used, we have to resort to trial and error. The transformation that yields the "best" straight line when the cumulative frequency of the transformed variable is plotted on normal probability paper is considered most suitable.

As an illustration of rectification by transformation, Fig. 10-8 shows a histogram for fatigue tests on 100 metal members.[3] This is markedly skewed,

3. J. Pope and N. Bloomer, "Statistics as Applied to Fatigue Testing," Metal Fatigue Symposium, Nottingham University, England, 1955.

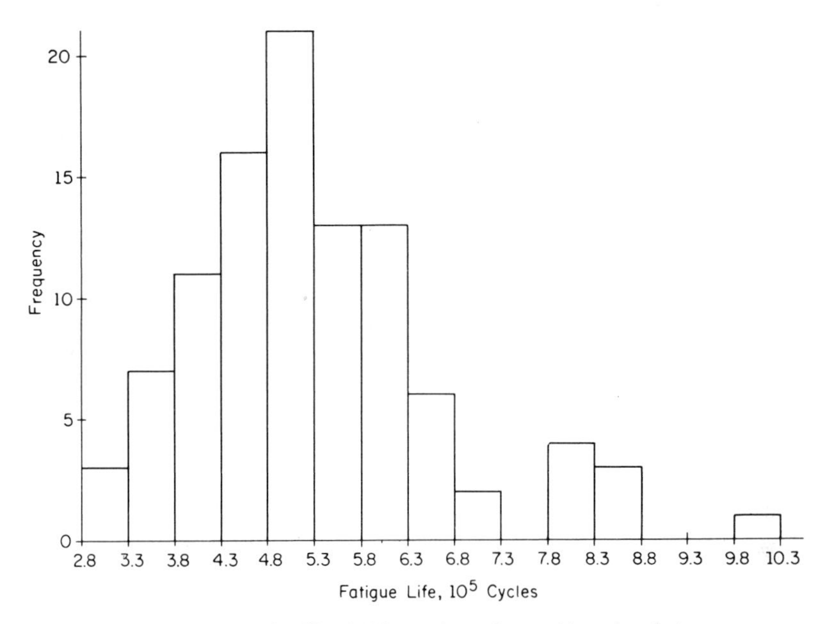

Fig. 10-8. Histogram for life of 100 metal members subjected to fatigue tests.

and the skewness is confirmed by the cumulative frequency plot in Fig. 10-9. Log transformation was applied to the data and resulted in an approximately normal distribution, as shown in Figs. 10-10 and 10-11.

We may note that the log transformation has to be applied to the original data before grouping; for this reason the number of class intervals and the class frequency after transformation are different from those in the original data.

In general terms, a random variable y whose logarithms x are normally distributed is said to have a *lognormal distribution.* As in any case of normal distribution, the range of x is $-\infty$ to $+\infty$, but the range of y is 0 to $+\infty$. Since

$$x = \log_e y$$

(although logarithms to any base can be used, the difference being only in a constant coefficient), when $y = 1$,

$$x = 0$$

when $y > 1$,

$$x > 0$$

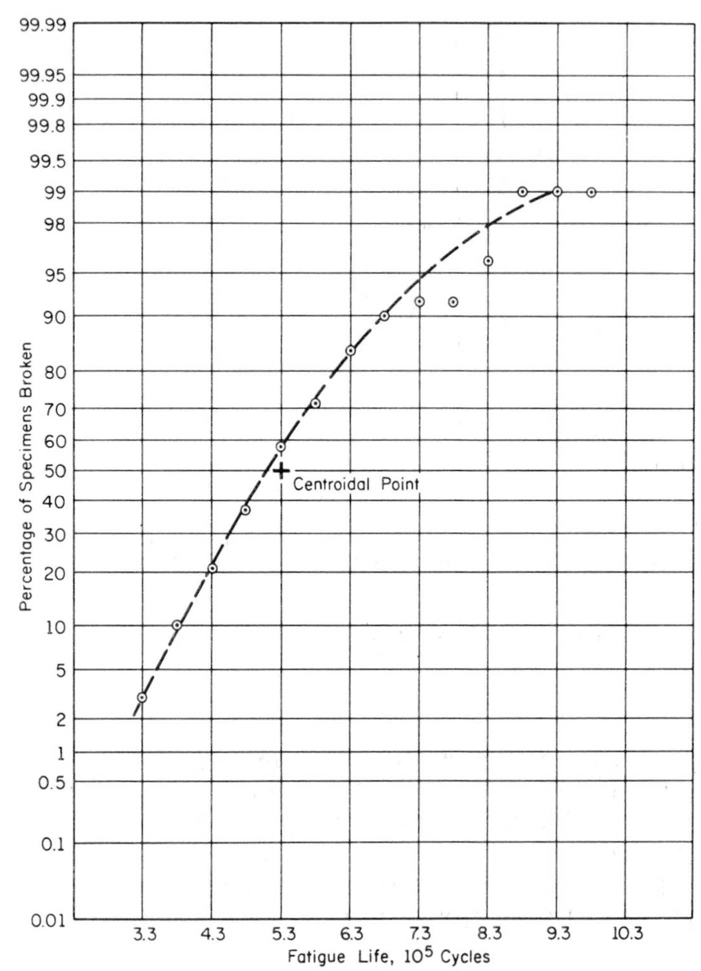

Fig. 10-9. Cumulative frequency diagram for data of Fig. 10-8 plotted on normal probability paper.

and when $0 \leq y < 1$,

$$-\infty \leq x < 0$$

It is evident that y cannot be negative, since the logarithm of a negative number is not defined. This fact has contributed, albeit probably in a small measure, to the use of the lognormal distribution when a finite probability of a negative value is physically absurd. However, the main reason for the use of a lognormal distribution is that the behavior of some variates is well described by a lognormal distribution. This applies to some hydrological

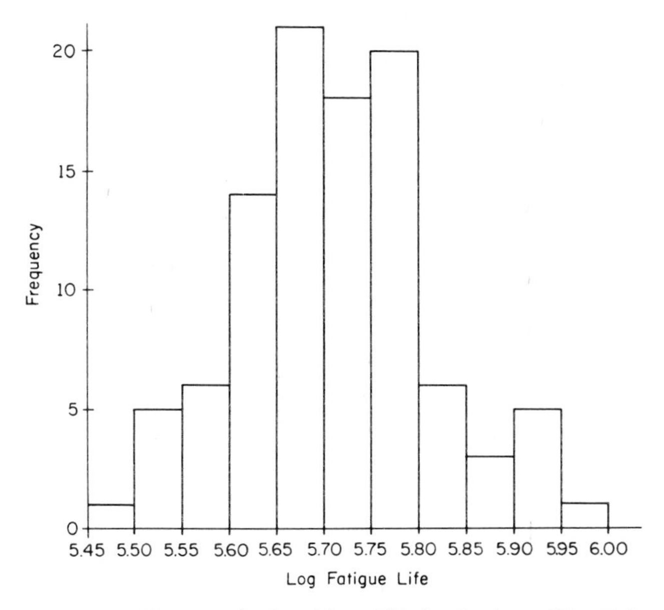

Fig. 10-10. Histogram for logarithm of life for the data of Fig. 10-8.

events, such as daily water flow, rainfall, or flood discharge, to the distribution of small particle sizes, to the strength of some materials, or to fatigue life. For dealing with such problems, lognormal probability paper, in which the abscissa is plotted to a logarithmic scale, is useful.

The use of lognormal distribution does not generally create any additional difficulties.

The use of lognormal distribution can be illustrated by its application in the field of soil testing. Let us assume that the "true" mean of the shear strength of a particular soil is τ_t. Using a small number of soil samples, n, we measure the shear strength τ. The mean value $\bar{\tau}[= (\sum \tau)/n]$ will deviate from the true value τ_t. The substructure is then designed on the basis of a reduced shear strength given by

$$\tau_d = \frac{\bar{\tau}}{L}$$

where L is a safety factor. Provided that a correct method of soil stability analysis is available, the probability of soil failure is expressed by

$$P\left[\left(\tau_t - \frac{\bar{\tau}}{L}\right) < 0\right] = \alpha$$

where α is the level of significance (see p. 158). Now, a normal distribution of the random variable τ is not likely, since τ cannot take negative values

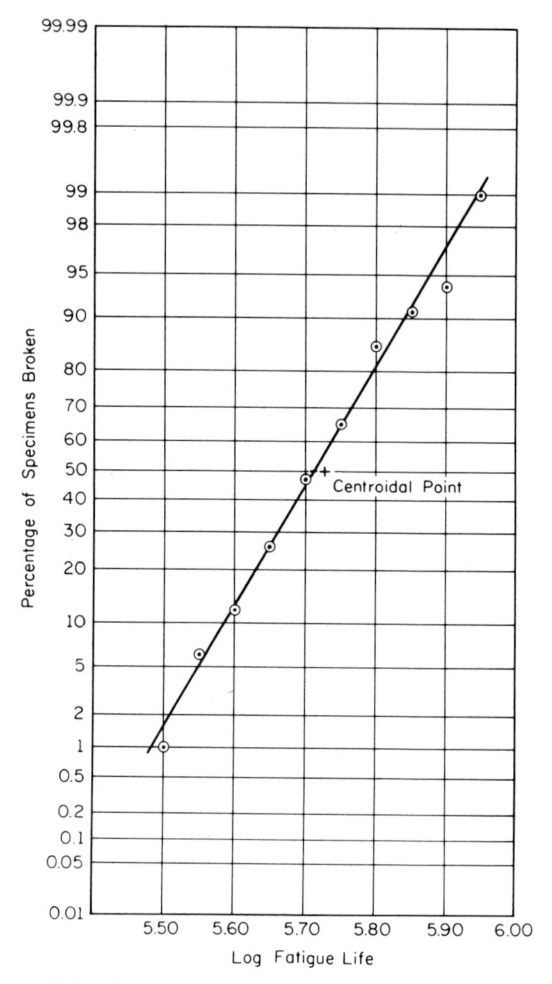

Fig. 10-11. Cumulative frequency diagram for data of Fig. 10-10, plotted on a normal probability paper.

(implied by the normal range $-\infty$ to $+\infty$). Instead, a lognormal distribution is assumed with a range from zero to ∞. The probability of soil failure can, therefore, be expressed as

$$P\left[\left(\log \tau_t - \overline{\log \frac{\tau}{L}}\right) < 0\right] = \alpha$$

where $\overline{\log (\tau/L)} = [\sum \log (\tau/L)]/n$. Since L is a constant, we can write the above probability statement as

$$P[(\log \tau_t - \overline{\log \tau} + \log L) < 0] = \alpha$$

Since we are dealing with a small sample, we can apply the t distribution (see Chapter 13). Thus

$$t = \frac{\overline{\log \tau} - \log \tau_t}{s/\sqrt{n}}$$

Substituting in the equation for probability, we obtain

$$P\left[\left(\log L - \frac{s}{\sqrt{n}}t\right) < 0\right] = \alpha$$

or

$$P\left[\left(t > \frac{\sqrt{n}}{s} \log L\right)\right] = \alpha$$

The standard deviation, s, is estimated from the sample data by

$$s = \sqrt{\frac{\sum (\log \tau - \overline{\log \tau})^2}{n - 1}}$$

This calculation is necessary, since τ_t is generally not available. In a given case the values of n, s, and L are known; hence the probability of failure P can be found readily corresponding to the t values from Table A-8. Other combinations of the various factors are also possible.

As an example let us consider the case when $n = 3$, $s = 0.02$ kN/m^2 and $L = 1.2$; we want to find the probability α. We obtain

$$t = \frac{\sqrt{3}}{0.02} \log_{10} 1.2 = \left[\frac{1.732}{0.02}\right](0.07918) = 6.85$$

The corresponding probability α from Table A-8 (for the one-sided test) is less than 0.025.

BINOMIAL APPROXIMATION

Although we have fitted different distributions to different types of sets of data, it is important to be aware of the relation between the various distributions. In Chapter 8 we saw how a markedly asymmetrical binomial distribution can be approximated by a Poisson distribution. Later we used a symmetrical binomial distribution in obtaining a normal distribution. The agreement between the terms of an expansion of $(q + p)^n$ (where $p = q = \frac{1}{2}$) and a normal distribution is better the larger the value of n. It is interesting to note that when n is very large, the approximation of the binomial distribution by the normal distribution is good, even if p differs considerably from q, as the binomial distribution loses a great deal of its skewness. For example, for $q = 0.8$, $p = 0.2$, and $n = 50$ the binomial distribution approximates closely to normal. The greater the difference between p and q, the larger n has to be for a given closeness of approximation.

LEVEL OF SIGNIFICANCE

When observations are normally distributed, it follows that $[1 - 2F(z)]$ represents the probability of a value's falling outside the range $\mu \pm z\sigma$. This probability is called the level of significance of a statistical test and is denoted by α. (See Fig. 10-12; also refer to p. 208 and Fig. 13-3.) Thus, when

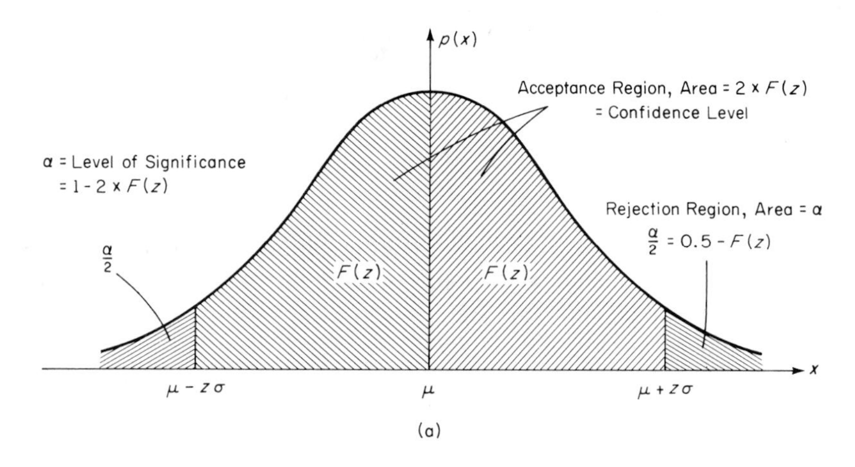

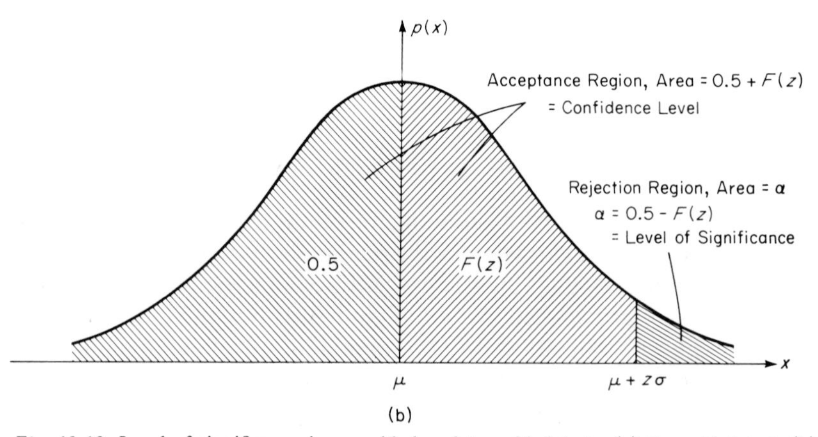

Fig. 10-12. Level of significance in one-sided and two-sided tests: (a) Two-sided test; (b) one-sided test.

$z = 1.96$, $[1 - 2F(z)] = 0.05$, and we say that the level of significance α is 5 percent. This means that if we obtain an observation that deviates from the mean by at least $\pm 1.96\sigma$, we can say that the observation is *significantly* different from the body of the data described by the given normal distribution, and the probability of our being in error is 5 percent. In other words, if we draw such a conclusion a large number of times, we shall be wrong in 5 percent of all the cases.

The term *significant* is used in the statistical sense of the word and means that the probability of the observed difference being due to chance alone is equal to the level of significance. A difference may be *statistically significant* but quite unimportant and not significant from the practical point of view.

From the fact that there is a 5 percent probability of an observation's having a deviation from the mean greater than $|1.96\sigma|$, it follows that there is a 95 percent probability that an observation will fall within the range $\mu \pm 1.96\sigma$, and this degree of confidence is referred to as the 95 percent confidence level. Thus, the *confidence level* (or confidence coefficient) is described by $2F(z)$ expressed as a percentage (see Fig. 10-12). The limits $(\mu - z\sigma)$ and $(\mu + z\sigma)$ are called *confidence limits*, and they describe between them the *confidence interval*. Thus we can say that if α measures our lack of confidence that a certain value x is not exceeded, then $(1 - \alpha)$ is the measure of our confidence or ability to prove that x is not exceeded. We can also think of the confidence level as the area of the acceptance region, whereas the significance level would be the area of the rejection region.

The values of z corresponding to the more commonly used values of the level of significance are presented in Table 10-2.

TABLE 10-2

Values of z for a Specified Percentage of Results to Lie Within the Range $\mu \pm z\sigma$

Level of significance (percentage of results outside the range), percent	z	*Confidence level (percentage of results within the range), percent*
10	1.645	90
5	1.960	95
2	2.326	98
1	2.576	99
0.1	3.291	99.9
0.01	3.891	99.99

The confidence interval, which can be calculated for any statistic, is of considerable importance as it expresses the reliability of our estimate of a parameter: the narrower the interval, the more precise the estimate.

If we know μ and σ, then we can say that, for example, the 99.74 percent confidence interval for the mean of a sample of size n is $\mu \pm (3\sigma/\sqrt{n})$. In the converse and more common case when μ is unknown, we can express the 99.74 percent confidence interval for μ as $\bar{x} \pm (3\sigma/\sqrt{n})$. This asserts that the true mean lies within the interval with a 99.74 percent probability of our being right.

When σ is not known but only estimated, the confidence limits must perforce be wider. This is discussed in Chapter 13; Chapter 14 deals with the confidence limits for variance.

We should stress the fact that all our statements are in terms of probability and it is not possible to *prove* whether or not an observation belongs to a population.

Example. Imagine that a coin was tossed 576 times and 256 heads were obtained. Are we justified in suspecting that the coin is biased or the experimenter dishonest? Use a level of significant of 1 percent.

Since $p = q = \frac{1}{2}$ and n is large, we can use the normal distribution as an approximation to binomial. We have

$$n = 576$$

$$\mu = np = 288 \text{ (a value obviously expected)}$$

$$\sigma = \sqrt{npq} = 12$$

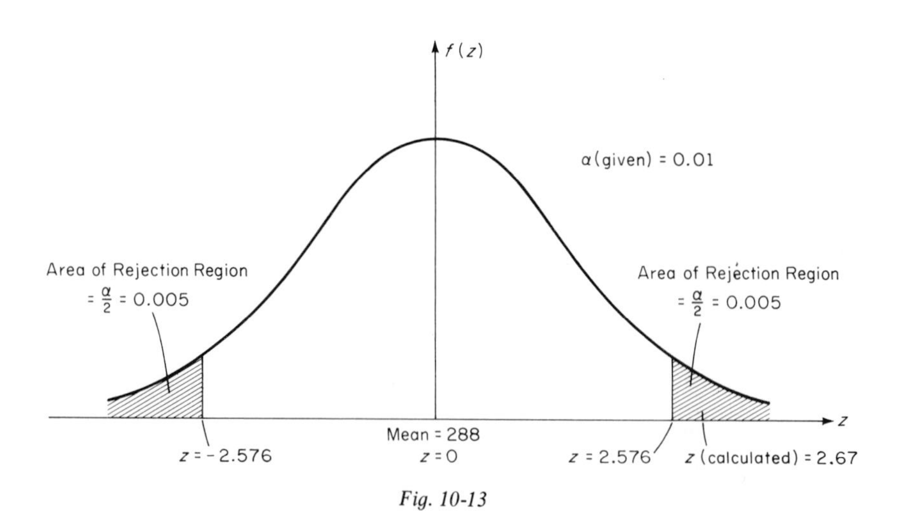

Fig. 10-13

The observed value is

$$x = 256$$

Hence

$$z = \frac{|x - \mu|}{\sigma} = 2.67$$

From Table 10-2 we find that a deviation of 2.576σ is exceeded only in 1 percent of all cases and the observed deviation is larger than the tabulated value. Figure 10-13 shows that the calculated z falls in the rejection region. We conclude, therefore, at the 1 percent level of significance, that the observed value does not belong to the population (of all the possible sets of 576 throws of a coin); the probability of our making a wrong accusation of dishonesty is 1 percent.

SPECIFYING A MINIMUM VALUE

When material such as concrete is used in construction, it is common to specify a certain minimum strength. Because of the statistical probability of encountering a test result falling below any specified minimum, the word is generally not taken to mean an absolute minimum. We require a specified confidence level; i.e., we stipulate that not less than a prescribed percentage of test results falls above the "minimum." If this is, say, 99 percent, the area of *one* tail is 1 percent,[4] i.e., $z = 2.326$. See Fig. 10-12b.

If the material used, for example, is concrete, the mix is proportioned so as to give a certain *mean* strength. This mean has to be chosen so that with the variance that is characteristic of the process of manufacture, the "minimum" has a value exceeded by 99 percent of the test results.

We can state the problem as follows: Let x_m be the "minimum" strength, μ the mean strength, and σ the standard deviation. Also, $z_m = (x_m - \mu)/\sigma$. Then,

$$F(z_m) + 0.5 \geq 0.99 \tag{10-3}$$

Since x_m is specified by the designer, the value of μ that satisfies Eq. (10-3) depends on σ. This is illustrated in Fig. 10-14 for $x_m = 20.7$ MN/m^2 and three different values of σ. In all cases the area under the normal curve to the left of the abscissa $x_m = 20.7$ MN/m^2 is the same.

It is clear that the higher the value of σ, the higher the necessary value of μ for the specified x_m, and if the cost of manufacture is related to μ, which is generally the case, then a higher value of σ requires the use of a more expensive material. On the other hand, a reduction in σ demands a closer

4. Note that Table 10-2 has been prepared for results falling within a range, i.e., with two tails taken into account.

control of manufacture and, therefore, a higher cost so that in practice the choice of σ (and therefore μ) is a result of a compromise.

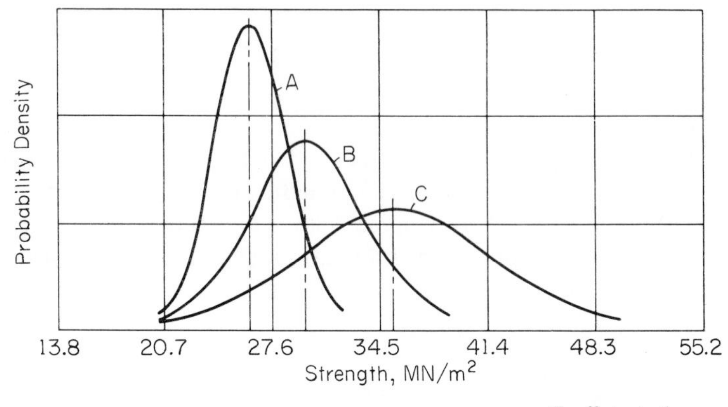

	Mean, MN/m²	*Standard deviation, MN/m²*	*Coefficient of variation, percent*
A	26.2	2.4	9.2
B	29.7	3.9	13.0
C	35.2	6.2	17.6

Fig. 10-14. Normal distribution curves for concrete with a minimum strength (exceeded by 99 percent of results) of 20.7 MN/m^2 and different values of standard deviation. [From A. M. Neville, *Properties of Concrete* (London: Pitman Publishing Corporation; New York: Wiley Interscience, 1973.)]

Solved Problems

10-1. Foil strain gages are produced with a mean resistance of 120.0 ohms. If the specification limits are 120 ± 0.5 ohms, what is the maximum allowable standard deviation that will permit no more than one gage in 1000 to be defective? It is assumed that the resistance of the gages is normally distributed.

SOLUTION

Given $\mu = 120.0$ ohms, $x - \mu = 0.5$ ohm. Refer to Fig. 10-15. The area under the normal curve inside the specification limits must be

$$1 - \frac{1}{1000} = 0.999$$

$$\text{half this area} = \frac{0.999}{2} = 0.4995$$

For this value of $F(z)$, Table A-4 gives $z = 3.27$. Hence

$$\sigma = \frac{x - \mu}{z} = \frac{0.5}{3.27}$$

$$= 0.153 \text{ ohm}$$

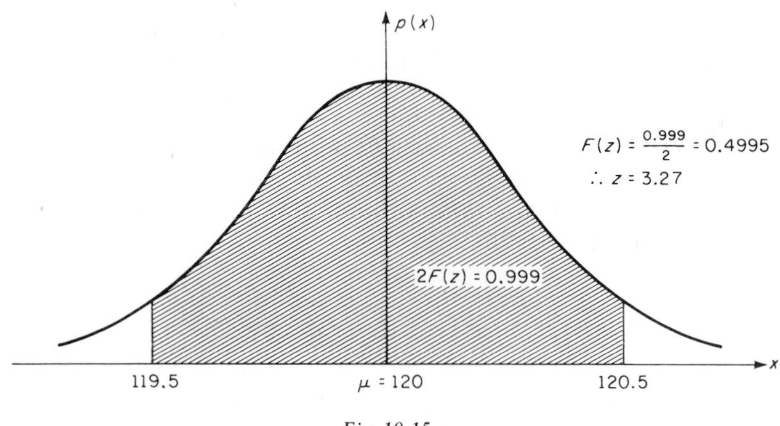

Fig. 10-15

10-2. A certain process of manufacture produces records whose mass is normally distributed with a standard deviation of 5 g. What must be the mean mass if the probability of obtaining a mass exceeding 210 g is to be 0.01?

SOLUTION

Given $\sigma = 5$ g and $x = 210$ g. The area under the normal curve is

$$F(z) = 0.5000 - 0.01$$
$$= 0.4900$$

The corresponding z from Table A-4 is

$$z = 2.33$$

But

$$z = \frac{x - \mu}{\sigma}$$

Therefore,

$$210 - \mu = z\sigma = 2.33 \times 5 = 11.65$$

Hence

$$\mu = 198.35 \text{ g} \simeq 198 \text{ g}$$

10-3. Results from a tensile test on 36 aluminum alloy specimens chosen at random from a certain mill were grouped as follows:

Tensile strength, MN/m^2	155—165	165—175	175—185	185—195	195—205	205—215	215—225
Frequency	1	4	7	11	9	3	1

a. Find the mean and standard deviation of the tensile strength of the specimens.
b. Plot on normal probability paper the tensile strength versus fractional cumulative frequency. Draw "by eye" a straight line through the points and obtain the mean and standard deviation from the graph; observe whether the results appear to follow the normal distribution.
c. Estimate the probability of obtaining a random measurement that has a deviation from the mean of between -10 MN/m^2 and 20 MN/m^2.
d. Calculate the range in which we would expect the mean of the population to fall with a probability of 95 percent.

SOLUTION

a. Take origin at 160 MN/m^2 and a class width $w = 10$ MN/m^2. See Table 10-3.

$$\text{mean} = \bar{x} = 160 + w \times \frac{\sum f_i X_i'}{\sum f_i} = 160 + 10 \times \frac{108}{36} = 190 \text{ MN/m}^2$$

standard deviation s

$$= w\sqrt{\frac{\sum f_i X_i'^2 - (\sum f_i X_i')^2/n}{n-1}} = w\sqrt{\frac{386 - (108)^2/36}{35}}$$

$$= 1.33 \times w = 1.33 \times 10$$

$$= 13.3 \text{ MN/m}^2$$

b. The cumulative frequency is plotted on normal probability paper in Fig. 10-16 using the upper boundary of each class interval. We find that the mean is 189.7 MN/m^2 and the standard deviation 12.7 MN/m^2. The graph indicates that the results follow a normal distribution very closely.
c. Given $\mu - x_1 = 10$ MN/m^2, and $x_2 - \mu = 20$ MN/m^2

$$\mu \text{ (estimated from the sample mean)} = 190 \text{ MN/m}^2$$

$$\sigma \text{ (estimated from the sample)} = 13.3 \text{ MN/m}^2$$

TABLE 10-3

Class interval MN/m^2	*Class midpoint* x_i	*Frequency* f_i	*Cumulative frequency* F	*Fractional cumulative frequency* F/n	*Deviation from origin in terms of class width* X_i'	$f_i X_i'$	$f_i X_i'^2$
155–165	160	1	1	0.028	0	0	0
165–175	170	4	5	0.139	1	4	4
175–185	180	7	12	0.333	2	14	28
185–195	190	11	23	0.639	3	33	99
195–205	200	9	32	0.889	4	36	144
205–215	210	3	35	0.972	5	15	75
215–225	220	1	36	1.000	6	6	36
Totals		$\sum f_i = 36$				$\sum f_i X_i' = 108$	$\sum f_i X_i'^2 = 386$

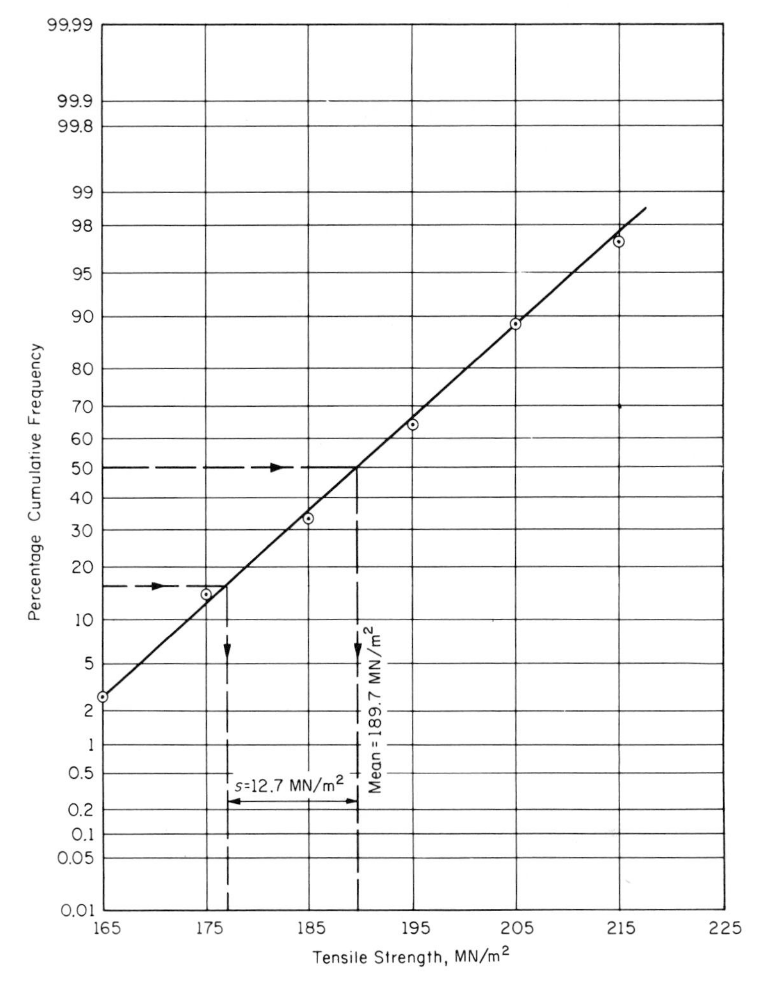

Fig. 10-16

(Refer to Fig. 10-17.) Then

$$z_1 = \frac{\mu - x_1}{\sigma} = \frac{10}{13.3} = 0.752$$

From Table A-4,

$$F(z_1) = 0.2740$$

Also,

$$z_2 = \frac{x_2 - \mu}{\sigma} = \frac{20}{13.3} = 1.504$$

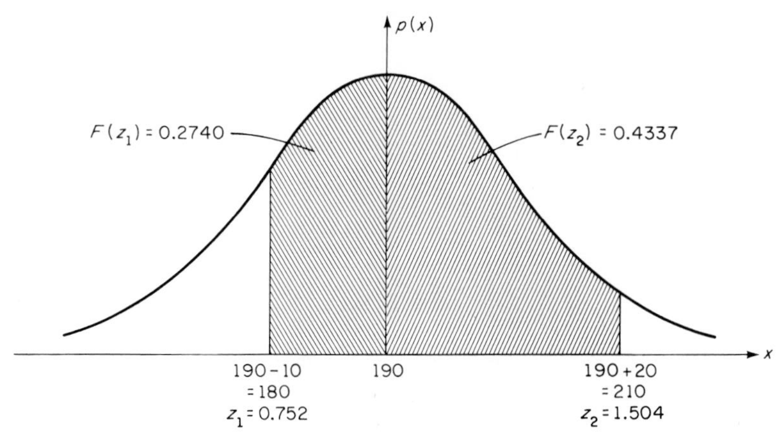

Fig. 10-17

Therefore,

$$F(z_2) = 0.4337$$

The probability that a random measurement will fall within the limits (190 − 10) MN/m^2 and (190 + 20) MN/m^2, i.e., between 180 and 210 MN/m^2, is

$$F(z_1) + F(z_2) = 0.2740 + 0.4337 = 0.7077$$

$$= 70.77 \text{ percent}$$

d. Table A-4 gives the area under the normal curve on one side of the mean. Thus for a probability of 95 percent the area required = 0.95/2 = 0.475. (See

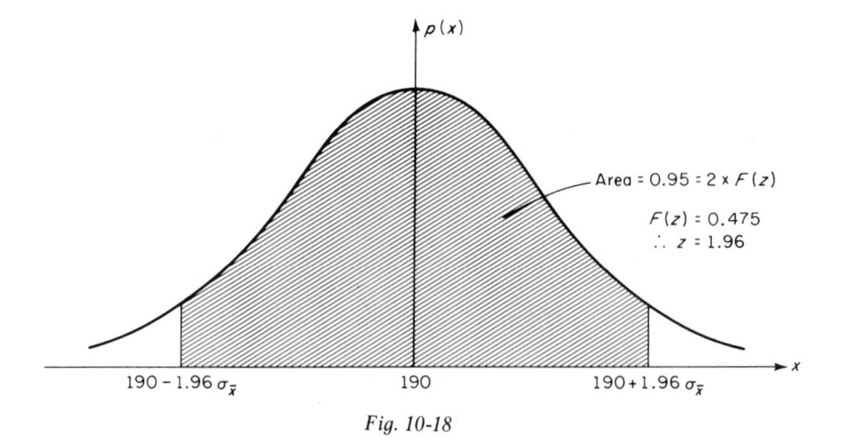

Fig. 10-18

Fig. 10-18.) The corresponding z from Table A-4 is $z = 1.96$. The range within which we would expect the population mean to fall is

$$190 \pm z\sigma_{\bar{x}} = 190 \pm 1.96 \times 13.3/\sqrt{36}$$
$$= 190 \pm 4.3 \text{ MN/m}^2$$

10-4. A steel company made numerous studies on the life span of their plate products immersed in water of a particular city. The results showed that the life span of such plate products was normally distributed, with a mean $\mu = 2160$ days and standard deviation $\sigma = 252$ days.

a. What is the probability that the mean of a sample of 121 plates will not differ from μ by more than 60 days?
b. Find the 95 percent confidence limits for the mean of a sample of 121 plates and for a single plate.
c. What should be the size of a sample in future studies if it is required that the probability of the sample mean being in error by more than 90 days is 5 percent?

SOLUTION

a. Given $\mu = 2160$ days and $\sigma = 252$ days, we find

$$\sigma_{\bar{x}} = \frac{\sigma}{\sqrt{n}} = \frac{252}{\sqrt{121}} = 22.9$$

$$z = \frac{\bar{x} - \mu}{\sigma_{\bar{x}}} = \frac{60}{22.9} = 2.62$$

From Table A-4 for such a value of z, $F(z) = 0.4956$. This is the area to one side of the mean of the normal curve. For a deviation in either direction, the area is $2 \times 0.4956 = 0.9912$. Therefore, the probability that the mean life span of 121 plates will not differ from $\mu = 2160$ days by more than 60 days is 99.12 percent.

b. For 95 percent confidence limits the area under the normal curve is 0.95. Therefore, the area to either side of the mean is $0.95/2 = 0.475$, for which (from Table A-4) $z = 1.96$. Hence the deviation of the mean life span of 121 plates is

$$z\sigma_{\bar{x}} = \frac{z\sigma}{\sqrt{n}} = \frac{1.96 \times 252}{\sqrt{121}} = 44.8 \text{ or } 45 \text{ days}$$

Thus the 95 percent confidence limits for the mean life span of 121 plates are

$$\mu \pm z\sigma_{\bar{x}} = 2{,}160 \pm 45 \text{ days}$$

For a single plate, the deviation of its life span from μ is $z\sigma = 1.96 \times 252 = 494$ days. Thus the range in which a single observation will fall with a probability of error of 5 percent is 2160 ± 494 days.

c. Table A-4 uses the area to one side of the mean. Thus

$$\text{required area} = \frac{1 - 0.05}{2} = 0.475$$

$$\text{corresponding } z = 1.96$$

The given deviation is $\bar{x} - \mu = 90$ days. From

$$z = \frac{\bar{x} - \mu}{\sigma_{\bar{x}}} = \frac{\bar{x} - \mu}{\sigma/\sqrt{n}} = \frac{90}{252/\sqrt{n}} = 1.96$$

we find

$$\sqrt{n} = 5.49$$

or

$$n = 30$$

Hence the sample size to be used is 30.

10-5. On a construction project the shear strength of 50 soil specimens was measured, and the following values (in kN/m^2) were observed:

2450	3300	3400	3650	3800
2650	3150	3100	3500	2850
3050	4300	3300	3300	3150
2100	3300	3650	3150	3550
2900	3250	3000	3400	3750
3900	3600	3150	3600	3000
4200	3700	3050	3300	2350
4150	2950	3200	3900	3200
3200	3450	2500	3050	2650
3050	2800	2700	3450	3400

a. Group these strengths into a frequency distribution with class width $w = 250$ kN/m^2, starting with 2000 kN/m^2.
b. Draw a histogram and a frequency polygon.
c. Calculate the mean, an estimate of the standard deviation, and the coefficient of variation from the grouped data; indicate the mean on the histogram.
d. What is the range of the given data? Find also the mode and the median of the grouped data.
e. By plotting the fractional cumulative frequency on normal probability paper, check whether the distribution of the data is approximately normal.
f. Assuming normal distribution, find the minimum shear strength that can be used for design, accepting a definite risk that 1 percent of test samples will have a strength less than this minimum. Indicate whether any of the 50 samples fall below this strength.
g. Calculate the standard error of the mean and explain its significance.
h. Obtain the equations to the normal probability curve and the normal frequency curve for the given data. Draw the normal probability curve.
i. Estimate the probability that a random test result will have a deviation from the mean lying between -200 and $+500$ kN/m^2.
j. Estimate the probability that a mean of a sample of 36 soil specimens from the same site will exceed 3500 kN/m^2.
k. Find the limits for the mean of a sample of 36 soil specimens at the 1 percent and 5 percent level of significance.

1. What should be the size of a sample in future tests in order that the probability of the sample mean being in error by more than 400 kN/m² be not more than 0.1?

SOLUTION

a. The required grouping of the test results is shown in Table 10-4.

TABLE 10-4

Class interval kN/m²	Class midpoint x_i	Frequency f_i	Cumulative frequency F	Fractional cumulative frequency F/n	Deviation from origin in terms of class width X'_i	$f_i X'_i$	$f_i X'^2_i$
2000–2250	2125	1	1	0.02	0	0	0
2250–2500	2375	2	3	0.06	1	2	2
2500–2750	2625	4	7	0.14	2	8	16
2750–3000	2875	4	11	0.22	3	12	36
3000–3250	3125	14	25	0.50	4	56	224
3250–3500	3375	11	36	0.72	5	55	275
3500–3750	3625	7	43	0.86	6	42	252
3750–4000	3875	4	47	0.94	7	28	196
4000–4250	4125	2	49	0.98	8	16	128
4250–4500	4375	1	50	1.00	9	9	81
Totals		$\sum f_i = 50$				$\sum f_i X'_i = 228$	$\sum f_i X'^2_i = 1210$

b. The histogram and the frequency polygon are shown in Fig. 10-19.

c. If the arbitrary origin is taken as 2125 kN/m², then the deviation of the class midpoint from this origin, X'_i, is calculated as shown in Table 10-4, with a class width $w = 250$ kN/m². Thus from this table we get the mean:

$$\bar{x} = 2125 + \frac{w \sum f_i X'_i}{\sum f_i} = 2125 + \frac{250 \times 228}{50}$$

$$= 3265 \text{ kN/m}^2$$

The estimate of the standard deviation is

$$s = w\sqrt{\frac{\sum f_i X'^2_i - [(\sum f_i X'_i)^2/\sum f_i]}{n-1}}$$

$$= 250\sqrt{\frac{1210 - [(228)^2/50]}{49}}$$

$$= 466 \text{ kN/m}^2$$

The coefficient of variation $V = s/\bar{x} \times 100 = 46{,}600/3265 = 14.27$ percent.

d. Range $= 4300 - 2100 = 2200$ kN/m² (from the raw data). For the grouped data, mode is the midpoint of the class interval with the highest frequency =

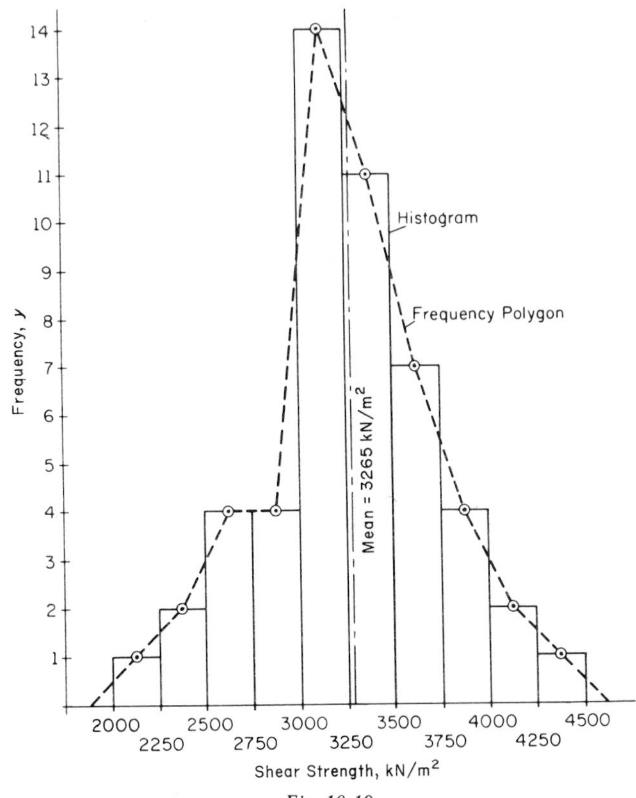

Fig. 10-19

3125 kN/m^2. Median is the value that divides the histogram into two equal areas = 3250 kN/m^2 (by inspection).

e. The fractional cumulative frequency (in percent) is plotted against the upper-class boundary on normal probability paper, Fig. 10-20. It can be seen that the plotted points closely follow a straight line drawn "by eye." From this graph, the mean is 3250 kN/m^2 and the standard deviation $s = 475$ kN/m^2. Both these values are close to those calculated from the grouped data, and we conclude that the distribution of the given values is sensibly normal.

f. For a 1 percent risk the tail area of the normal probability curve (and we are interested in one tail only) is 0.01. Therefore, the area $F(z)$ is $0.50 - 0.01 = 0.49$. For this value of $F(z)$ Table A-4 gives $z = 2.33$. But

$$z = \frac{\text{deviation from the mean}}{s}$$

Hence

$$\text{deviation from the mean} = zs = 2.33 \times 466$$
$$= 1086 \text{ kN/m}^2$$

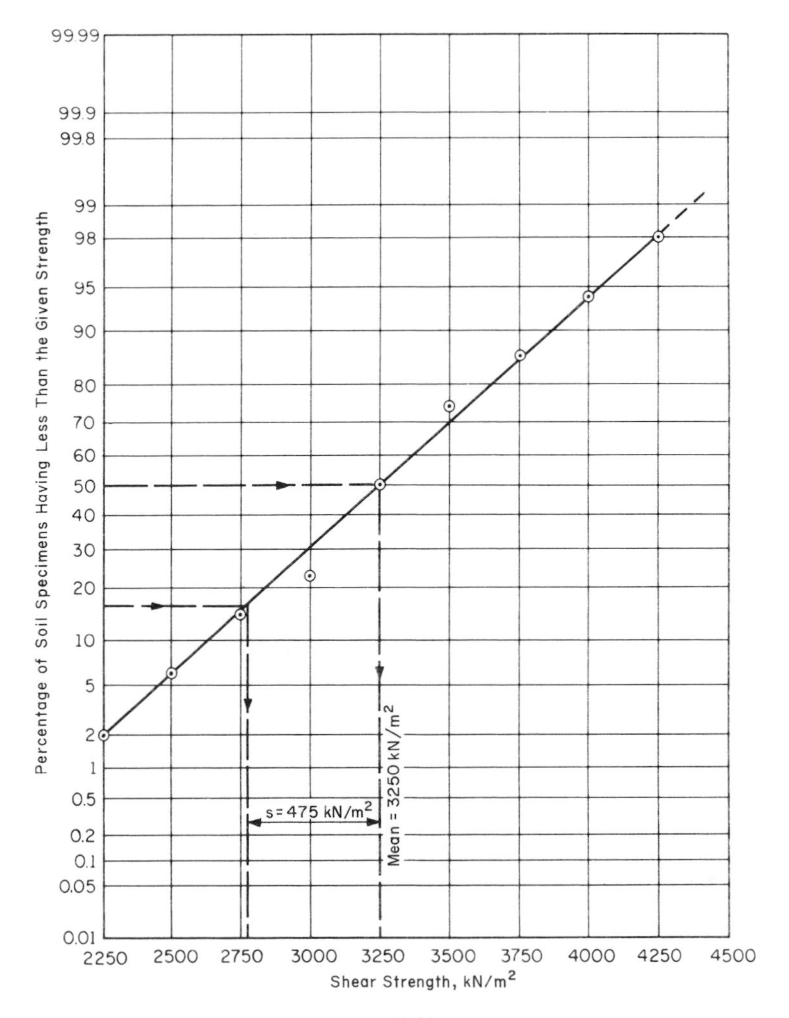

Fig. 10-20

Therefore, the minimum shear strength used in design is

$$3265 - 1086 = 2179 \text{ kN/m}^2$$

Among the data in hand only 1 specimen has a strength that falls below this minimum.

g. The standard error of the mean is $s/\sqrt{n} = 466/\sqrt{50} = 66$ kN/m^2. Thus there is a 68.26 percent probability that the mean of the population (i.e., of all the soil represented by the test specimens) lies within the range 3265 ± 66, i.e., between, say, 3200 and 3330 kN/m^2.

h. Equation (9-18) gives the normal probability curve as

$$f(z) = \frac{1}{\sqrt{2\pi}} e^{-z^2/2}$$
$$= 0.3989e^{-z^2/2}$$

where $z = X/\sigma$ is the deviation from the mean of the grouped data in terms of standard deviation.

From Eq. (9-14) the equation to the normal frequency curve is

$$y = \frac{N}{\sigma\sqrt{2\pi}} e^{-z^2/2}$$

where N is the area under the histogram.

$$N = w \sum f_i = 250 \times 50$$

Hence

$$y = \frac{250 \times 50}{466\sqrt{2\pi}} e^{-z^2/2}$$
$$= 10.7012e^{-z^2/2}$$

where z is as defined in (f).

To plot the normal *probability* curve, the values of the ordinate $f(z)$ (the probability density) are obtained from Table A-3 for selected deviations $\pm z$ from the mean 3265 kN/m². The normal probability curve is shown in Fig. 10-21.

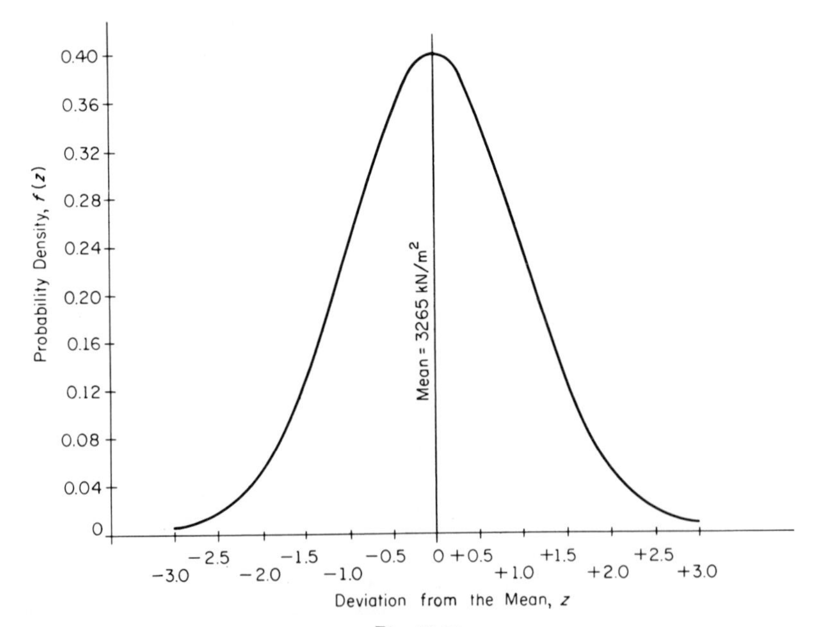

Fig. 10-21

i. To find the probability that a random measurement will have a deviation from the mean lying between −200 and +500 kN/m², we calculate

$$z_1 = \frac{|\text{deviation}|}{\sigma} = \frac{200}{466} = 0.43$$

The corresponding area under the normal curve is given in Table A-4 as $A_1 = 0.1664$. (Refer to Fig. 10-22.)

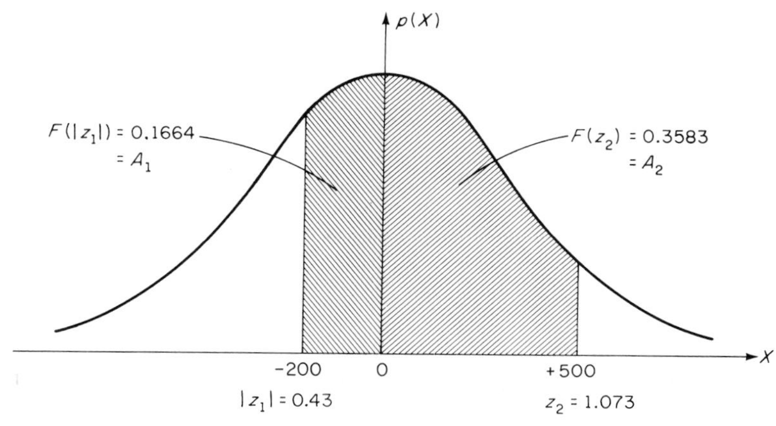

Fig. 10-22

Also,

$$z_2 = \frac{500}{466} = 1.073$$

From Table A-4, $A_2 = 0.3583$;

$$\begin{aligned}\text{total area under normal probability curve} &= A_1 + A_2 \\ &= 0.1664 + 0.3583 \\ &= 0.5247\end{aligned}$$

Thus there is a 52.47 percent probability that a single test has a deviation from the mean of 3265 kN/m², falling between −200 and +500 kN/m².

j. From Eq. (6-7), after taking $s = \sigma$ (assuming that we have available σ from the finite population with size $N = 50$), we obtain

$$\sigma_{\bar{x}} = s\sqrt{\frac{N-n}{n(N-1)}} = 466\sqrt{\frac{50-36}{36 \times 49}} = 41.5 \text{ kN/m}^2$$

$$\text{deviation } X = 3500 - 3265 = 235 \text{ kN/m}^2$$

Therefore,

$$z = \frac{X}{\sigma_{\bar{x}}} = \frac{235}{41.5} = 5.66$$

From Table A-4, the area under the normal curve for such a value of z can be taken as 0.5. Therefore, the area under the curve for a value of $z > 5.66$ is considered to be zero. Thus the required probability is zero.

k. For the 5 percent level of significance the total area under the normal curve outside the appropriate $\pm$ deviations from the mean is 0.05. Therefore,

$$\text{area inside the appropriate} \pm \text{deviations} = 1 - 0.05 = 0.95$$

(See Fig. 10-23.) The area on either side of the mean is 0.95/2. Hence, from

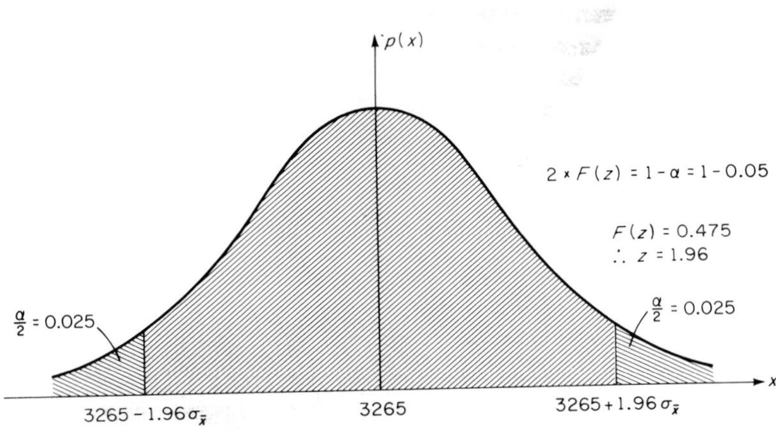

Fig. 10-23. Limits for the mean at $\alpha = 5$ percent.

Table A-4, $z = 1.96$. Therefore,

$$\text{deviation } X = z\sigma_{\bar{x}} = 1.96 \times 41.5 = 81 \text{ kN/m}^2$$

Thus the limits of the mean at the 5 percent level of significance, or the 95 percent confidence limits, are 3265 ± 81 kN/m^2.

Similarly, for the 1 percent level of significance $z = 2.575$.

$$\text{deviation } X = z\sigma_{\bar{x}} = 2.575 \times 41.5 = 107 \text{ kN/m}^2$$

Thus the limits at the 1 percent level of significance are 3265 ± 107 kN/m^2.

l. The value of z for a 10 percent probability of the mean $\bar{x}$ being in error by more than 400 kN/m^2 is the value corresponding to the area under the normal curve of $(1.00 - 0.10)/2 = 0.45$; from Table A-4, $z = 1.645$. To find the size of the sample n, we use Eq. (6-9), where (again assuming $s = \sigma$):

$$\sigma_{\bar{x}} = \frac{s}{\sqrt{n}}$$

Now

$$z = \frac{\text{deviation}}{\sigma_{\bar{x}}} = \frac{400}{466/\sqrt{n}}$$
$$= 1.645$$

Hence

$$\sqrt{n} = 1.916$$

or

$$n = 3.67$$

Therefore use a sample size of $n = 4$.

Problems

10-1. Tests on 16 tubes drawn at random from a normal population have yielded the mean value of resistance $\bar{y} = 20$ microohms and $\sum (y_i - \bar{y})^2 = 260$.

a. Estimate the population mean and variance.
b. Calculate the standard deviation of the sample mean.
c. Find the 99 percent confidence interval for the mean.

10-2. Of 10,000 children entering grade 1 in a given year, 5170 are male. Using the approximation of normal distribution, establish whether the figures suggest that the numbers of males and females are not equally distributed. Use $\alpha = 1$ percent.

10-3. In a true-false examination, find the probability that a student can guess the answers to (a) 14 or more out of 27, (b) 22 or more out of 40 questions.

10-4. From past experience, it has been found that a particular machine produces piston rings of which 10 percent are defective. Find the probability that in a random sample of 500 rings the following will be defective (a) at most 50, (b) between 50 and 60, (c) 65 or more.

10-5. The mass in milligrams of a certain pharmaceutical product is distributed normally with a variance of 0.0025 $(\text{mg})^2$. Find the size of the sample necessary to estimate, at the 0.5 percent level of significance, the mean mass within 0.025 mg.

10-6. (a) The diameter of a certain shaft is normally distributed with a mean of 2.79 cm and a standard deviation of 0.01 cm. The specification limits are 2.77 ± 0.03. If 1000 shafts were produced, how many would be unacceptable? What are the limits for the 1 percent level of significance about the mean. Using these limits, how many of the 1000 shafts would one expect to be rejects? (b) What is the probability that a diameter measurement would deviate from the true mean by ± 0.02 cm? (c) Estimate the size of a sample in future measurements in order that the probability will not be greater than 0.05 of the sample mean being in error by more than ± 0.01 cm.

10-7. Fit a normal frequency curve to the data given in Problem 2-1. Draw both the histogram and the fitted curve on the same set of coordinate axes. Compare the two plots and comment.

10-8. Find the equations for the normal probability and frequency curves representing the data in Problem 2-4. From a plot on normal probability paper of the cumulative frequency distribution, estimate the mean and standard deviation of the population distribution. Compare these values with those obtained previously and comment.

10-9. From the 64 values of hardness given in the first table of Problem 4-5,

a. Estimate the 90 percent confidence limits for hardness (estimated by a mean of 10 readings).
b. Fit a normal frequency curve to the data.
c. Test for normality of the distribution of hardness number by means of a plot on normal probability paper. Comment on the result.

10-10. Fit a normal frequency curve to the data of Problem 4-6 and calculate what proportion of results will in the long run fall below 17.2 MN/m^2. Check this proportion from a plot of the cumulative frequency distribution on normal probability paper.

10-11. The "minimum" strength of concrete on a certain job is specified to be 28.0 MN/m^2, the minimum being defined as a value exceeded by $\frac{9}{10}$ of all tests. If the coefficient of variation is 11.8 percent, find the mean strength of the concrete. What would the mean strength have to be if the coefficient of variation increased to 19.5 percent?

10-12. A fixed flow is known to have a bedload of 10 kg/s. An electromagnetic flowmeter was used to take a large number of readings one-half of which fell between 9.5 and 10.5 kg/s; a plot of the results indicated that they follow a normal distribution.

a. Determine the precision index h for the flowmeter.
b. An alarm will ring if the flowmeter reads 9 kg/s or less; how many false alarms will ring in a 14-day period knowing that the flow is checked twice daily? If one wishes to reduce the number of false alarms by a factor of 3 what should be the value of the flowmeter precision index h?

Chapter 11

Rejection of Outliers

Probably every engineer or scientist has encountered observations which, he is convinced, are incorrect. Specifically, in a group of readings or in a supposedly homogeneous sample, one of the observations may be very different from all the others. Such an observation is called an *outlier*. The question arises: Is the experimenter justified in discarding the outlier and in treating the data as if the "faulty" observation did not exist?

REASONS FOR OUTLIERS

The cause of a faulty observation may be a mistake, but we generally assume that all mistakes have been eliminated. A second possibility is that an additional variable has entered the picture; for example, an overload of the grid may have appreciably reduced the voltage which had previously been reasonably constant, or a changed wind direction may have caused an excessive amount of solid matter in the air. We thus have an observation from a different population. Such a population may have a different mean or the same mean but a larger variance. In the former case, we are dealing with a location error, in the latter with a scalar error.

Under such circumstances we may encounter an observation that will not fall within the expected range, i.e., one whose deviation from the mean will be greater than expected. What is it that we expect, however, or are justified in expecting? Table 10-2 showed that, when the variate is normally distributed, 95 percent of all observations are expected to fall within 1.96σ of the mean, i.e., the chance of an observation's falling outside the range $(\mu - 1.96\sigma, \mu + 1.96\sigma)$ is 1 in 20. For a deviation of $\pm 2.58\sigma$ the chance is 1 in 100, for $\pm 3.29\sigma$ it is 1 in 1000, and for $\pm 4.89\sigma$, one in a million.[1]

It is thus reasonable under normal circumstances to reject an observation which differs from the mean by more than a specified $z\sigma$; the value of z depends on size of the sample. Such a rejection is justified either because we were probably wrong to assume that the sample containing the extreme observation came from the specific population that we are testing (as the occurrence of such a large deviation is unlikely in a sample from the population in question), or because we are interested in testing representative samples, and a sample containing such a large deviation is not representative. It is important to appreciate this reasoning and not simply to reject indiscriminately observations that appear to be more widely scattered than we would like. In particular, if large deviations occur in a number of samples, we should suspect the presence of additional factors, probably of intermittent character; the experiment must then be carefully examined.

Under industrial conditions it is important to distinguish between the two cases mentioned in the last paragraph, namely, whether the outlier is caused by a real factor or is just an improbable result. If the outlier is caused by a real factor, we may want to identify it. The factor causing the different product may then be eliminated or, in some cases, deliberately used to obtain a better or new product.

A number of criteria for the rejection of outliers have been proposed, but we shall discuss only two of them here.

REJECTION ON BASIS OF ESTIMATED VARIANCE

In Nair's method the maximum deviation from the sample mean $\bar{x}$ that can be expected for single values in samples of size n is related to the estimated variance of the population; the estimate must be based on a larger sample than the one containing the outlier.

1. These are values for both tails and should be distinguished from the case where we are interested only in a value smaller *or* greater than the mean by a certain number of standard deviations.

Let x_m be the greatest or smallest value of x that can be expected in a sample of size n at a given significance (probability) level; the levels considered are 10 percent, 5 percent, 1 percent, and 0.1 percent, but data for other levels are also available. Let s_e be the estimated standard deviation, the estimate being based on data with v degrees of freedom. Table A-5 gives the values of $|x_m - \bar{x}|/s_e$ that are not normally expected to be exceeded in samples of size n between 3 and 9. If a deviate larger than the tabulated value is observed, then the observation may be rejected as being significantly different from the remainder of the sample. The level of significance at which the rejection is decided upon represents the risk of error; for example, a 5 percent level of significance means that there is a 5 percent probability that we reject a value which does correctly belong to the sample in question.

Example. Twenty-five observations on the masses of material from a packing machine have given an estimate of standard deviation as 0.40 kg. Assume that masses, in kilograms, taken by a customer on 5 packages from his consignments are 100.4, 100.2, 100.5, 100.2, and 99.2. The question is whether the value of $x_m = 99.2$ can be discarded. Use a 5 percent level of significance.

Since $\bar{x} = 100.1$ and $s_e = 0.40$,

$$\frac{|x_m - \bar{x}|}{s_e} = \frac{0.9}{0.40} = 2.25$$

We now enter Table A-5 with the value of 2.25 for $n = 5$ and the number of degrees of freedom[2] $v = 25 - 1 = 24$. The extreme deviates in Table A-5 are 2.23 at the 5 percent level of significance, and 2.85 at the 1 percent level. Thus the value $x_m = 99.2$ can be rejected with a 5 percent risk of error of wrong rejection.

Once the extreme deviate has been rejected, the sample mean becomes 100.32 kg and the standard deviation estimated from the sample is $s = \sqrt{0.0225}$; this is based on the number of degrees of freedom $v = 4 - 1 = 3$. It may be of interest to compare the standard deviation from this sample with the value of 0.40 estimated for 24 degrees of freedom. This is done by means of the F test (see Chapter 14). We find

$$F = \frac{(0.4)^2}{0.0225} = 7.11$$

For $v_1 = 24$ and $v_2 = 3$, F at the 5 percent level of significance is 8.64. We conclude, therefore, that the variation in the estimates of variance given by the two samples is due to chance alone.

2. See Chapter 4.

CHAUVENET'S CRITERION

Another criterion of rejection of outliers is due to Chauvenet: An observation in a sample of size n is rejected if it has a deviation from the mean greater than that corresponding to a $1/(2n)$ probability. The probability is calculated on the assumption of a normal distribution, using an estimate of variance on the basis of the sample considered. For example, if $n = 10$, then $1/(2n) = 0.05$, which is the probability of a deviate of at least 1.96σ. Thus an outlier which deviates from the mean by at least $1.96s$ would be rejected.[3] The mean and standard deviation of the remaining 9 observations are then calculated and used in further work.

To simplify calculations, Table A-6 gives the maximum values of $|x_m - \bar{x}|/s$ for different values of n; an outlier exceeding the tabulated value can be rejected.

For example, for the 5 masses given in the preceding example, $s = 0.52$, and

$$\frac{|x_m - \bar{x}|}{s} = \frac{0.9}{0.52} = 1.73$$

For $n = 5$, Table A-6 gives the maximum value of $|x_m - \bar{x}|/s$ as 1.64. We are justified, therefore, in rejecting the outlier, and both Chauvenet's and Nair's methods lead to the same conclusion, but this need not always be the case.

We should also note that the rejection of an outlier decreases s. When Chauvenet's criterion is used, this could easily lead to successive rejection of extreme observations—a procedure that must never be used.

The decision to reject an observation should be based on experience and must not be made lightly. It is important to realize that in rejecting an observation we *may* be in our ignorance, throwing away vital information which *could* lead to the discovery of a hitherto unrecognized factor.

Problems

11-1. Ten measurements of the diameter of a shaft were as follows: 6.06, 5.92, 6.01, 6.01, 5.99, 5.99, 6.02, 6.03, 6.02, and 5.97 (cm). Using Chauvenet's criterion, determine whether any one of the observations can be considered as an outlier.

11-2. Twenty observations of deflection of similar beams have given an estimate of the standard deviation as 0.5 cm. If the deflections of 6 beams are 3.4, 3.2, 3.5, 3.2, 4.6, and 3.4 cm, determine whether the deflection of any beam can be considered abnormal, using the following methods. (a) Nair's method. (b) Chauvenet's criterion. After discarding the appropriate measurement, recalculate the mean and compare it with the original mean deflection.

3. We use s because we only estimate σ.

11-3. A student using a cathetometer was measuring the height of a mercury column in a manometer. The results in centimeters were as follows: 8.92, 8.98, 9.01, 8.99, 9.02, 8.97, 9.02, 9.03, 9.00, 9.06.

a. Determine whether any measurement should be discarded as an outlier using Chauvenet's criterion.
b. After rejecting the outlier, recompute the mean and compare it with the original mean.

Chapter 12

The Chi-Squared Test

We have stated repeatedly that experimental observations are subject to scatter, and all experimental results must be viewed in relation to this scatter. This statement is fundamental to *tests of significance*, which enable us to judge whether any observed differences are real or are due to chance alone. The differences studied may be those between two comparable sets of observations or between an experimental distribution and a hypothetical distribution.

NULL HYPOTHESIS

In this chapter we consider the χ^2 test,[1] which enables us to find whether observed frequencies differ *significantly* from frequencies expected from an assumed model. The test requires, in general, the use of frequencies and not percentages.

1. Sometimes written chi, pronounced *kī* as in kite.

The derivation of the χ^2 distribution will not be given here. In the χ^2 test, as in all tests of significance, we always postulate that there is no significant difference between the distributions being compared, i.e., that they are drawn from the same population. This is known as the *null hypothesis.* We measure the probability of the actual difference occurring due to chance alone, and if this probability is very small, we reject the null hypothesis and infer that a real difference exists. However, we can never formally prove the null hypothesis to be correct. This may seem an unsatisfactory state of affairs, but statistical inference is not an end in itself; it is only a tool that enables us to fit a hypothesis to observed physical facts or, alternatively, makes us reject it and seek another pattern.

SIGNIFICANCE OF TEST

To perform the χ^2 test, we compare each expected class frequency E with the observed frequency O, and compute for each class the term

$$\frac{(O - E)^2}{E}$$

The statistic χ^2 is then defined as

$$\chi^2 = \sum \frac{(O - E)^2}{E} \tag{12-1}$$

the summation extending over all classes and with the constraint condition that

$$\sum_{i=1}^{k} O_i = \sum_{i=1}^{k} E_i = N \tag{12-2}$$

where N is the total frequency and k is the number of classes.

The calculated value is then compared with tabulated values of χ^2; the latter are values that cannot be exceeded by the calculated value *when there is no real difference* at a specified level of significance, i.e., with a given probability. Table A-7 gives values of χ^2 for a range of probabilities from 0.001 to 0.99.

To find the requisite probability, we locate on the table the appropriate number of degrees of freedom (see the following section) and find the highest tabulated value that is exceeded by the calculated value. At the head of the column containing this tabulated value we read off the probability of the null hypothesis' not being true. It is usual to reject the null hypothesis at the 5 percent or the 1 percent level of significance (i.e., a probability of 0.05 and 0.01, respectively).

If the level of significance is higher, we generally do not reject the null hypothesis, but this does not necessarily mean that the observed distribution is the same as the hypothetical one. It is possible that we simply do not have adequate data in hand, and further tests should be made before a reliable conclusion can be drawn.

We should note that the χ^2 test determines the probability of obtaining the values of $|O - E|$ of *at least* the magnitude observed and not of exactly that magnitude.

Equation (12-1) defines χ^2 as a criterion to measure the discrepancies between the observed and the expected values. We shall gain a better understanding of the χ^2 test if we examine how reasonable it is as a criterion. Consider any class; the discrepancy between the observed and expected values is $(O - E)$. The degree of this discrepancy can be better judged if it is relative; hence we divide the term $(O - E)$ by E (since E is a better basis for comparison than O) to yield the quotient $(O - E)/E$. Such quotients will be either positive or negative and their sum might be close to zero, thus impairing its adequacy. It is better, therefore, to square the quotients and then sum them up. However, in situations involving relative values it is judicious to consider the sum of weighted squared quotients. For example, in the case when the values are $O = 11$ and $E = 10$, the discrepancy quotient is 10 percent and this is far less significant (since it can happen by chance) than the 10 percent discrepancy quotient in a case when $O = 330$ and $E = 300$. Thus we multiply each squared quotient by its expected value, E, and then sum up all classes, yielding

$$\sum_{\text{all classes}} E\left(\frac{O - E}{E}\right)^2$$

which we call χ^2. Thus

$$\chi^2 = \sum_{\text{all classes}} \frac{(O - E)^2}{E} \qquad [12\text{-}1]$$

DEGREES OF FREEDOM IN THE χ^2 TEST

The concept of degrees of freedom, denoted by v, was introduced and defined in Chapter 4. Applying the definition to problems involving the use of the χ^2 test, we find that

a. The degree of freedom $v = k - 1$ if the expected frequencies can be calculated without having to estimate any population parameters from the sample. We subtract one from k, the number of classes because of the constraint of Eq. (12-2). In other words, since the total is fixed, we can arbitrarily assign expected frequencies to only $(k - 1)$ classes.

b. The degree of freedom $v = k - 1 - m$ if the expected frequencies can be calculated only after estimating m number of population parameters from the sample (e.g., the mean μ, standard deviation σ). For instance, if we are determining the goodness of fit of observed data having six classes to a theoretical binominal distribution, we need to calculate the parameter p from the observed sample. This would then yield the expected frequencies. Thus, in this case, $v = k - 1 - m$, with $k = 6$, $m = 1$, i.e., $v = 4$. If we were comparing the same observed data to a normal distribution, we would have to estimate the parameters $\bar{x}$ and $s(m = 2)$ from the sample before being able to calculate the expected frequencies. Hence the degrees of freedom $v = 6 - 1 - 2 = 3$ in this case.

Example. Three shifts, A, B, and C are in competition to produce similar units that must pass a standard test before being declared acceptable. In the first week they produced 2, 9, and 10 acceptable units, respectively. It is assumed that the total number of units produced by each of the three shifts is identical and very large. Can we conclude that there is no difference in the quality of workmanship between the three shifts? Use a level of significance of 5 percent.

We adopt the null hypothesis that there is no significant difference between shifts so that the expected number of acceptable units per shift is $(2 + 9 + 10)/3 = 7$. We can set out the data as in Table 12-1.

TABLE 12-1

Shift	O	E	$\lvert O - E \rvert$
A	2	7	5
B	9	7	2
C	10	7	3

Hence

$$\chi^2 = \frac{5^2}{7} + \frac{2^2}{7} + \frac{3^2}{7} = 5.428$$

The number of degrees of freedom is 2 since, given the total number of acceptable units, we can assign arbitrarily only two classes. Table A-7 gives $\chi^2 = 5.991$ at the 5 percent level of significance, and we cannot conclude, therefore, that there is a difference in workmanship between the three shifts.

If the foreman of shift C is not satisfied with this conclusion, we answer that we do not deny that a difference *may* exist, but the evidence available is inadequate to regard the difference as established.

The same foreman, still unconvinced, may collect further evidence. Let us assume that at the end of the second week, all conditions having remained unaltered, the numbers of acceptable units have increased to 3, 13, and 14. Now the expected number of acceptable units per shift is $(3 + 13 + 14)/3 = 10$. Hence

$$\chi^2 = \frac{7^2}{10} + \frac{3^2}{10} + \frac{4^2}{10} = 7.4$$

With 2 degrees of freedom ($\nu = k - 1 = 3 - 1$) this value is significant at the 5 percent level, and we consider our suspicions confirmed: The workmanship in the three shifts is not of uniform quality.

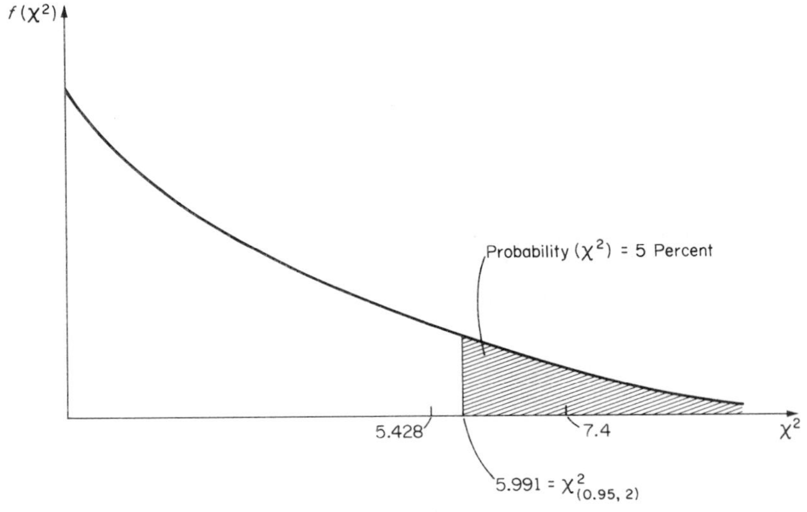

Fig. 12-1. Probability density function of χ^2 variable ($\nu = 2$) for the example on p. 185

MINIMUM CLASS FREQUENCY

The preceding example illustrates the point that the χ^2 test is sensitive to the size of the samples used. In general, the test should not be used when an *expected* class frequency is less than 5, as the relative frequency for such a class is very small. However, it is usually possible to combine adjacent classes with a frequency below 5 in order to reach or exceed this value.

Example. Five pressures are used in extruding a difficult shape, and the numbers of rejects are given in Table 12-2. From numerous past records the expected numbers of rejects are also listed. Using $\alpha = 5$ percent, check whether there is a significant difference between the observed and expected numbers of rejects produced by the five pressures.

TABLE 12-2

Pressure, pascals	2.0	2.2	2.4	2.6	2.8
Class	A	B	C	D	E
Observed number of rejects O	2	3	6	6	3
Expected number of rejects E	$1\frac{1}{2}$	$5\frac{1}{2}$	6	$5\frac{1}{2}$	$1\frac{1}{2}$

To satisfy the requirement of an expected class frequency of not less than 5, we combine A and B, and D and E:

Class	*A and B*	*C*	*D and E*
O	5	6	9
E	7	6	7

Hence

$$\chi^2 = \frac{2^2}{7} + 0 + \frac{2^2}{7}$$

That is, $\chi^2 = 1.143$ with 2 degrees of freedom ($\nu = k - 1 = 3 - 1$). A value of 1.386 can be obtained with a probability of 50 percent; therefore, we conclude that there is no significant difference between the observed and expected numbers of rejects for the pressures: (A and B), C, and (D and E).

A close examination of Table A-7 reveals that if the calculated value of χ^2 is smaller than the number of degrees of freedom ν, we can conclude that χ^2 is not significant.

CONTINUITY CORRECTION

We should note that χ^2 is a continuous variable and, if the actual distribution is discontinuous, we have to apply a *continuity correction*. This can be done only when the number of degrees of freedom is 1. The correction consists of reducing by 0.5 the values of observed frequency that are greater than the expected frequency, and increasing those that are smaller. Failure to apply the correction leads to too high a value of calculated χ^2. Thus, without the correction, we may wrongly reject the null hypothesis at a specified level of significance. If, however, the null hypothesis is not rejected without the correction being applied, the correction would not affect our conclusion. (See also p. 145.)

Example. We toss a coin 50 times and obtain 17 heads. Are we justified in suspecting that the coin is biased? Use $\alpha = 1$ percent. The number of degrees of freedom is 1.

We tabulate the data as follows (Table 12-3).

TABLE 12-3

	Heads	*Tails*
Observed	17	33
Expected E	25	25
Observed, corrected O	17.5	32.5
$\|O - E\|$	7.5	7.5

Hence

$$\chi^2 = \frac{7.5^2}{25} + \frac{7.5^2}{25} = 4.5$$

This is significant at the 5 percent level but not at the 1 percent level; we have, therefore, a good, but not an overwhelming, reason to suspect the coin.

CONTINGENCY TABLES

The χ^2 tests can also be written down in a somewhat different form. For example, if we compare two methods of treatment in order to establish whether or not there is a significant difference between them, we can set out the results in the form of a contingency table (Table 12-4).

TABLE 12-4

Item	*Number of successes*	*Number of failures*	*Totals*
Method A	A_1	A_2	A_t
Method B	B_1	B_2	B_t
Totals	T_1	T_2	T_t

Instead of the χ^2 calculations given in Eq. (12-1), we can use a mechanical formula:

$$\chi^2 = \frac{(B_1 A_2 - A_1 B_2)^2 \times T_t}{A_t B_t T_1 T_2} \qquad (12\text{-}3)$$

The number of degrees of freedom is 1, as only one of the values of A_1, A_2, B_1, and B_2 can be assigned arbitrarily, all the others being governed by the totals T_1, T_2, A_t, and B_t.

It can easily be shown that Eq. (12-3) gives the same result as the calculation of χ^2 by Eq. (12-1). It should be noted that the tabulated values must be adjusted *before* applying Eq. (12-3).

Let us consider a more general case: to test the hypothesis that rows and columns in Table 12-5 represent an independent classification. We do this by computing an expected number E_{ij} for each cell and using the χ^2 test.

TABLE 12-5

Observed Table

	1	*2*	*3*	...	*c*	*Row totals*
1	O_{11}	O_{12}	O_{13}		O_{1c}	R_1
2	O_{21}	O_{22}	O_{23}		O_{2c}	R_2
⋮	⋮	⋮	⋮	⋮	⋮	⋮
r	O_{r1}	O_{r2}	O_{r3}		O_{rc}	R_r
Column totals	C_1	C_2	C_3	...	C_c	N

By independence we mean that the proportion of each row total to that belonging in, say, the jth column, is the same for all rows, and the same when the words column and row are interchanged in the preceding statement. This requirement can be expressed mathematically by saying that the probability of any random element's belonging in the (i, j)th cell is equal to the product of the probability of its belonging to the ith row (R_i/N) multiplied by the probability of its belonging to the jth column (C_j/N), i.e., $R_i C_j/N^2$. Hence the expected number is the sample size N multiplied by the estimated probability, i.e.,

$$E_{ij} = \frac{R_i C_j}{N^2} \times N = \frac{R_i C_j}{N}$$

On this basis we construct the expected table and proceed to compute the χ^2.

In such cases the number of degrees of freedom, v, is given by

a. The degree of freedom $v = (r - 1)(c - 1)$ if the expected frequencies can be computed without estimating population parameters from the observed sample (i.e., when the population parameters required in calculating the expected frequencies are *known*). This value can be readily deduced in the following way: If in the expected table we leave out a single number in each row and column, such numbers can then be determined from the known fixed totals of each row and column. Thus we have freedom in assigning numbers to only $(r - 1)(c - 1)$ classes in the table, the others being uniquely determined. Another way of deducing the expression for v is as follows. We have a total of rc cells in all with the total N being fixed, and hence there is one constraint. Also, $(r-1)+(c-1)=(r+c-2)$ independent parameters have been estimated from the sample. Thus $v = rc - 1 - (r+c-2) = rc - r - c + 1 = (r-1)(c-1)$.

b. The degree of freedom $v = (r-1)(c-1) - m$ when the population parameters are *not known* but are estimated from the observed sample to compute the expected frequencies. Here we are imposing m additional constraints by assuming that m *population* parameters are equal to the corresponding *sample* statistics.

If the calculated χ^2 exceeds the tabulated value for a probability α, we reject the hypothesis and conclude that rows and columns do not represent an independent classification. If this conclusion is indeed wrong, then we are committing a Type I error (see Chapter 13) and the measure of the risk involved is the probability α, the level of significance.

TOO GOOD A FIT

Table A-7 gives the values of χ^2 for probabilities of 0.001 to 0.10, i.e., for levels of significance of 0.1 to 10 percent which are the usual levels at which the rejection of the null hypothesis is considered. The table also contains, however, values for probabilities of 0.50 to 0.99. The occurrence of χ^2 corresponding to a probability higher than about 0.99 makes us suspect that the data have been "rigged." This has been shown to have been the case, for instance, with some of the test results reported by Mendel's disciples. Too high a value of χ^2 may also occur when, for example, spurious pulses of uniform frequency are mixed with pulses being observed, or, in general, when there is a lack of randomness.

Example. In the statistics laboratory we ask students to draw samples of 4 from a bowl containing red and black balls in equal proportions, the drawn balls being returned into the bowl after each test; 160 samples are drawn and the students report their results. We compare the observed distribution with that expected, i.e., with the binomial distribution. Are the results suspicious? Use a 5 percent level of significance.

TABLE 12-6

Number of red balls in sample r	0	1	2	3	4
Number of samples with r red balls observed O	9	40	59	41	11
Number of samples with r red balls expected E	10	40	60	40	10
$\lvert O - E \rvert$	1	0	1	1	1

Hence

$$\chi^2 = \frac{1}{10} + \frac{1}{60} + \frac{1}{40} + \frac{1}{10} = 0.24$$

The number of degrees of freedom is $\nu = k - 1 = 5 - 1 = 4$.

Table A-7 gives $\chi^2 = 0.297$ at the 99 percent level of significance. Therefore, the probability of obtaining χ^2 *as small* as calculated is less than 1 percent, and we are justified, therefore, in suspecting that the experiment was not performed but that the "results" were arbitrarily written down so as to appear plausible in the students' eyes.

χ^2 AS A MEASURE OF GOODNESS OF FIT

As mentioned earlier, the χ^2 test is used to test the goodness of fit; in this case the null hypothesis states that there is no significant difference between the observed distribution and a postulated standard distribution. The preceding example involved the binomial distribution, but perhaps a better illustration is offered by the data in Chapter 8 on the number of deaths caused by a horse-kick. On the assumption of Poisson distribution, we calculated there the expected number of deaths per army corps per annum, and we can now apply the χ^2 test to find how well the observed data fit the assumed distribution. The data for the example are repeated in Table 12-7.

TABLE 12-7

Number of deaths per corps per annum r	0	1	2	3	4
Observed number of corps with r deaths O	109	65	22	3	1
Expected number of corps with r deaths E	109	66	20	4	$0.6 \simeq 1$
$\lvert O - E \rvert$	0	1	2		1

Hence

$$\chi^2 = \frac{1^2}{66} + \frac{2^2}{20} + \frac{1^2}{5} = 0.415$$

Since we deduced the expected frequencies by first estimating np, the mean number of deaths/corps/annum, from the observed sample (see p. 117), the number m is 1. Here we are imposing a restriction by assuming that the mean of the population (expected frequencies) is equal to the mean of the sample (observed frequencies). Thus the number of degrees of freedom in this case is $\nu = k - 1 - m = 4 - 1 - 1 = 2$. With $\nu = 2$, Table A-7 gives $\chi^2 = 0.446$ at the 80 percent level of significance. Thus the goodness of fit is good and we are satisfied that the observed distribution is a Poisson distribution.

Solved Problems

12-1. An experiment was conducted to test the effect of the rate of loading on the type of failure in steel rods. Twenty-four specimens were tested to failure, 12 at a fast-loading rate and the other 12 at a slow rate. It was found that 11 of the slow specimens showed a complete cone failure, whereas only 5 of the fast specimens showed this type of failure. Establish whether the rate of loading influences the shape of the failure zone for the given steel. Use $\alpha = 5$ percent.

SOLUTION

There are two criteria here and hence two-way classification. Let a complete cone failure be denoted as A failure and all other types of failure as B. Then the *observed* 2×2 contingency table, Table 12-8, follows.

TABLE 12-8

	Failure A	*Failure B*	*Total*
Slow method	11	1	12
Fast method	5	7	12
Total	16	8	24

The *expected* values are shown in Table 12-9.

TABLE 12-9

	Failure A	*Failure B*	*Total*
Slow method	8	4	12
Fast method	8	4	12
Total	16	8	24

The number of degrees of freedom, $v = (r - 1)(c - 1) = (2 - 1)(2 - 1) = 1$. The value of χ^2 *without the continuity correction* is

$$\chi^2 = \frac{(3)^2}{8} + \frac{(3)^2}{8} + \frac{(3)^2}{4} + \frac{(3)^2}{4} = 6.75$$

The value of χ^2 *with the continuity correction* is

$$\chi^2 = \frac{(2.5)^2}{8} + \frac{(2.5)^2}{8} + \frac{(2.5)^2}{4} + \frac{(2.5)^2}{4} = 4.69$$

This is significant at the 5 percent level, and hence the rate of loading does influence the shape of the failure zone at the above level of significance.

12-2. Five machines, *A*, *B*, *C*, *D*, and *E* are experimentally used to make precision tools. The following are the numbers of tools made by the five machines and the numbers rejected:

Machine	*A*	*B*	*C*	*D*	*E*	*Total*
Number made	20	18	16	24	22	100
Number rejected	12	16	10	14	18	70

Test the hypothesis that there is no difference in the performance of the machines. Use a level of significance of $\alpha = 5$ percent.

SOLUTION

The observed number of tools in each category is listed in Table 12-10.

TABLE 12-10

Machine	*A*	*B*	*C*	*D*	*E*	*Total*
Rejected	12	16	10	14	18	70
Accepted	8	2	6	10	4	30
Total	20	18	16	24	22	100

SOLUTION

$$\text{Expected number of rejects from any machine} = \frac{\text{total number of rejects}}{\text{total number of tools made}} \times \text{number of tools made by that particular machine}$$

The expected numbers are shown in Table 12-11.

TABLE 12-11

Machine	*A*	*B*	*C*	*D*	*E*	*Total*
Rejected	14.0	12.6	11.2	16.8	15.4	70
Accepted	6.0	5.4	4.8	7.2	6.6	30
Total	20	18	16	24	22	100

$$\chi^2 = \sum \frac{(O-E)^2}{E} = \frac{(12-14)^2}{14} + \frac{(16-12.6)^2}{12.6} + \frac{(10-11.2)^2}{11.2}$$

$$+ \frac{(14-16.8)^2}{16.8} + \frac{(18-15.4)^2}{15.4} + \frac{(8-6)^2}{6} + \frac{(2-5.4)^2}{5.4}$$

$$+ \frac{(6-4.8)^2}{4.8} + \frac{(10-7.2)^2}{7.2} + \frac{(4-6.6)^2}{6.6} = 7.458$$

For $v = (r-1)(c-1) = (2-1)(5-1) = (1)(4) = 4$ degrees of freedom, Table A-7 gives the probability of such a value of χ^2 as between 0.10 and 0.20. We have, therefore, no reason to reject the null hypothesis, and we conclude that there is no difference in the performance of the five machines.

12-3. Test the null hypothesis in Solved Problem 12-2 using the following data:

Machine	*A*	*B*	*C*	*D*	*E*	*Total*
Number made	12	18	24	24	22	100
Number rejected	4	16	18	14	18	70

Use $\alpha = 10$ percent.

SOLUTION

We calculate the observed and expected numbers as before (Table 12-12).

TABLE 12-12

Machine	*A*	*B*	*C*	*D*	*E*	*Total*
Observed rejected	4	16	18	14	18	70
Observed accepted	8	2	6	10	4	30
Expected rejected	8.4	12.6	16.8	16.8	15.4	70
Expected accepted	3.6	5.4	7.2	7.2	6.6	30

Since the expected number in A is smaller than 5, we pool the expected numbers of A and B and, at the same time, pool the observed rejects of A and B. Thus

$$\chi^2 = \frac{(20-21)^2}{21} + \frac{(18-16.8)^2}{16.8} + \frac{(14-16.8)^2}{16.8} + \frac{(18-15.4)^2}{15.4}$$

$$+ \frac{(10-9)^2}{9} + \frac{(6-7.2)^2}{7.2} + \frac{(10-7.2)^2}{7.2} + \frac{(4-6.6)^2}{6.6} = 3.463$$

The number of degrees of freedom is $v = (r-1)(c-1) = (2-1)(4-1) = 3$, since we pooled the expected numbers in A and B. From Table A-7, for $v = 3$, the probability is approximately 50 percent. Therefore, there are no grounds for rejection of the null hypothesis. However, nothing can be said about the difference between the machines A and B until further tests have been made.

12-4. The manufacturer of a particular casting kept the following record of the number of defective units produced in 50 shifts:

Number of defectives per shift r	0	1	2	3	4	5	6	7	8
Shifts with r defectives O	2	6	10	10	7	6	4	3	2

Determine whether the data support the hypothesis that the number of defective castings is completely random, i.e., we are not justified in suspecting that different shifts produce significantly different numbers of defective castings. Use $\alpha = 10$ percent.

SOLUTION

If the distribution is random, it will not differ significantly from a Poisson distribution. Therefore, we start by fitting a Poisson distribution to the observed data (Table 12-13).

TABLE 12-13

Number of defectives per shift r	*Shifts with r defectives O*	$r \times O$
0	2	0
1	6	6
2	10	20
3	10	30
4	7	28
5	6	30
6	4	24
7	3	21
8	2	16
Totals	$\sum O = 50$	$\sum (r \times O) = 175$

$$\text{mean number of defectives per shift} = np = \frac{175}{50} = 3.5$$

Using the cumulative Poisson probability graph of Fig. 8-1, we obtain the data given in Table 12-14.

TABLE 12-14

r	0	1	2	3	4	5	6	7	8
P cumulative	0.03	0.13	0.32	0.535	0.725	0.86	0.93	0.97	0.99
P_r	0.03	0.10	0.19	0.215	0.19	0.135	0.07	0.04	0.02
E	1.5	5.0	9.5	11.0	9.5	7.0	3.5	2.0	1.0

where $E = 50 \times P_r$ = expected number of shifts with r defective castings.

Since the expected number of shifts producing 0 defectives and those producing 6, 7, and 8 defectives are smaller than 5, we pool the expected numbers of shifts of 0 and 1 defectives, and 6, 7, and 8 defectives, respectively; at the same time the corresponding numbers of observed shifts O are pooled, as shown in Table 12-15.

TABLE 12-15

Number of defectives	0 and 1	2	3	4	5	6, 7, and 8	$\sum$
Observed numbers of shifts O	8	10	10	7	6	9	50
Expected number of shifts E	6.5	9.5	11	9.5	7.0	6.5	50

$$\chi^2 = \sum \frac{(O - E)^2}{E} = \frac{(8 - 6.5)^2}{6.5} + \frac{(10 - 9.5)^2}{9.5} + \frac{(10 - 11)^2}{11}$$

$$+ \frac{(7 - 9.5)^2}{9.5} + \frac{(6 - 7)^2}{7} + \frac{(9 - 6.5)^2}{6.5}$$

$$= 0.346 + 0.026 + 0.091 + 0.658 + 0.143 + 0.961$$

$$= 2.225$$

The number of degrees of freedom is $\nu = k - 1 - m = 6 - 1 - 1 = 4$. The value of k is 6 after pooling the classes, and $m = 1$, since we have one added constraint in assuming that the means (np) of the observed and expected distributions are the same. Thus, from Table A-7, for $\nu = 4$ the probability of such a value of χ^2 is approximately 0.70. Hence there is no justification in claiming that the identity of the shift affects the number of defective castings produced.

Problems

12-1. A product is supposed to contain 5 percent of defective items. We take a sample of 100 items and find it to contain 12 defectives. Are we justified in suspecting that the consignment is not up to specification? Use $\alpha = 1$ percent.

12-2. For the data given in Problem 8-1, use the χ^2 test to determine whether the distribution differs significantly from a Poisson distribution. Use $\alpha = 10$ percent.

12-3. For the data given in Problem 8-3, use the χ^2 test to check on the goodness of fit of the assumed Poisson distribution. Use $\alpha = 5$ percent.

12-4. A course is taught in two classes. In class A there are 27 failures out of 202 students taking the course. In class B there are 9 failures out of 199 students. Can we conclude that class B receives better instruction? Use $\alpha = 1$ percent.

12-5. A number of machines of two types were used over a period of time; the records of their serviceability are as follows:

Type	*Broken down*	*Temporarily out of order*	*Always serviceable*
A	11	132	212
B	58	29	13

Can we conclude that type A is superior insofar as it leads to "less trouble"? Use $\alpha = 5$ percent.

12-6. Of 10,000 children entering grade 1 in a given year, 5170 are male. Use the χ^2 test to determine whether the figures suggest that the numbers of males and females differ significantly. Use $\alpha = 1$ percent.

12-7. Samples of 100 items each were taken from two machines producing the same product. Among those from machine A there were 17 defectives, but there were only 3 defectives among those from machine B. Should we conclude that there is a significant difference between the two machines? Use $\alpha = 5$ percent.

12-8. At the end of the first semester the number of failures in three sections of a class, the sections being chosen at random and each with 30 students, was 2, 9, 10, respectively. Can we conclude that the three instructors differ in their marking? Use (a) 5 percent level of significance. (b) 1 percent level of significance.

12-9. In order to improve the performance of a certain engine a new carburetor nozzle was designed. A lot of one hundred identical engines was divided randomly into two sublots of fifty engines. One sublot was fitted with the new nozzle and the other with the regular nozzle. The engine test results are shown in Table 12-16.

TABLE 12-16

	Improvement	*No Improvement*
New nozzle	30	20
Regular nozzle	19	31

Test the hypothesis that the new nozzle has no effect in improving engine performance. Use a 5 percent level of significance.

Chapter 13

Comparison of Means

In the present chapter we are concerned with comparing means of samples for the purpose of determining whether the observed difference is due to chance only, or whether we should suspect some real cause to be responsible and hence consider the difference to be statistically significant.

Suppose that we have determined the compressive strength of concrete supplied by two ready-mix concrete manufacturers and found that the mean strength of the concrete from supplier A was 48.3 MN/m^2 and the standard deviation was 2.8 MN/m^2, the corresponding values for the concrete from supplier B being 63.5 and 2.8 MN/m^2, respectively. In each case the result was obtained from a sample of 10 specimens. We have no doubt that the latter concrete has a higher strength because the difference between the means (15.2 MN/m^2) is more than five times the value of the standard deviation of the values in either sample. It is, therefore, highly improbable that the two samples have been drawn from the same population and that their difference is due to chance. This conclusion is intuitively obvious and statistical proof is not necessary.

INFERENCE ERRORS

When, however, the difference is smaller—for example, if concrete *B* has a mean strength of 50.4 MN/m^2 with the standard deviation remaining at 2.8 MN/m^2—it is far from obvious that this concrete is *really* superior to concrete *A*. We can use statistical methods in an attempt to infer whether or not there is a real difference between the strengths of the two concretes; however, our answer cannot be *guaranteed* to be correct, but it has only a specified probability of being correct. It is important at this stage to know the type of inference error that we may commit. The usual nomenclature is as follows.

A Type I error is said to have been made if we infer that there is a real difference between the two samples, while in fact the observed difference is due to chance only. To reduce the risk of a Type I error, we may insist on a higher level of significance of the difference being studied before we accept the difference as real; for example, we may require a 1 percent level of significance, rather than 5 percent, i.e., a probability of only 1 percent that such a difference may occur by chance and not be due to real causes. (The choice in a practical case depends on judgment and experience.)

However, a decrease in the risk of a Type I error increases the chances of committing a Type II error. This type of error is said to occur if we conclude that there is no real difference between two samples, while the difference does in fact exist. Thus the errors of inference can be summarized as in Table 13-1.

TABLE 13-1

	Actual situation	
Decision	*Null hypothesis true*	*Null hypothesis false*
Accept null hypothesis	Correct	Type II error
Reject null hypothesis	Type I error	Correct

The situation is illustrated in Fig. 13-1. Sample (1) may belong to the same population as sample (2). Conversely, it is possible that the two samples belong to two distinct populations which overlap at their tails, so that there is always a small chance that a sample (1) belonging to population *B* is so near the mean of population *A* that sample (1) can be erroneously believed to belong to *A*.

Let us consider the distribution curves of Fig. 13-2 and assume that we want the probability of Type I error, α, to be 1 percent. This means that the

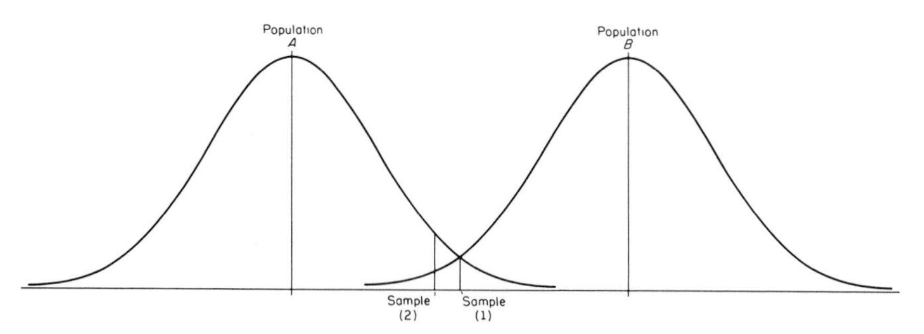

Fig. 13-1. Normal distribution curves for two populations. Which population does sample (1) belong to? If we infer that it does not belong to A while, in fact, it does, we have committed a Type I error. If we infer that (1) does belong to A while, in fact, it does not, we have committed a Type II error. (In the diagram, the area under the normal distribution curve for population A to the right of sample (1) represents 2.5 percent of the total area under the curve.)

criterion for the rejection of the null hypothesis is a value of the abscissa at least equal to the value indicated by X such that the area under curve A to the right of X is 1 percent of the total area under the curve. The probability of Type II error, β, is then equal to the shaded area under curve B. This probability can be calculated from the knowledge of the mean of the population B, μ_B, the standard deviation of the means of the sample drawn from this population, $s_B/\sqrt{n}$, and from the value of X. This probability is

$$P(\bar{x} < X) = P\left(\frac{\bar{x} - \mu_B}{s_B/\sqrt{n}} < \frac{X - \mu_B}{s_B/\sqrt{n}}\right) = \beta$$

where $\bar{x}$ is the observed sample mean.

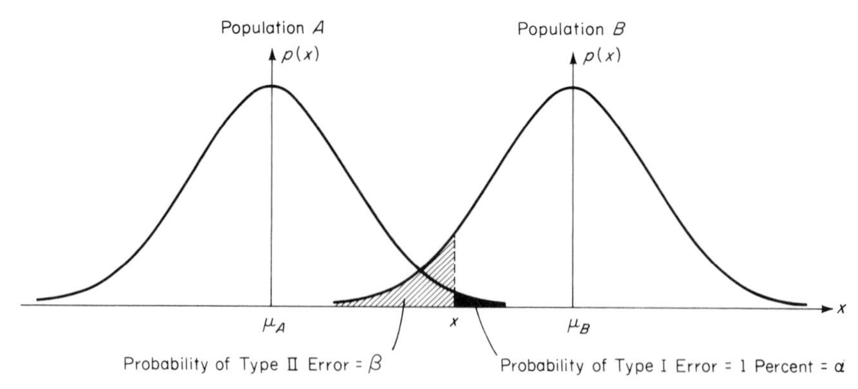

Fig. 13-2. Graphical representation of the probability of obtaining a Type II error, assuming that a Type I error of 1 percent is acceptable.

The value of the probability β can be found with the aid of Table A-4. This is illustrated in Solved Problem 13-2 at the end of the chapter.

It is not possible to reduce both Types I and II errors in a given test without changing the sample size. For a given sample size, the smaller the Type I error, the larger the Type II error. See Solved Problem 13-2. Usually, the tests of significance are arranged so that there is a specified risk of committing a Type I error (this risk is expressed as the level of significance of the test) without a provision for controlling the risk of a Type II error, which depends on the true unknown value of the parameter in question. However, if the resulting Type II error is unacceptably high, the probability of committing it can be decreased by increasing the sample size, but this, of course, means a higher cost. The choice of sample size for specified risks of committing errors of both types is discussed in Chapter 21. Here it suffices to say that the level of acceptability of the error of either type depends on the consequences of its occurrence: We must err on the safe side; this, indeed, is common engineering sense. (See Table 13-2.)

TABLE 13-2

Examples of Type I and Type II Errors.

Engineering problem	*Type I error*	*Type II error*
	Selection of a more expensive device that *seems* to behave more dependably than others but does not really do so	Failure to select the truly best device
Parachute opening device	Not serious	Extremely serious (safety)
Electric relay in a cheap toy	Serious (economics)	Not very serious

NORMAL DISTRIBUTION TEST

We frequently want to determine whether a set of observations (i.e., a sample) accords with the hypothesis that the population mean has a specific value. The standard deviation of the population may be known (as in quality control work when a large amount of previous data is available) or may have to be estimated from the standard deviation of the actual observations (as is usually the case in experimental work). In the latter case the sample must be sufficiently large (say, $n \geq 30$) for a close estimate of the population standard deviation σ to be possible.

To answer our question, we apply the normal distribution test, provided, of course, that the underlying distribution is normal. The test consists of calculating

$$u = \frac{|\mu - \bar{x}|}{\sigma/\sqrt{n}} \tag{13-1}$$

where μ is the population mean, $\bar{x}$ the sample mean, n the sample size, and $\sigma/\sqrt{n}$ the standard deviation of the mean [from Eq. (6-8)].

We can now find from Table A-4 the probability of obtaining a value of u at least this large. (Note that u is essentially the same as z, but the variate is now the mean; for this reason the standard deviation of the variate is $\sigma/\sqrt{n}$.) This probability is represented by $1 - 2 \times F(z)$. The factor 2 is appropriate here, since we are generally interested in testing the significance of the difference $|\mu - \bar{x}|$, and are not concerned with whether $\mu > \bar{x}$ or $\mu < \bar{x}$. When the probability given by Table A-4 is small (below a specified minimum), we reject the hypothesis.

If the underlying population is not infinite but consists of N items, then u is modified to

$$u = \frac{|\mu - \bar{x}|}{(\sigma/\sqrt{n})\sqrt{(N-n)/(N-1)}} \tag{13-2}$$

Example. The mean ultimate strength of a certain aluminum alloy wire is stated by the manufacturer to be 250 MN/m^2. A contractor buys a consignment of wire and tests specimens from 35 coils. The mean value from these tests is 247.4 MN/m^2 with a standard deviation of 11.2 MN/m^2. Is the contractor justified in concluding that the consignment does not accord with the manufacturer's statement? Use $\alpha = 5$ percent.

We have $\mu = 250$, $\bar{x} = 247.4$, and $s = 11.2$.

Because the sample size is large ($n = 35$), we can use the normal distribution test and take the standard deviation of the mean as

$$\frac{11.2}{\sqrt{35}} = 1.89$$

From Eq. (13-1),

$$u = \frac{250 - 247.4}{1.89} = 1.37$$

From Table A-4 the probability of obtaining a value at least this large is greater than 5 percent. We conclude, therefore, that the consignment belongs to the manufacturer's stated population and advise the contractor to accept the wire.

The normal distribution test can also be used to compare the means of two large samples, provided that they are *independent* quantities. The variance of the difference of the means σ_d^2 of two such quantities is the sum of variances of the two sample means [Eq. (14-14)], i.e.,

$$\sigma_d^2 \simeq \frac{s_1^2}{n_1} + \frac{s_2^2}{n_2} \tag{13-3}$$

where s_1^2 and s_2^2 are the variance estimates from the two samples of size n_1 and n_2, respectively. It should be noted that since the two samples are large, the variances of the two sample means will be approximately normally distributed (see the Central Limit Theorem, p. 151). Therefore, the estimated variance σ_d^2 is, in fact, quite precise under such conditions. A sample size is considered large when n is greater than about 30.

As in Eq. (13-1), the significance of the difference of means is measured by the ratio of the difference to its standard deviation, namely,

$$u = \frac{|\bar{x}_1 - \bar{x}_2|}{\sigma_d} \tag{13-4}$$

or

$$u = \frac{|\bar{x}_1 - \bar{x}_2|}{\sqrt{\dfrac{\sum (x_1 - \bar{x}_1)^2}{n_1(n_1 - 1)} + \dfrac{\sum (x_2 - \bar{x}_2)^2}{n_2(n_2 - 1)}}} \tag{13-5}$$

where x is the sample mean and n the sample size. The subscripts 1 and 2 refer to the two samples being compared.

The probability of $|\bar{x}_1 - \bar{x}_2|$ being smaller than $u\sigma_d$ (if drawn by chance from the same population) is given by the area under the normal curve between the limits $\mu \pm u\sigma_d$. Since we are testing the difference between the two means regardless of sign (i.e., the absolute value of $\bar{x}_1 - \bar{x}_2$), we require double the area given in Table A-4; this is the so-called two-sided (or two-tail) test. For convenience the more common values of the areas between $\mu - u\sigma_d$ and $\mu + u\sigma_d$ are given in Table 13-3. It should be stressed that the test is applicable only if the standard deviation σ_d is closely estimated and if the distribution within samples is approximately normal. The normal distribution test is thus a particular case of the t test (see p. 205) when the sample size is large, i.e., $n > 30$. This can be seen from a comparison of Table 13-3 and the last line of Table A-8.

Example. Traffic studies before and after traffic control improvements were made at a certain intersection in a city. The condensed data are given in Table 13-4.

TABLE 13-3

Probability for Normal Distribution

u	*Probability of* $\lvert \bar{x}_1 - \bar{x}_2 \rvert$ *being smaller than* $u\sigma_d$, *percent*
0.524	40
0.674	50
1.036	70
1.282	80
1.645	90
1.960	95
2.326	98
2.576	99
3.291	99.9

TABLE 13-4

Time	*Mean speed* $\bar{x}$, *km/h*	*Number of speed observations* n	*Standard deviation* s, *km/h*
Before	29.3	140	5.65
After	30.9	160	5.07

Does the difference between the two mean speeds represent a significant increase? Use $\alpha = 1$ percent.

The standard deviations of the two means are

$$\frac{s_1}{\sqrt{n_1}} = \frac{5.65}{\sqrt{140}} = 0.478 \qquad \text{and} \qquad \frac{s_2}{\sqrt{n_2}} = \frac{5.07}{\sqrt{160}} = 0.401$$

From Eq. (13-3) the standard deviation of the difference of means is

$$\sigma_d = \sqrt{(0.478)^2 + (0.401)^2} = 0.624$$

We now calculate u [given by Eq. (13-4)]:

$$u = \frac{\lvert \bar{x}_1 - \bar{x}_2 \rvert}{\sigma_d} = \frac{30.9 - 29.3}{0.624} = 2.56$$

This is a one-sided test, since we are checking the null hypothesis $\bar{x}_1 = \bar{x}_2$ against the alternative hypothesis $\bar{x}_1 > \bar{x}_2$. From Table A-4 we find the probability of obtaining by chance a value of u of at least this magnitude as $(0.5000 - 0.4948) = 0.0052 = 0.52$ percent.

We conclude, therefore, that the observed increase in the mean speed is significant and not due to chance error.

THE t TEST

When the conditions stated at the end of the preceding section are not satisfied, we apply Student's t test. (Student was the pseudonym of W. S. Gosset, a chemist at Guinness' Brewery in Dublin.) Specifically, we may find in practice that the true standard deviation σ in Eq. (13-1) is unknown. In such cases, σ is estimated by s, which is computed from the sample. Then s replaces σ and the random variable t replaces u in Eq. (13-1) to yield

$$t = \frac{|\mu - \bar{x}|}{s/\sqrt{n}} \tag{13-6}$$

where t has a sampling distribution that is related to the normal and χ^2 distributions (see p. 237). Values of the statistic t are given in Table A-8 in terms of the number of degrees of freedom v. This number is the same as the one for the estimate s. As we have seen in Chapter 4, for a sample of size n, v for s is $(n - 1)$. For a large-sample size $(n \geq 30)$ the t distribution closely approximates the normal distribution.

The test is applied to the null hypothesis that the two samples being compared are drawn from the same population, and we calculate the probability of the difference $|\bar{x}_1 - \bar{x}_2|$ having a value as large as, or greater than, observed. If the samples belong to the same population, then the sample means (being means of random samples) are normally distributed about the population mean, even if the distribution within the samples is not normal. The combined (population) variance s_c^2 is estimated by pooling the sums of squares of the residuals $(x - \bar{x})$ of both samples and dividing by the total number of degrees of freedom $v = (n_1 - 1) + (n_2 - 1)$. Thus

$$s_c^2 = \frac{\sum (x_1 - \bar{x}_1)^2 + \sum (x_2 - \bar{x}_2)^2}{(n_1 - 1) + (n_2 - 1)} \tag{13-7}$$

If the data in hand give estimates of the standard deviation of the two samples, s_1 and s_2, rather than the sums of squares, Eq. (13-7) can be written as

$$s_c^2 = \frac{s_1^2(n_1 - 1) + s_2^2(n_2 - 1)}{(n_1 - 1) + (n_2 - 1)} \tag{13-8}$$

Thus each estimated variance is weighted by the number of degrees of freedom available for its calculation. This is the only method of obtaining a combined variance. Averaging of variances without considering the numbers of degrees of freedom involved is incorrect. Averaging of standard deviations is also incorrect.

The standard deviations of the two means are given in the usual manner by $s_c/\sqrt{n_1}$ and $s_c/\sqrt{n_2}$, respectively. The standard deviation of the difference of means is thus:

$$s_d = \sqrt{\frac{s_c^2}{n_1} + \frac{s_c^2}{n_2}}$$

or

$$s_d = s_c\sqrt{\frac{n_1 + n_2}{n_1 n_2}} \tag{13-9}$$

The significance of the difference is measured by the ratio of the difference to its standard deviation, and is denoted by t, so that

$$t = \frac{|\bar{x}_1 - \bar{x}_2|}{s_d} \tag{13-10}$$

i.e.,

$$t = \frac{|\bar{x}_1 - \bar{x}_2|}{\sqrt{\dfrac{\sum (x_1 - \bar{x}_1)^2 + \sum (x_2 - \bar{x}_2)^2}{(n_1 - 1) + (n_2 - 1)} \times \left(\dfrac{n_1 + n_2}{n_1 n_2}\right)}} \tag{13-11}$$

the number of degrees of freedom being the number of observations less two (which were used in determining the variances), i.e., $(n_1 - 1) + (n_2 - 1)$.

Since the null hypothesis being examined by the t test assumes that the two samples belong to the same population, the two variance estimates must be consistent with this hypothesis, i.e., the two variances must not be significantly different. This should be verified by means of the F test (see Chapter 14) before the t test is applied. When $n_1 = n_2 = n$, Eq. (13-11) is simplified to

$$t = \frac{|\bar{x}_1 - \bar{x}_2|}{\sqrt{[\sum (x_1 - \bar{x}_1)^2 + \sum (x_2 - \bar{x}_2)^2]/n(n-1)}} \tag{13.12}$$

The probability of $\bar{x}_1 - \bar{x}_2$ exceeding ts_d, if drawn by chance from the same population, represents the odds *against* the null hypothesis and, similarly to the case in Chapter 10, is known as the level of significance. Values of t for various levels of significance and degrees of freedom are given in Table A-8.

We usually specify the level of significance at which we are prepared to reject the null hypothesis as 5 percent or 1 percent. If the calculated t is greater than the tabulated value at the specified level of significance, we reject the null hypothesis and conclude that the difference is significant. If the calculated t is not greater than the tabulated t at, say, the 5 percent level of significance, the null hypothesis is accepted, but we cannot tell whether there is no difference between the means being compared or whether the data are inadequate to establish whether or not there is a difference. As previously stated, there is no question of ever proving the null hypothesis. We must remember also that statistical considerations are not the sole basis for drawing inferences; a physical appreciation of the problem, judgment, and experience should also be brought into the picture.

We can now answer the question posed at the beginning of this chapter. The difference between the mean strengths of the two concretes is $50.4 - 48.3 = 2.1$ MN/m^2; and

$$s_d = 2.8\sqrt{\frac{20}{100}} = 1.25 \text{ MN/m}^2$$

Hence

$$t = \frac{2.1}{1.25} = 1.68$$

For $20 - 2 = 18$ degrees of freedom, Table A-8 gives $t = 1.734$ at the 10 percent level of significance, and we conclude, therefore, that there is no significant difference between the strengths of the two concretes.

We should note that for a given probability (i.e., a specified level of significance) the size of the sample required to make a decision possible increases as the difference between the means decreases.

Example. The slopes of two types of valley walls were measured. Sample A comprised slopes at whose base talus and slope wash have accumulated, indicating that considerable time has elapsed since stream erosion was active against the slope base. Sample B comprised slopes at whose base stream erosion has recently been active. We wish to determine whether the slopes of Sample A differ significantly from those of B. In other words, do the data indicate that a slope, if left to weather and waste without basal cutting, tends to decline in angle rather than retreat in parallel planes?[1] Use $\alpha = 1$ percent.

Sample	*Mean slope* $\bar{x}$	*Standard deviation* s	*Sample size* n
A	38.23°	2.70°	34
B	44.82°	3.27°	172

1. A. N. Strahler, "Statistical Analysis in Geomorphic Research," *Journal of Geology*, Jan. 1954, p. 12.

The F test (see Chapter 14) shows that the variances of the two samples do not differ significantly, and, of course, the two samples are independent of one another.

From Eq. (13-8) the pooled estimate of variance is

$$s_c^2 = \frac{33 \times 2.7^2 + 171 \times 3.27^2}{33 + 171} = 10.142$$

and

$$s_c = 3.2°$$

Using Eq. (13-9), we obtain

$$s_d = 3.2\sqrt{\frac{1}{34} + \frac{1}{172}} = 0.60$$

and from Eq. (13-10),

$$t = \frac{44.82 - 38.23}{0.60} = 10.98$$

The number of degrees of freedom is $172 + 34 - 2 = 204$, and Table A-8 gives the probability of less than 0.1 percent of obtaining a value of t equal to or greater than 10.98. Therefore, the null hypothesis can be rejected, and we conclude that the absence of erosion at the base of a slope leads to a decline in the angle of the slope, but when erosion takes place, the slope retreats parallel to itself.

ONE-SIDED AND TWO-SIDED TESTS

There are two general questions that we may seek to answer by means of the t test. The first one is: Is $\bar{x}_1$ significantly different from $\bar{x}_2$? Here we are not interested in whether $\bar{x}_1 > \bar{x}_2$ or $\bar{x}_1 < \bar{x}_2$; the null hypothesis can be wrongly rejected in favor of either of these possibilities, and the test is therefore two-sided. The null hypothesis is rejected if the proportional area under *two* tails is greater than that corresponding to the specified level of significance, i.e., when calculated $|t| >$ tabulated t.

The second question is of the type: Is $\bar{x}_1 \geq \bar{x}_2$? If $\bar{x}_1 \ngtr \bar{x}_2$, we accept the null hypothesis regardless of how much smaller $\bar{x}_1$ is than $\bar{x}_2$. This is a one-sided test, and the sign of $\bar{x}_1 - \bar{x}_2$ is material. In a one-sided test we are interested in one tail of the distribution only, and the probability given at the top of Table A-8 should therefore be halved. This follows from the fact that the distribution of t is symmetrical, and, for example, a 5 percent level of significance means that there is a $2\frac{1}{2}$ percent probability of obtaining t

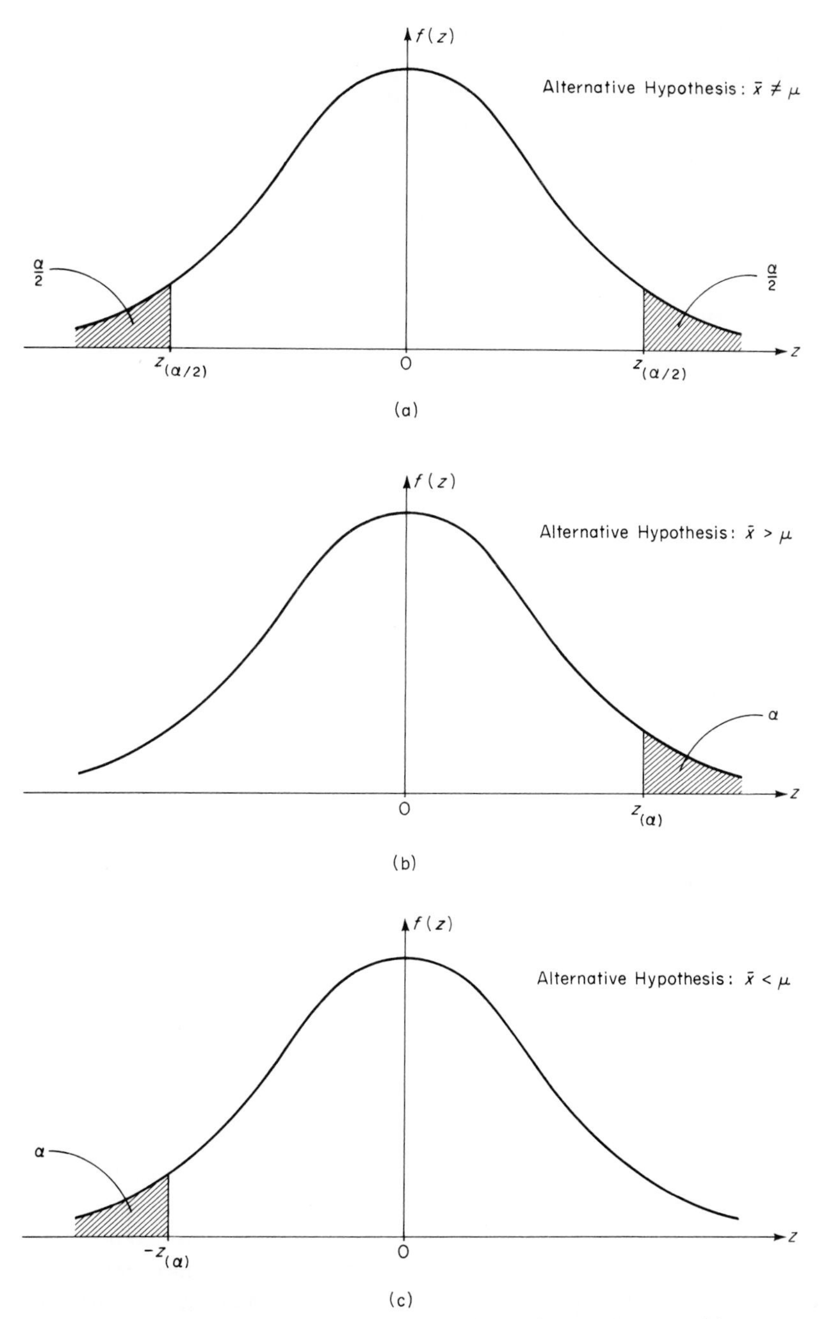

Fig. 13-3. One-sided and two-sided normal tests with null hypothesis $\bar{x} = \mu$: (a) two-sided test; (b) one-sided test with positive z; and (c) one-sided test with negative z. (Here $z = u$.)

greater than tabulated and a $2\frac{1}{2}$ percent probability of obtaining t smaller than the negative tabulated value [i.e., the tabulated value times (-1)].

It is advisable to decide whether we should apply a one-sided or a two-sided test before commencing the calculations so as to avoid the tendency to lean toward the test that will give a more "convenient" result. As a general rule, a one-sided test is appropriate if a deviation in a direction opposite to that postulated would have no practical significance. For example, if we require fuel with a certain minimum octane rating level, we would test whether the actual rating is significantly lower than that specified, but we would not be concerned if it were higher.

An example of a one-sided test is afforded by the traffic speed problem, considered earlier in this chapter, since the question asked concerned an increase in speed.

The concepts of one-sided and two-sided normal distribution tests are illustrated in Fig. 13-3, and these are similar to the t tests. See also Solved Problems 13-1 and 13-2.

We should stress the fact, mentioned earlier, that the t test can be used for any underlying distribution while the "normal distribution test" of Eq. (13-5) is limited to normal distribution with a close estimate of population variance available. Thus, although Eq. (13-4) and (13-10) are similar, they are applied in different cases. We may note that t itself is not normally distributed but becomes so when the number of degrees of freedom ν approaches infinity. This can be seen from a comparison of the values of t for $\nu = \infty$ (Table A-8) and the values of z for the corresponding $2F(z)$ (Table A-4). For example, at the 10 percent level of significance $t = 1.645$. Now for $z = 1.645$, $2F(z) = 0.90$, which corresponds to the 10 percent level of significance.

THE t TEST FOR PAIRED DATA

The comparison of means discussed in the preceding pages is applicable when the two samples have been drawn independently of one another; for example, they may represent products of two factories, and we may want to compare these products in order to determine whether there is a significant difference between them. If, however, the samples are drawn from one source and then subjected to two different treatments whose effects are being studied, we can use a somewhat different technique, although the t test on the difference of means is still permissible.

In the case of paired variables we consider the mean difference between two samples as the variate and compare it with *its* standard deviation, i.e., we apply the t test to the paired data. In this manner, the effect of the test-to-test variation is eliminated, but this advantage is offset by a loss in precision due to the standard deviation's being based on fewer degrees of freedom (the number of *pairs* less one, instead of the total number of observations less two).

It should be stressed that this technique can be used only when the pairs of samples are truly correlated; pairing by random choice or by arrangement of samples in each group by rank is not permitted. Because of the somewhat confusing terminology, it is important to distinguish between a test for the difference between means and a test for the mean difference between pairs of observations.

INTERPRETATION OF RESULTS

When the data are truly paired, we can apply either form of the t test. Which one is more convenient to use will depend on whether the variation between tests, which may be made under different conditions, obscures the difference between two members of a pair forming one test. This is illustrated by the next example.

Having a choice of tests of significance, the reader may wonder what to do when one test indicates a significant difference between two sets of values while the other does not. We should remember that failure to detect a significant difference does not mean that there is none, but only that we cannot say with a sufficiently high probability of being right that a difference is present. Thus if *any* method of test shows a definitely significant difference, its testimony is vital, even though another method fails to show a similar result.

Example. The influence of the size of the test specimen on the tensile strength of briquettes was tested as shown in Table 13-5. Seven mixes were made and from each one large and one small concrete specimen were prepared and tested.

TABLE 13-5

Strength of specimen, kN/m²		
Small x_2	*Large* x_1	*Difference, kN/m²* $x_2 - x_1$
4404	4140	264
4326	3984	342
3788	3842	−54
3475	3053	422
3418	3145	273
2262	1813	449
7415	6867	548
$\sum x_2 = 29{,}088$	$\sum x_1 = 26{,}844$	$\sum (x_2 - x_1) = 2244$

Test whether there is a significant difference between the strengths of the two types of specimens. Use $\alpha = 5$ percent. Hence

$$\bar{x}_2 = \frac{\sum x_2}{n_2} = \frac{29{,}088}{7} = 4155.43 \text{ kN/m}^2$$

and

$$\bar{x}_1 = \frac{26{,}844}{7} = 3834.86 \text{ kN/m}^2$$

Since the data are paired, we consider $x_2 - x_1$ as the variable. Let $y = x_2 - x_1$. Then

$$\bar{y} = \frac{\sum y}{n} = \frac{2244}{7} = 320.57 \text{ kN/m}^2$$

Now the standard deviation of y is

$$s = \sqrt{\frac{\sum (y - \bar{y})^2}{n - 1}}$$

$$= \sqrt{\frac{224{,}731.71}{6}} = 193.53 \text{ kN/m}^2$$

The standard deviation of $\bar{y}$ is

$$s_{\bar{y}} = \frac{s}{\sqrt{n}}$$

$$= \frac{193.53}{\sqrt{7}} = 73.15 \text{ kN/m}^2$$

We now apply the t test to $\bar{y}$ by comparing it to a zero mean difference:

$$t = \frac{|\bar{y} - 0|}{s_{\bar{y}}} = 4.38$$

The number of degrees of freedom is 6 ($\nu = n - 1 = 7 - 1$), and Table A-8 gives $t = 3.707$ at the 1 percent level of significance. The difference in strength is thus significant.

Let us now ignore the pairing and apply the t test to the mean difference of strengths.

Using Eq. (13-7), we compute the combined population variance s_c^2:

$$s_c^2 = \frac{\sum (x_1 - \bar{x}_1)^2 + \sum (x_2 - \bar{x}_2)^2}{2n - 2}$$

$$= \frac{14{,}484{,}406.90 + 15{,}442{,}547.80}{12}$$

$$= 2{,}493{,}912.89$$

The standard deviation of the difference of means is given by Eq. (13-9):

$$s_d = s_c\sqrt{\frac{2}{n}}$$

$$= 1579.21\sqrt{\frac{2}{7}} = 844.12 \text{ kN/m}^2$$

The mean difference of strengths is

$$\bar{x}_2 - \bar{x}_1 = 4155.43 - 3834.86 = 320.57 \text{ kN/m}^2$$

We now apply the t test:

$$t = \frac{|\bar{x}_2 - \bar{x}_1|}{s_d}$$

$$= \frac{320.57}{844.12} = 0.380$$

The number of degrees of freedom is $(2n - 2) = 12$. Table A-8 gives $t = 2.179$ at the 5 percent level. The difference thus appears to be not significant.

It is clear that the conclusion from the last test is due to the considerable difference in the level of strength of the different mixes. The paired test is more discriminating, and we conclude that there is a significant difference between the tensile strengths of the specimens of the two sizes.

CASE OF NONHOMOGENEOUS VARIANCES

Equation (13-11) was obtained on the assumption that the variances of the two samples being compared are not significantly different (i.e., are homogeneous), this being implicit in the hypothesis that the two samples belong to the same population. The homogeneity of variances is examined by the F test (see Chapter 14).

However, in some cases the condition of homogeneity of variances may not be satisfied, but we may still want to test the significance of the difference of two means. This is the case, for example, when errors of measurement in the two samples are due to different causes so that the estimates of variance cannot correctly be pooled. The t test cannot, therefore, be applied, and we use a test in which the ratio of the standard deviations of the two sample means, $s_{\bar{x}1}/s_{\bar{x}2} = \tan\theta$, is considered in determining the significance of the difference of the means. The difference $\bar{x}_1 - \bar{x}_2$ is considered significant if

$$\frac{|\bar{x}_1 - \bar{x}_2|}{\sqrt{s^2_{\bar{x}1} + s^2_{\bar{x}2}}} > d \tag{13-13}$$

where d is given in Table A-9 for two levels of significance and for different values of θ, ν_1, and ν_2. The numbers of degrees of freedom in the two samples, ν_1 and ν_2, are equal to $n_1 - 1$ and $n_2 - 1$, respectively (n_1 and n_2 are the sample sizes).

Example. Imagine that we have used two different methods to determine a physical constant. Each method yields a mean value, and we want to determine at $\alpha = 1$ percent whether or not there is a real discrepancy between the two results. Because different methods have been used, the variances may differ and cannot be pooled. Let the results be

$$\bar{x}_1 = 5.289 \qquad \bar{x}_2 = 5.261$$

$$s_{\bar{x}1}^2 = 0.00008 \qquad s_{\bar{x}2}^2 = 0.00001$$

for $n_1 = 13$ and $n_2 = 60$.

We may check that the F test yields

$$F = \frac{s_{\bar{x}1}^2}{s_{\bar{x}2}^2} = 8$$

which, for the degrees of freedom $\nu_1 = 12$ and $\nu_2 = 59$, shows a significant difference at the 1 percent level (F from Table A-10 is 2.50). We calculate

$$\tan\theta = \sqrt{\frac{0.00008}{0.00001}} = 2.828$$

whence $\theta = 70.5°$. Now

$$\sqrt{s_{\bar{x}1}^2 + s_{\bar{x}2}^2} = 0.0095$$

Therefore,

$$d = \frac{|\bar{x}_1 - \bar{x}_2|}{\sqrt{s_{\bar{x}1}^2 + s_{\bar{x}2}^2}} = \frac{0.028}{0.0095} = 2.95$$

Table A-9 gives the values of θ of 60° and 75°; therefore, we have to interpolate. For the 5 percent level of significance the values of d are for $\nu_1 = 12$:

$$\nu_2 = 24 \quad \begin{cases} \theta = 60° & d = 2.142 \\ \theta = 75° & d = 2.168 \end{cases}$$

$$\nu_2 = \infty \quad \begin{cases} \theta = 60° & d = 2.120 \\ \theta = 75° & d = 2.163 \end{cases}$$

As all these values are close to one another, we shall take simply the arithmetic mean of the four values of d; this is $d = 2.15$.

For the 1 percent level of significance the four values of d at the same points are: 2.938, 3.020, 2.909, and 3.014. The arithmetic mean is thus 2.97.

We can, therefore, conclude that the difference between the means $\bar{x}_1$ and $\bar{x}_2$ is significant at the 5 percent level but not quite at the 1 percent level.

A special case arises when the standard deviation is not constant for either group but varies with the level of x, the variation being the same in both groups. For example, in testing the strength of concrete it has been found that the standard deviation is proportional to strength.[2] In such a case logarithmic transformation has to be applied to the variate. The variance of the natural logarithm of the original variate is approximately equal to the square of the coefficient of variation V^2. We can, therefore, use Eq. (13-10), substituting $\log_e x$ for x, and V^2 for s_c^2. Then

$$t = \frac{|\log_e \bar{x}_1 - \log_e \bar{x}_2|}{\sqrt{\dfrac{V_1^2(n_1 - 1) + V_2^2(n_2 - 1)}{n_1 + n_2 - 2} \times \dfrac{n_1 + n_2}{n_1 n_2}}} \tag{13-14}$$

where $\bar{x}_1$ and $\bar{x}_2$ are the sample means, n_1 and n_2 the sample sizes, and V_1 and V_2 the coefficients of variation. Note that t has $(n_1 + n_2 - 2)$ degrees of freedom.

If n is large, we can consider V^2 to be the variance of $\log_e \bar{x}$ and not an estimate, and Eq. (13-14) reduces to

$$t = \frac{|\log_e \bar{x}_1 - \log_e \bar{x}_2|}{\sqrt{\left(\dfrac{V_1^2}{n_2} + \dfrac{V_2^2}{n_1}\right) \times \dfrac{n_1 + n_2}{n_1 + n_2 - 2}}} \tag{13-15}$$

CHOICE OF APPROACH

The various tests of the null hypothesis of no difference between two means have been described in the preceding pages, and it may be convenient to have them summarized in one table. This is done in Table 13-6.

DISTRIBUTION-FREE OR NONPARAMETRIC TESTS

In certain applications, one is confronted with sampling distributions arising from populations with unknown parameters. In such cases one should not make assumptions about such parameters other than those of

2. A. M. Neville, "The Relation Between Standard Deviation and Mean Strength of Concrete Test Cubes," *Magazine of Concrete Research* vol. 10, no. 31, July 1959.

TABLE 13-6

Summary of Test Methods for Comparison of Means[a]

Null hypothesis[b]	*Conditions of applicability*	*Equation to be used*
$\bar{x} = \mu$	μ and σ are known, and the population is normally distributed and is { infinite / finite.	Eq. (13-1) (normal distribution test) / Eq. (13-2) (normal distribution test)
$\bar{x} = \mu$	μ is known but σ is unknown; s, estimated from a small sample, is used.	Eq. (13-6) (t test)
$\bar{x}_1 = \bar{x}_2$	$\bar{x}_1$ and $\bar{x}_2$ are means of large samples (independent of one another). The means can be considered as normally distributed.	Eq. (13-4) and (13-5) (normal distribution test)
$\bar{x}_1 = \bar{x}_2$	$\bar{x}_1$ and $\bar{x}_2$ are means of small samples whose variances are homogeneous.	Eq. (13-10) (t test)
$\bar{x}_1 = \bar{x}_2$	$\bar{x}_1$ and $\bar{x}_2$ are means of samples whose variances are not homogeneous.	Eq. (13-13)

[a] μ = population mean, $\bar{x}$ = sample mean, σ = standard deviation of population, and s = estimate of standard deviation from the sample used.

[b] See Fig. 13-3 for the alternative hypothesis.

random sampling and continuity of the distribution. Tests that deal with such distributions are, therefore, called distribution-free tests; since they do not involve the use of parameters, they are known as nonparametric tests.

SIGN TEST

This is the simplest of all nonparametric tests. Let us consider an example (see Table 13-7). The degree of corrosion of a pipe made of a new alloy is to be investigated. Pipe pieces made from the new alloy and from the standard alloy are paired. The pairs are placed at the same depth of the same soil, positioned in the same manner, and for the same length of time.

Is there a significant difference in the degree of corrosion in the pipe due to change in alloy? Use $\alpha = 5$ percent.

If we apply the null hypothesis that the two alloys are affected by corrosion to the same degree, the number of plus signs should be approximately equal to the number of minus signs. Such a number in a sample of size n is a random variable with a binomial distribution, with $n = 12$, $p = q = \frac{1}{2}$. The expected number of minus signs is the mean $np = (12)(\frac{1}{2}) = 6$. However, the

TABLE 13-7

Depth of Maximum Pits (in 10^{-2} mm)

Pair	*Standard Pipe, A*	*New Alloy Pipe, B*	*Sign of difference (A − B)*
1	20	25	−
2	26	29	−
3	31	28	+
4	42	37	+
5	35	40	−
6	19	29	−
7	33	41	−
8	38	43	−
9	29	21	+
10	27	35	−
11	40	47	−
12	37	41	−

observed number of minus signs is 9. The probability that 9 or more minus signs are observed is

$$\sum_{r=9}^{r=12} P_r = P_9 + P_{10} + P_{11} + P_{12}$$

We find from Table A-15 that for $p = 0.5$, $q = 0.5$, $n = 12$, $r = 9$, and $r' = 4$, the corresponding value from the table is 0.9270. Therefore, the required probability is $1 - 0.9270 = 0.0730$. Bearing in mind that this is a two-sided test, the probability of 0.0730 is compared to 0.5α, i.e., 0.025. Since $0.0730 > 0.025$, there is no strong evidence that there is a difference in the degree of corrosion in the pipes made from the two alloys.

Applying the normal test as an approximation to the binomial distribution, we find:[3]

$$\text{mean} = np = 6, \qquad \sigma = \sqrt{npq} = \sqrt{(12)(\tfrac{1}{2})(\tfrac{1}{2})} = 1.732$$

$$z = \frac{8.5 - 6}{1.732} = \frac{2.5}{1.732} = 1.442$$

and from Table A-4,

$$F(z) = 0.4254$$

3. Since we need to approximate the area of the histogram to the right of $x = 9$, we decrease 9 by $\frac{1}{2}$ to better fit a continuous normal distribution to a discrete binomial distribution.

The required probability = 0.5 − 0.4254 = 0.0746, which is greater than $\frac{1}{2}\alpha$, i.e., 0.025; therefore, we arrive at the same conclusion.

It is evident from this example, that the sign test does not take account of the magnitude of the observed differences and is, therefore, not very sensitive. This deficiency is rectified by the Wilcoxon signed rank test.

WILCOXON SIGNED RANK TEST

In this test the absolute values of the differences are first ranked and then these ranks are affixed with the associated sign of the difference. For no significant difference between the means of the two samples, the total of the ranks associated with positive differences and that associated with negative differences should be about the same. Otherwise, there may be a significant difference between the means of the two samples. This is so if the probability of obtaining by chance alone a signed rank total less than or equal to the smaller of the signed rank totals is less than a critical value at the appropriate significance level. Let us apply this test to the previous example (see Table 13-8).

TABLE 13-8

Depth of Maximum Pits (in 10^{-2} mm)

Pair	*Pipe A*	*Pipe B*	*Difference*	*Rank*	*Signed rank*
1	20	25	−5	$5\frac{1}{2}$	$-5\frac{1}{2}$
2	26	29	−3	$1\frac{1}{2}$	$-1\frac{1}{2}$
3	31	28	3	$1\frac{1}{2}$	$1\frac{1}{2}$
4	42	37	5	$5\frac{1}{2}$	$5\frac{1}{2}$
5	35	40	−5	$5\frac{1}{2}$	$-5\frac{1}{2}$
6	19	29	−10	12	−12
7	33	41	−8	10	−10
8	38	43	−5	$5\frac{1}{2}$	$-5\frac{1}{2}$
9	29	21	8	10	10
10	27	35	−8	10	−10
11	40	47	−7	8	−8
12	37	41	−4	3	−3

We notice that some of the observed differences are tied for rank. In such cases Wilcoxon suggested that the tied values be assigned the mean of the respective ranks. For example, pairs 2 and 3 have the same differences corresponding to ranks 1 and 2. The mean rank of $(1 + 2)/2 = 1.5$ is assigned to each of the two differences.

If there is no difference in the degree of corrosion in the pipes made from the two alloys, one would expect the sum of the positive ranks to be nearly equal numerically to the sum of the negative ranks.

$$\text{sum of all ranks} = 1 + 2 + \cdots + n$$
$$= \frac{n(n+1)}{2} = \frac{(12)(13)}{2} = 78$$

and hence the expected sum of either positive or negative signed ranks is

$$\frac{78}{2} = 39$$

Now the observed sum of negative ranks is

$$5.5 + 1.5 + 5.5 + 12 + 10 + 5.5 + 10 + 8 + 3 = 61$$

and the observed sum of positive ranks is

$$1.5 + 5.5 + 10 = 17$$

To check for a significant difference, we compare the absolute value of the smaller sum of ranks, i.e., 17 to the critical value of R given in Table A-16; this is $R = 14$ for $n = 12$, at the 5 percent level of significance. Since $17 > 14$, the null hypothesis cannot be rejected and we can conclude that there is no difference in the degree of corrosion in the pipes made from the two different alloys.

Note that the values given in Table A-16 are for a two-sided test. If a one-sided test is performed, the levels of significance given at the top of the table must be halved.

RANK TEST FOR TWO INDEPENDENT SAMPLES

When the samples are not paired and their probability distributions are assumed to differ only in their means, one can use a modified rank test to check for significant difference. We shall illustrate the test by considering the aforementioned data on pipe corrosion but assuming that the data are not paired. (See Table 13-9.)

We arrange all observations in order of magnitude and rank them. If no significant difference exists, one would expect the sum of ranks for both samples to be nearly the same. For the above data the sum of ranks in sample $A = 1 + 2 + 5 + \cdots + 22 = 132.5$; the sum of ranks in sample $B = 3 + 4 + 7 + \cdots + 24 = 167.5$. Let $T_1 =$ the sum of ranks of smaller sample (in this case $T_1 = 132.5$), and $T_2 = n_1(n_1 + n_2 + 1) - T_1$ (in this case $T_2 = 167.5$). Then let T be the smaller sum of ranks T_1 or T_2.

Entering Table A-17, with n_1 (size of sample A) = 12 and n_2 (size of sample B) = 12, we find the critical value T as 115. For a two-sided test,

TABLE 13-9

Sample	*Depth of Maximum Pits* (*in* 10^{-2} mm)	*Rank*
A	19	1
A	20	2
B	21	3
B	25	4
A	26	5
A	26	6
B	28	7
A	29	9
B	29	9
B	29	9
A	31	11
A	33	12
A	35	$13\frac{1}{2}$
B	35	$13\frac{1}{2}$
A	37	$15\frac{1}{2}$
B	37	$15\frac{1}{2}$
A	38	17
A	40	$18\frac{1}{2}$
B	40	$18\frac{1}{2}$
B	41	$20\frac{1}{2}$
B	41	$20\frac{1}{2}$
A	42	22
B	43	23
B	47	24

since $T_2 > T$ the difference in corrosion is not significant and the null hypothesis is accepted at the 5 percent level. If T_2 had been below the critical value T, we would have had to reject the null hypothesis at the 5 percent level. It should be noted that for unequal sample sizes n_1 applies to smaller samples. Again, for a one-sided test, the level of significance equals 0.5α, i.e., $2\frac{1}{2}$ percent in this case.

From the foregoing it appears that nonparametric tests are easily applied. Because of this and other advantages mentioned earlier, their use has increased considerably. However, when justifiable assumptions about the form and/or parameters of the population sampled can be made, nonparametric tests become less efficient than tests utilizing correct assumptions.

Solved Problems

13-1. It is suspected that the state of stress in a steel wire affects the percentage loss in the ultimate tensile strength in the wire when such a wire has been immersed in a calcium chloride solution. Twelve lengths of wire were obtained, and each one was cut into four specimens. Two of these were stressed, one in solution, one in air; the other two were unstressed, likewise one in solution, one in air.

Table 13-10 gives the percentage loss in the ultimate tensile strength for the appropriate pairs.

TABLE 13-10

Test number	*Stressed*	*Unstressed*
1	10.4	7.1
2	8.1	8.4
3	8.5	7.2
4	9.7	8.3
5	8.2	6.8
6	10.1	8.5
7	7.9	7.9
8	9.8	8.2
9	8.4	8.4
10	8.7	6.5
11	9.3	7.0
12	8.6	8.8

Establish whether stressing affects the percentage loss in strength caused by immersion in the solution. Use $\alpha = 1$ percent.

SOLUTION

Since the samples are paired, we shall apply the t test to find the significance of the mean difference between the unstressed and stressed specimens. (See Table 13-11.)

TABLE 13-11

Test number	*Stressed*	*Unstressed*	*Difference d*	d^2
1	10.4	7.1	3.3	10.89
2	8.1	8.4	−0.3	0.09
3	8.5	7.2	1.3	1.69
4	9.7	8.3	1.4	1.96
5	8.2	6.8	1.4	1.96
6	10.1	8.5	1.6	2.56
7	7.9	7.9	0	0
8	9.8	8.2	1.6	2.56
9	8.4	8.4	0	0
10	8.7	6.5	2.2	4.84
11	9.3	7.0	2.3	5.29
12	8.6	8.8	−0.2	0.04
Totals			$\sum d = 15.1 - 0.5 = 14.6$	$\sum d^2 = 31.88$

$$\text{mean difference } \bar{d} = \frac{14.6}{12} = 1.217$$

$$\text{estimated standard deviation } s = \sqrt{\frac{\sum d^2 - [(\sum d)^2/n]}{n-1}} = \sqrt{\frac{31.88 - [(14.6)^2/12]}{11}}$$

$$= \sqrt{1.283} = 1.13$$

Therefore,

$$s_{\bar{d}} = \frac{s}{\sqrt{n}} = \frac{1.13}{\sqrt{12}}$$

Hence

$$t = \frac{|\bar{d} - 0|}{s_{\bar{d}}} = \frac{1.217}{1.13} \times \sqrt{12} = 3.731$$

This is a two-sided test, since we are checking the null hypothesis that $\bar{d} = 0$ against the alternative hypothesis $\bar{d} \neq 0$. For $\nu = n - 1 = 12 - 1 = 11$, Table A-8 gives for the 1 percent level of significance $t = 3.106$. Since the calculated $t > 3.106$, the difference is significant at the 1 percent level. (See Fig. 13-4.) In other words, there is

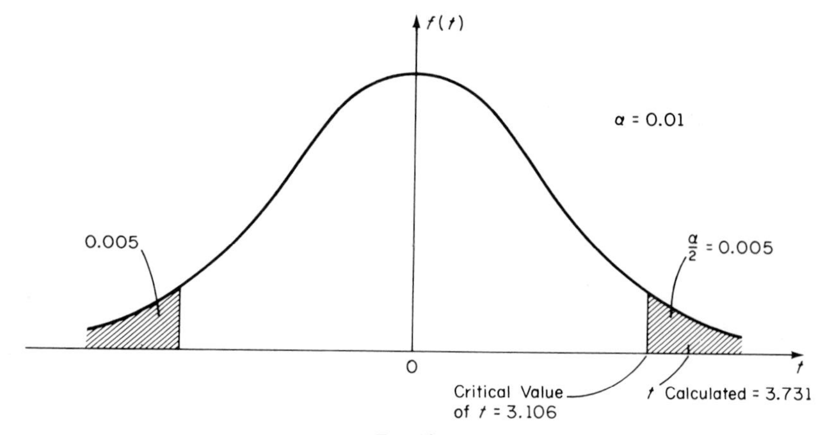

Fig. 13-4

strong evidence to support the contention that the state of stress affects the percentage loss in ultimate strength after immersion in a calcium chloride solution.

The χ^2 test can be used to give a quick, albeit rough, check on the significance of the difference. We calculate the expected numbers of positive and negative signs of the difference on the basis of the null hypothesis and compare these with the observed numbers of positive and negative signs. In tabular form:

	+ *ve* sign	− *ve* sign
observed number of signs O	8	2
expected number of signs E	5	5

$$\chi^2 = \sum \frac{(O - E)^2}{E} = \frac{(8-5)^2}{5} + \frac{(2-5)^2}{5} = \frac{9}{5} + \frac{9}{5} = 3.6$$

For $\nu = 1$, from Table A-7, the probability of obtaining such a value of χ^2 by chance is between 0.10 and 0.05. Therefore, we suspect the null hypothesis and should test the significance by the t test (as already done).

As a somewhat more accurate alternative, the binomial distribution can be used to calculate the possibility of obtaining the distribution of positive and negative signs actually observed.

If the specimens had not been paired; i.e., if we had measured the loss in strength on immersion in the solution for 12 consignments of wire using stressed specimens and for another 12 consignments using unstressed specimens, pairing would have not been possible. In such a case the t test must be applied to the difference of means, but the procedure is permissible even when the data are paired. Thus we have

a. For stressed specimens (Table 13-12):

TABLE 13-12

Strength loss x_1	*Deviation,* $x_1 - \bar{x}_1$		$(x_1 - \bar{x}_1)^2$
10.4	1.425		2.031
8.1		−0.875	0.766
8.5		−0.475	0.226
9.7	0.725		0.526
8.2		−0.775	0.601
10.1	1.125		1.266
7.9		−1.075	1.156
9.8	0.825		0.681
8.4		−0.575	0.331
8.7		−0.275	0.076
9.3	0.325		0.106
8.6		−0.375	0.141
$\sum x_1 = 107.7$	$\sum (x_1 - \bar{x}_1) = +4.425$	$-4.425 = 0$	$\sum (x_1 - \bar{x}_1)^2 = 7.907$

$$\bar{x}_1 = \frac{107.7}{12} = 8.975$$

b. For unstressed specimens (Table 13-13):

$$\bar{x}_2 = \frac{\sum x_2}{n} = \frac{93.1}{12} = 7.758$$

The pooled estimate of variance is

$$s_c^2 = \frac{\sum (x_1 - \bar{x}_1)^2 + \sum (x_2 - \bar{x}_2)^2}{(n_1 - 1) + (n_2 - 1)}$$

$$= \frac{7.907 + 6.788}{11 + 11} = \frac{14.695}{22} = 0.668$$

TABLE 13-13

Strength loss x_2	*Deviation*, $(x_2 - \bar{x}_2)$		$(x_2 - \bar{x}_2)^2$
7.1		−0.6583	0.433
8.4	0.6417		0.412
7.2		−0.5583	0.312
8.3	0.5417		0.293
6.8		−0.9583	0.918
8.5	0.7417		0.550
7.9	0.1417		0.020
8.2	0.4417		0.195
8.4	0.6417		0.412
6.5		−1.2583	1.583
7.0		−0.7583	0.575
8.8	1.0417		1.085
$\sum x_2 = 93.1$	$\sum (x_2 - \bar{x}_2) = +4.1916$	$-4.1915 \simeq 0$	$\sum (x_2 - \bar{x}_2)^2 = 6.788$

The standard deviation of the difference of means is

$$s_d = s_c\sqrt{\frac{n_1 + n_2}{n_1 n_2}}$$

$$= \sqrt{\frac{0.668}{6}} = 0.3337$$

The difference of means $= \bar{x}_1 - \bar{x}_2 = 8.975 - 7.758 = 1.217$. Hence

$$t = \frac{1.217}{0.3337} = 3.647$$

The number of degrees of freedom $= 24 - 2 = 22$.

From Table A-8, $t = 2.819$ at the 1 percent level of significance and 3.792 at the 0.1 percent level for a two-sided test. The difference is, therefore, highly significant.

13-2. The ultimate strengths of prestressing wires manufactured by a steel company have a mean of 2000 N and a standard deviation of 100 N. By employing a new manufacturing technique, the company is claiming that the ultimate strength is now increased. To verify this claim, a builder tests a sample of 50 wires produced by the new process and finds that the mean ultimate strength is 2050 N.

a. Do the data support the manufacturer's claim at the 1 percent level of significance?
b. What is the probability of accepting the old process when, in fact, the new process has increased the mean ultimate strength to 2050 N, assuming still $\sigma = 100$ N?

SOLUTION

a. We have to decide between two hypotheses: either $\mu = 2000$ N, and there is really no change in ultimate strength; or $\mu > 2000$ N, and there is a change in ultimate strength.

Since we are testing whether the new process is better than the old one (and not better *or* worse), a one-sided test is appropriate. Thus, for a one-sided test at $\alpha = 0.01$, assuming a normal distribution, $F(u) = 0.5 - 0.01 = 0.49$. The corresponding critical value of u from Table A-4 is $u = 2.33$. (Note: We are using the notation of u instead of z, since we are dealing with the mean as a variate.) The decision is: If the calculated $u > 2.33$, reject the hypothesis $\mu = 2000$ N; accept the hypothesis $\mu = 2000$ N. (See Fig. 13-5.) Otherwise, from Eq. (13-1) the actual u

$$= \frac{\bar{x} - \mu}{\sigma/\sqrt{n}} = \frac{2050 - 2000}{100/\sqrt{50}} = \frac{50}{100/\sqrt{50}} = 3.54$$

Since $3.54 > 2.33$, the results are highly significant and we conclude that the hypothesis $\mu = 2000$ N should be rejected and the manufacturer's claim should be accepted.

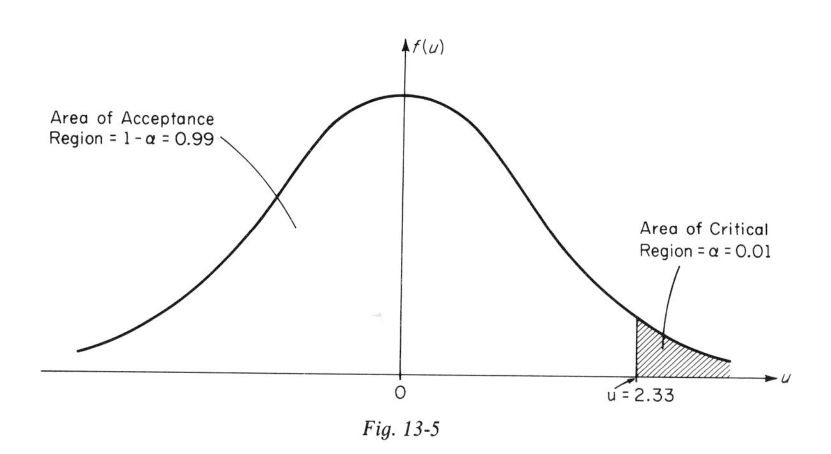

Fig. 13-5

To calculate the critical mean strength for rejecting the old process at $\alpha = 0.01$ with $n = 50$, we have the critical value $u = 2.33$, $\sigma = 100$ N, $\mu = 2000$ N; thus, from Eq. (13-1),

$$u = \frac{\bar{X} - \mu}{\sigma/\sqrt{n}} = \frac{\bar{X} - 2000}{100/\sqrt{50}} = 2.33$$

Therefore, $\bar{X} = 2033$ N.

We can rephrase the above conclusion by saying that we reject the hypothesis $\mu = 2000$ N, since $2050 > 2033$.

b. The two hypotheses are now: $\mu = 2000$ N against the alternative of $\mu = 2050$ N. The distributions of the mean ultimate strengths corresponding to the above hypotheses are shown in Fig. 13-6. The probability of committing a Type II

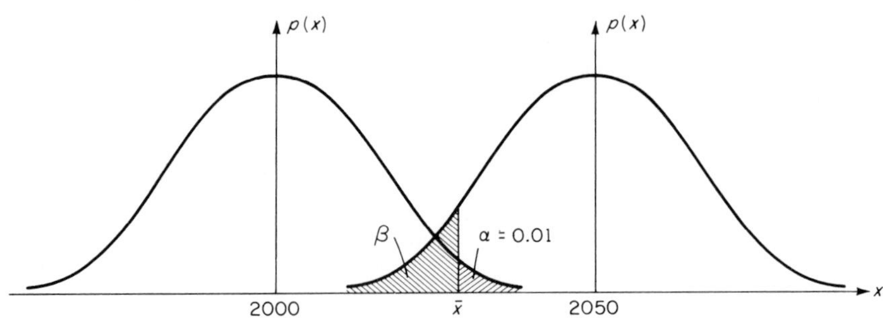

Fig. 13-6

error is β, which is the area under the right-hand normal curve corresponding to

$$u = \frac{2033 - 2050}{100/\sqrt{50}} = \frac{-17}{(100/7.08)} = -1.21$$

Therefore,

$$\beta = 0.5 - F(1.21) = 0.5 - 0.3869 = 0.1131 \simeq 0.11$$

The decision in this case becomes: Reject the hypothesis $\mu = 2000$ N if $\bar{X} \geq 2033$ N (the probability of Type I error, α, is 0.01). Accept the hypothesis that $\mu = 2000$ N if $\bar{X} < 2033$ N (the probability of Type II error, β, is 0.11).

If the probability for Type II error is too high and we wish to limit it, e.g., to $\beta = 0.05$, what should the size of the sample n be? Assume the same standard deviation.

$$\text{critical } u = \frac{\bar{X} - 2000}{100/\sqrt{n}} = 2.33 \qquad \text{for } \alpha = 0.01$$

Hence $\bar{X} = 2000 + 233/\sqrt{n}$.

$$\beta = 0.05 = 0.5 - F\left(\frac{2050 - \bar{X}}{100/\sqrt{n}}\right)$$

Thus

$$0.45 = F\left(\frac{2050 - 2000 - 233/\sqrt{n}}{100/\sqrt{n}}\right)$$

$$= F\left(\frac{\sqrt{n}}{2} - 2.33\right)$$

From Table A-4, for $F(u) = 0.45$ the corresponding u is 1.645. Therefore,

$$1.645 = \sqrt{n}/2 - 2.33, \qquad \sqrt{n} = 7.95, \qquad \text{or } n = 62$$

Then

$$\text{critical } \bar{X} = 2000 + \frac{233}{\sqrt{n}} = 2000 + \frac{233}{\sqrt{62}} = 2029 \text{ N}$$

The decision changes in this case to

1. Reject the hypothesis that $\mu = 2000$ N if $\bar{X} \geq 2029$ N ($\alpha = 0.01$).
2. Accept the hypothesis that $\mu = 2000$ N if $\bar{X} < 2029$ N (now $\beta = 0.05$) and it is agreed to test 62 wires.

In this problem we were given α and n and computed β in the first part, while in the second part we were required to compute n for specified values of α and β. In a similar manner we can calculate α, given n and β.

Although in the present case we have applied the normal distribution, the procedure would be the same when we use other distributions, e.g., the t distribution, as applicable.

These computations can be much simplified by the use of operating characteristic curves, which are discussed in Chapter 21, p. 391.

Problems

13-1. A special brand of cement is sold in bags containing 50 kg. We choose 11 bags at random and find their masses in kilograms: 49.2, 50.1, 49.8, 49.7, 50.1, 50.5, 49.6, 49.9, 50.4, 50.2, and 49.7. Are these results consistent with the assumption that the bags belong to a population with a mean of 50 kg? Use α = 10 percent.

13-2. To check two weighing machines, 8 samples were weighed on each machine. Do the results suggest that there is a significant difference between the two machines? Use α = 5 percent.

Machine A	10.063	8.051	9.036	9.067	3.056	5.076	5.074	2.006
Machine B	10.063	8.050	9.033	9.062	3.060	5.070	5.070	2.000

13-3. To test his laboratory a manufacturer took 13 samples of his product, halved each of them, and had one-half tested in his laboratory (A) and the other in an independent laboratory (B). Is there a significant difference between the test results of the two laboratories? Use α = 5 percent.

Sample no.	1	2	3	4	5	6	7	8	9	10	11	12	13
Laboratory A	17.2	17.0	17.4	18.0	18.3	15.2	13.2	18.7	16.7	19.4	14.7	17.7	16.9
Laboratory B	19.1	19.8	17.9	18.0	18.3	15.0	14.3	18.2	17.7	19.3	16.7	16.8	16.7

13-4. The mean resistance of a box has been established by the manufacturer to be 250 ohms. A purchaser tests 10 boxes and finds the following values: 246, 261, 249, 254,

235, 242, 235, 231, 266, and 239. Does the consignment meet the specification if values within the 10 percent level of significance are tolerated?

13-5. In a laboratory it was suspected that the measurements of viscosity obtained in the morning were lower than in the afternoon. Ten samples were, therefore, split in half, one-half of each being tested in the morning, the other in the afternoon. Do the data suggest that the "afternoon viscosity" is higher? Use $\alpha = 5$ percent. (See Table 13-14.)

TABLE 13-14

	Viscosity (coded)	
Sample no.	*Morning*	*Afternoon*
1	43	45
2	48	48
3	48	50
4	50	53
5	55	54
6	50	52
7	72	73
8	75	75
9	73	72
10	54	56

13-6. Results of chemical analyses for the content of A in materials from two sources are as follows:

	Content of A, percent					
Source 1	93.12	93.57	92.81	94.32	93.77	93.52
Source 2	92.54	92.38	93.21	92.06	92.55	

Test the hypothesis that there is no difference in the content of A between the two sources. Use $\alpha = 1$ percent.

13-7. The strength of two alloys was compared, 10 samples of each being tested. Alloy A had a mean strength of 31,400 kN/m^2 with a coefficient of variation of 19 percent; the corresponding values for alloy B were 27,100 kN/m^2 and 15 percent. Can we conclude that the strengths of the two alloys do not differ at the 1 percent level of significance?

13-8. Measurements of a certain angle in castings over two periods yielded the following data:

$$n_1 = 154 \qquad n_2 = 149$$
$$\theta_1 = 90.0^\circ \qquad \theta_2 = 89.7^\circ$$
$$s_1 = 3.6^\circ \qquad s_2 = 3.5^\circ$$

Does this mean that the angle is becoming smaller? (If the angle were increasing, we would not worry.) Use $\alpha = 5$ percent.

13-9. In order to determine whether the use of rubber packing between the concrete specimen and the platen of the testing machine affects the observed strength, two specimens were made from each of six batches of concrete. (See Table 13-15.) Of each pair of specimens one was tested with the packing, the other without. Is there a significant difference between the strengths obtained by the two test methods? Use $\alpha = 1$ percent.

TABLE 13-15

	Tensile strength, MN/m^2	
Batch no.	*With packing*	*Without packing*
1	2.76	2.48
2	2.72	2.00
3	2.65	2.28
4	2.62	2.10
5	2.96	2.38
6	2.48	2.18

Source: P. J. F. Wright, "Statistical Methods in Concrete Research," *Magazine of Concrete Research*, vol. 5, no. 15, March 1954, p. 143. The original data was in psi units.

13-10. To compare the tensile strength of two cements, six mortar briquettes were made with each cement, and the following strengths (kN/m^2) were recorded:

Cement A: 4600, 4710, 4820, 4670, 4760, 4480
Cement B: 4400, 4450, 4700, 4400, 4170, 4100

Is there a significant difference between the tensile strengths of the two cements? Use $\alpha = 5$ percent.

13-11. The resistance (in ohms) of 40 boxes supplied from each of three manufacturers was found to be as listed in Table 13-16. Test whether there is a significant difference between the mean values of Groups A and B, and Groups B and C. Use $\alpha = 10$ percent.

TABLE 13-16

			Group A				
6040	7240	6160	7000	7160	8000	7360	6800
6040	5680	6320	6120	8240	8040	7760	7040
6720	6920	7600	6600	6920	7280	6400	6200
8120	8120	8000	7800	7320	7560	7200	7280
7560	7520	7520	6840	6640	7160	7280	6680
			Group B				
7040	7640	6480	6000	6640	6880	6200	6480
6160	6480	7320	6680	6440	6600	6280	7480
7320	6320	7880	7520	8760	8280	7880	7040
6720	6600	8080	7120	6600	7960	6440	5960
6680	6600	6600	6040	6080	6720	6640	6600
			Group C				
7240	7240	6840	7240	7320	7080	7320	7720
7280	7360	7320	7440	7240	7240	8400	8440
7800	7720	7640	7640	7520	7720	7640	7600
8600	8520	8880	8800	8440	8400	7320	8800
7520	7520	6320	5680	7440	7640	6960	8920

13-12. Apply the t test to the differences in strength of Problem 7-8, and hence establish whether there is a real difference between the strengths of the 10 and 20 cm cubes. Use $\alpha = 10$ percent.

13-13. Concrete is formed by pressurized compaction with a view to increasing its compressive strength. To test whether this is true, six sample cylinders are made by this new method and another six cylinders are made by the standard method. The results in MN/m^2 are as follows:

New method	33.1	31.0	34.5	33.8	35.9	29.0
Standard method	27.6	29.0	26.2	30.3	31.7	29.6

Check, at the 5 percent level, whether the new method has increased the compressive strength of concrete using:

1. The sign test, assuming the samples are paired.
2. The Wilcoxon signed rank test, assuming paired samples.
3. The test for two independent samples, assuming no pairing.

Chapter 14

Comparison of Variances and Their Properties

In the tests on the significance of means we either assumed that the two samples whose means were being compared had the same standard deviation (or, strictly speaking, deviations belonging to the same population of standard deviations) or we allowed for the inequality of the standard deviations (as in the text on p. 213). However, it is often important to know with some degree of certainty whether the standard deviations of two samples are the same, i.e., they do not differ significantly; such standard deviations are said to be *homogeneous*. For example, if we want to establish that two samples belong to the same population, we should test their means and determine that they do not differ significantly, and also test their standard deviations and determine that they do not differ significantly. It is, of course, possible for one of these conditions to be satisfied but not the other.

Although we refer to standard deviations, it is really variances that are the statistic being studied.

THE F TEST

If there are only two variances, we apply the variance ratio test, known as the F test. The ratio is that of two sample variances s_1^2 and s_2^2, the samples being drawn from the same population. If repeated pairs of samples are taken, it can be shown that the ratio s_1^2/s_2^2 follows a distribution called the F-distribution. It should be remembered that the variances s_1^2 and s_2^2 are really estimates of the population variance obtained from the two samples.

In other tests of significance, we adopt a null hypothesis, which in this case is that the variances of the two samples belong to the same population. The F value is calculated as

$$F = \frac{s_1^2}{s_2^2} \tag{14-1}$$

where $s_1 > s_2$. This condition must be satisfied as the F test is a one-sided test when the alternative to the null hypothesis is $s_1^2 > s_2^2$. If the alternative is simply $s_1^2 \neq s_2^2$, the test is two-sided and the probabilities in Table A-10 are doubled.

The associated degrees of freedom ν_1 and ν_2 are, in fact, those for the variances s_1^2 and s_2^2, respectively. If the two samples are of size n_1 and n_2, the corresponding numbers of degrees of freedom are $\nu_1 = n_1 - 1$ and $\nu_2 = n_2 - 1$, since, as explained in Chapter 3, by calculating the deviations $(x_i - \bar{x})$ from the sample mean $\bar{x}$, we lose 1 degree of freedom because of the constraint equation, namely,

$$\sum_{i=1}^{n_1} (x_i - \bar{x}_1) = 0 \qquad \text{for sample 1}$$

and

$$\sum_{i=1}^{n_2} (x_i - \bar{x}_2) = 0 \qquad \text{for sample 2}$$

It should be remarked that the null hypothesis implicit in Eq. (14-1) is actually $\sigma_1^2 = \sigma_2^2 = \sigma^2$, as explained in the following discussion.

The F_{ν_1,ν_2} statistic is defined as the ratio of two independent variates (each distributed as chi square), each divided by its number of degrees of freedom, or

$$F_{\nu_1,\nu_2} = \frac{\chi_1^2/\nu_1}{\chi_2^2/\nu_2}$$

(See also definition on p. 238.) Now

$$\chi_1^2 = \frac{\nu_1 s_1^2}{\sigma_1^2} \qquad \text{and} \qquad \chi_2^2 = \frac{\nu_2 s_2^2}{\sigma_2^2}$$

(For a distribution having a unit variance and zero mean, the expressions for χ^2 reduce to that on p. 237.) Therefore,

$$F_{\nu 1,\, \nu 2} = \frac{\sigma_2^2 s_1^2}{\sigma_1^2 s_2^2}$$

which reduces to Eq. (14-1) when the above null hypothesis is adopted.

Table A-10 gives values of F for various degrees of freedom of the two samples; if the calculated F exceeds the tabulated value, the probability that the difference between the two variances is caused by chance alone is smaller than the specified probability (5, 1, or 0.1 percent), and we are justified in rejecting the null hypothesis (with the given probability of committing a Type I error).

It has been shown that the larger the sample the more accurately the variance is determined. For this reason, the larger the samples being compared, the lower the value of F at which the null hypothesis is rejected with a given probability of a correct decision. This is illustrated by Table A-10.

Example. In tests on a plastic the following data were obtained for two samples from two sources.

Source	*Sample size* n	*Estimate of standard deviation* s
A	11	300
B	21	200

Can we say that the standard deviation of plastic from source A is greater than from source B at the 5 percent level of significance? We have:

$$s_1 = 300 \qquad \nu_1 = 11 - 1 = 10$$

$$s_2 = 200 \qquad \nu_2 = 21 - 1 = 20$$

This is a one-sided test. From Eq. (14-1),

$$F = \left(\frac{300}{200}\right)^2 = 2.25$$

From Table A-10, $F = 2.35$ at the 5 percent level of significance. Therefore, the difference between the variances is not significant, as shown in Fig. 14-1, but we would be wise to take further specimens and repeat the F test for the enlarged samples.

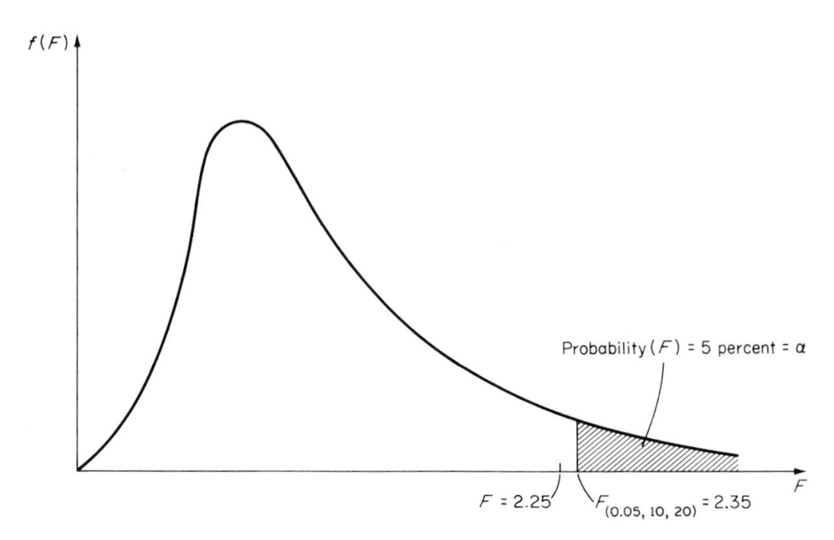

Fig. 14-1. Probability density function of F variable for $\nu_1 = 10$ and $\nu_2 = 20$ for the example on p. 233.

It may be of interest to mention that the value of F for $\nu_1 = 1$ and ν_2 degrees of freedom is equal to the value of t^2 with ν_2 degrees of freedom. For example, for a level of significance $\alpha = 0.05$, $\nu_1 = 1$ and $\nu_2 = 2$, $F = 18.51$ from Table A-10. Now, the corresponding value of t from Table A-8 with $\nu = \nu_2 = 2$ is 4.303. Thus $F = t^2$ (see p. 238).

If variance is a function of the level of the mean and the populations whose variability is being compared have different means, the F test cannot be applied directly to the variances. For example, if the standard deviation is directly proportional to the mean strength of the population,[1] then it is the squares of the coefficients of variations that have to be tested for homogeneity. The square of the coefficient of variation is approximately equal to the variance of the natural logarithm of the original variate; we can imagine thus that we are dealing with the variance of a log-transformed variable.

BARTLETT'S TEST

The F test can be used to compare two variances only. If more than two variances are involved, then a test of homogeneity of the variances, known as Bartlett's test, may be applied. This is a special application of the χ^2 test, in which we compare the difference between the total number of degrees of

1. See Chapter 13.

freedom times the natural logarithm of the pooled estimate of variance and the sum, extended over all samples, of the product of the degrees of freedom and the natural logarithm of the estimate of variance. Thus, if n_i is the sample size, s_i^2 is the estimate of variance from sample i, and $\bar{s}^2$ is the pooled estimate of variance, then Bartlett's test requires the calculation of

$$\chi^2 = 2.3026\{\log \bar{s}^2 \times \sum (n_i - 1) - \sum [(n_i - 1) \log s_i^2]\} \qquad (14\text{-}2)$$

The coefficient 2.3026 is introduced by conversion of natural logarithms to those to base 10, since

$$\log_e a = 2.3026 \log_{10} a$$

When all samples are of the same size n, Eq. (14-2) reduces to

$$\chi^2 = 2.3026(n - 1)(k \log \bar{s}^2 - \sum \log s_i^2) \qquad (14\text{-}3)$$

where k is the number of samples whose variances are being compared.

The computations are conveniently set out in Table 14-1.

TABLE 14-1

Sample number	*Sample size*	*Sum of squares of deviations*	*Degrees of freedom* v	*Estimated variance* s_i^2	*Logarithm of variance*	*Product*	*Reciprocal of* v
1							
.	.	.	.	.	.	.	.
.	.	.	.	.	.	.	.
.	.	.	.	.	.	.	.
i	n_i	$\sum_1^{n_i} (x - \bar{x}_i)^2$	$n_i - 1$	$\frac{\sum (x - \bar{x}_i)^2}{n_i - 1}$	$\log s_i^2$	$(n_i - 1) \log s_i^2$	$\frac{1}{n_i - 1}$
.	.	.	.	.	.	.	.
.	.	.	.	.	.	.	.
.	.	.	.	.	.	.	.
k							
		$\sum_1^k \sum_1^{n_i} (x - \bar{x}_i)^2$	$\sum (n_i - 1)$		$\sum \log s_i^2$	$\sum [(n_i - 1) \log s_i^2]$	$\sum \frac{1}{n_i - 1}$

Now,

$$\bar{s}^2 = \frac{\sum \sum (x - \bar{x}_i)^2}{\sum (n_i - 1)}$$

and, substituting the other summations in Eq. (14-2), χ^2 can be calculated. This enables us to test the hypothesis that all the variances are homogeneous, using Table A-7 with $(k - 1)$ degrees of freedom. If the calculated value of χ^2 is greater than the tabulated value at a specified level of significance, we conclude that the variances are not homogeneous. The level of significance represents, as always, the probability of our having reached the wrong conclusion.

The value of χ^2 as calculated from Eq. (14-2) is biased toward the high side so that we may wrongly reject the null hypothesis. If χ^2 indicates acceptance, then, after correction for bias, the hypothesis would be even more likely to be correct so that we need not worry about the bias. However, rejection at the 5 percent level has to be checked by calculating a corrected value of χ^2, say χ_c^2. This is given by

$$\chi_c^2 = \frac{\chi^2}{C} \tag{14-4}$$

where

$$C = 1 + \frac{1}{3(k-1)}\left[\sum\left(\frac{1}{n_i - 1}\right) - \frac{1}{\sum (n_i - 1)}\right] \tag{14-5}$$

Often a considerable computational effort can be saved by applying an F test to the largest and smallest variances before Bartlett's test. If the F test indicates that the largest variance is not significantly different from the smallest one, then one can reasonably assume that the variances lying in between do not differ significantly, and all the variances can be regarded as homogeneous.

Example. The performance of five testing machines was compared by testing four specimens, all from the same source, in each machine (see Table 14-2). Does the variability of the different machines differ significantly? Use $\alpha = 10$ percent.

TABLE 14-2

Machine	*Coded results*[a] x_i	$\bar{x}_i$	$\sum (x_i - \bar{x})^2$	s_i^2	$\log s_i^2$
1	2, 3, 5, 2,	3.0	6.00	2.00	0.30103
2	3, 4, 4, 1	3.0	6.00	2.00	0.30103
3	3, 3, 3, 4	3.25	0.75	0.25	−0.60206
4	2, 1, 3, 4	2.5	5.00	1.67	0.22272
5	5, 2, 2, 5	3.5	9.00	3.00	0.47712
Totals				$\sum s_i^2 = 8.92$	$\sum \log s_i^2 = 0.69984$

[a] "Coding" or simplifying the computations is explained in Chapters 3 and 18.

Hence

$$\bar{s}^2 = \frac{3 \times 8.92}{5 \times 3} = 1.784$$

$$\log \bar{s}^2 = 0.25139$$

From Eq. (14-3)

$$\chi^2 = 2.3026 \times 3(5 \times 0.25139 - 0.69984)$$
$$= 3.85$$

The number of degrees of freedom is 4, and Table A-7 gives $\chi^2 = 7.779$ at the 10 percent level of significance. Since the test indicates an acceptance of the null hypothesis, there is no need to apply the correction of Eq. (14-5), and we conclude that there is no significant difference between the variances of the test results of the five machines.

RELATION BETWEEN THE NORMAL, χ^2, t, AND F DISTRIBUTIONS

We recall that the *normal* distribution is that associated with a *random* variable $x(-\infty < x < \infty)$ whose density function is given by

$$p(x) = \frac{1}{\sqrt{2\pi}\,\sigma} e^{-(x-\mu)^2/2\sigma^2} \qquad [9\text{-}16]$$

where μ and σ, respectively, are the mean and the standard deviation of the random variable x. The sample mean, $\bar{x}$, is a random variable, which is normally distributed when the distribution of the underlying random variable is normal. (See also the Central Limit Theorem on p. 151.)

However, the distribution of other random variables that are functions of the normally distributed, underlying random variables is *not* necessarily normal. For example, the random variable x^2 is not normally distributed, even though x is.

Consider a set of random variables $X_1, X_2, \ldots X_\nu$ which are independent and normally distributed, each with a zero mean and a unit variance. Then the sum of the squares of X, denoted by χ^2, must also be a random variable, since it is a function of random variables. The quantity

$$\chi_\nu^2 = \sum_{i=1}^{i=\nu} X_i^2$$

is said to have a chi-square distribution with ν degrees of freedom.

It is obvious that the range of all possible values of χ^2 is between zero and $+\infty$, since χ^2 is a sum of squares. The distribution is always skewed to the right but for a large number of degrees of freedom the distribution tends toward normal.

Let us further consider X, which is a random variable, normally distributed with a zero mean and a unit variance. Assume that we have another random variable Y^2, independent of X, which has a χ^2 distribution with v degrees of freedom. Then the quantity t, given by

$$t = \frac{X}{\sqrt{Y^2/v}}$$

has a t distribution with v degrees of freedom. This distribution is symmetrical about zero (as is the case with the normal distribution) and its range extends from $-\infty$ to ∞, since the values of X lie in this interval and the values of Y^2 are nonnegative. With an increase in the number of degrees of freedom, the t distribution approaches the normal distribution. In fact, the two distributions will coincide when $v = \infty$.

Let us consider now two random variables X^2 and Y^2, independent of each other. If X^2 is distributed as χ^2 with v_1 degrees of freedom and Y^2 distributed as χ^2 with v_2 degrees of freedom, then the ratio

$$F_{v1,\,v2} = \frac{X^2/v_1}{Y^2/v_2}$$

is said to have an F distribution with v_1 and v_2 degrees of freedom. The range of all possible values of the random variable F is given by the interval zero to $+\infty$, since the values of X^2 and Y^2 are all nonnegative. By comparing the F distribution with the χ^2 and t distributions, it can readily be deduced that the density function of the F distribution approaches that of the χ^2 distribution as v_2 increases; also, for $v_1 = 1$,

$$F_{\alpha,\,1,\,v} = t^2_{\alpha,\,v}$$

It can be shown also that

$$F_{\alpha,\,v1,\,v2} = \frac{1}{F_{1-\alpha,\,v2,\,v1}}$$

STANDARD DEVIATIONS OF VARIOUS STATISTICS

It may be appropriate here to list the standard deviations of various statistics for a sample of size n and an estimate of variance of s^2. Equation (6-8) showed that

$$\text{standard deviation}^2 \text{ of the mean} = \frac{s}{\sqrt{n}} \tag{14-6}$$

2. This standard deviation and those that follow are known as standard errors (see Chapter 10).

We can now list the following without proof:

$$\text{standard deviation of variance} = s^2\sqrt{\frac{2}{n}} \tag{14-7}$$

$$\text{standard deviation of standard deviation} = \frac{s}{\sqrt{2n}} \tag{14-8}$$

$$\text{standard deviation of the coefficient of variation } V = \frac{V}{\sqrt{2n}} \tag{14-9}$$

Equations (14-6) to (14-9) are valid only when the underlying distribution is normal, but even then the standard deviations of variance, of standard deviation, or of the coefficient of variation are not normally distributed.[3] These standard deviations are useful as only an approximate guide to the precision of the estimate of the appropriate statistic and are not greatly used.

PROPAGATION OF ERRORS

If a quantity Q is a function of two or more measured quantities, it is clear that the errors in these measured values will affect the value of Q.

Suppose that we want to determine a quantity Q which is determined from two measured quantities x_1 and x_2; for example, let

$$Q = ax_1 \pm bx_2 \tag{14-10}$$

Assume that we have a large number of observations of x_1 and x_2, and hence of Q. We can compute the standard deviations of x_1 and x_2, and we want to determine the standard deviation of Q.

Let μ_Q, μ_1, and μ_2 be the true mean values of the three quantities. Then

$$\mu_Q = a\mu_1 \pm b\mu_2 \tag{14-11}$$

If ΔQ, Δx_1, and Δx_2 denote the deviations of observations from the appropriate means, then

$$\Delta Q = a\ \Delta x_1 \pm b\ \Delta x_2$$

and

$$(\Delta Q)^2 = a^2(\Delta x_1)^2 + b^2(\Delta x_2)^2 \pm 2ab\ \Delta x_1\ \Delta x_2 \tag{14-12}$$

We can write Eq. (14-12) for each set of observations. The mean value of $(\Delta Q)^2$ then represents the variance of Q, σ_Q^2. Similarly, the mean value of $(\Delta x_1)^2$ is the variance of x_1, σ_{x1}^2, and the mean value of $(\Delta x_2)^2$ is the variance of x_2, σ_{x2}^2.

3. Variance has a distribution close to χ^2 but as n increases, the distribution tends to normal.

The mean value of $\Delta x_1 \, \Delta x_2$ is the *covariance* of x_1 and x_2, and is a measure of correlation between x_1 and x_2.

If the variables x_1 and x_2 are independent, i.e., not correlated, the covariance is zero. Under such circumstances, the mean values of the various terms of Eq. (14-12) yield the relation

$$\sigma_Q^2 = a^2\sigma_{x_1}^2 + b^2\sigma_{x_2}^2 \tag{14-13}$$

When $a = b = 1$, i.e., for $Q = x_1 \pm x_2$, we have

$$\sigma_Q^2 = \sigma_{x_1}^2 + \sigma_{x_2}^2 \tag{14-14}$$

Thus the variance of a sum or of a difference of two *independent* variables is equal to the sum of the variances of the variables. The argument can be extended to any number of variables.

Equation (14-14) is of great importance in apportioning errors to various causes and forms the basis of the analysis of variance, which is the subject of Chapter 18.

In a more general way, Eq. (14-13) depicts the propagation of errors. For example, if Q is the perimeter of an isosceles triangle, x_1 the length of the base, and x_2 the length of the side, then

$$Q = x_1 + 2x_2$$

From Eq. (14-13), the variance of the perimeter is

$$\sigma_Q^2 = \sigma_{x_1}^2 + 4\sigma_{x_2}^2$$

and the standard deviation of the perimeter is

$$\sigma_Q = \sqrt{\sigma_{x_1}^2 + 4\sigma_{x_2}^2}$$

Equation (14-13) was derived for the case when Q is a sum or a difference of two quantities. Let us now consider the case when Q is a product of two quantities, for example,

$$Q = ax_1x_2 \tag{14-15}$$

where a is a constant assumed free from error.

Using the same notation as before, a typical measurement of Q inclusive of error ΔQ is

$$Q + \Delta Q = a(x_1 + \Delta x_1)(x_2 + \Delta x_2) \tag{14-16}$$

Expanding Eq. (14-16) and subtracting from it Eq. (14-15), we obtain for one set of observations the value of error

$$\Delta Q = a(x_1 \, \Delta x_2 + x_2 \, \Delta x_1 + \Delta x_1 \, \Delta x_2)$$

The error product term is a second-order quantity and can, therefore, be ignored, so that

$$\Delta Q = a(x_1 \ \Delta x_2 + x_2 \ \Delta x_1) \tag{14-17}$$

This expression can be written for any set of measurements. For n measurements we can write the variance σ_Q^2 as

$$\sigma_Q^2 = \frac{\sum (\Delta Q)^2}{n}$$

i.e.,

$$\sigma_Q^2 = \frac{a^2}{n}[x_1^2 \sum (\Delta x_2)^2 + x_2^2 \sum (\Delta x_1)^2 + 2x_1 x_2 \sum (\Delta x_1 \ \Delta x_2)]$$

Again ignoring the second-order term (which tends to zero[4]), we obtain

$$\sigma_Q^2 = a^2[x_1^2 \sigma_{x_2}^2 + x_2^2 \sigma_{x_1}^2] \tag{14-18}$$

Let us now turn to the more general case of Q, which is any function of two variables x_1 and x_2

$$Q = f(x_1, x_2) \tag{14-19}$$

Then a typical measurement is

$$Q + \Delta Q = f(x_1 + \Delta x_1, x_2 + \Delta x_2)$$

If this function is continuous, so that it has derivatives (and the majority of functions determined experimentally satisfy this requirement), we can expand it in a Taylor series. Using the first two terms only, we find that

$$Q + \Delta Q = f(x_1, x_2) + \left[\left(\frac{\partial Q}{\partial x_1}\right) \frac{x_1 + \Delta x_1 - x_1}{1!} + \left(\frac{\partial Q}{\partial x_2}\right) \frac{x_2 + \Delta x_2 - x_2}{1!} + \cdots\right] \tag{14-20}$$

Subtracting Eq. (14-19) from Eq. (14-20), we obtain as a first-order approximation

$$\Delta Q = \frac{\partial Q}{\partial x_1} \Delta x_1 + \frac{\partial Q}{\partial x_2} \Delta x_2$$

Hence

$$\sum (\Delta Q)^2 = \left(\frac{\partial Q}{\partial x_1}\right)^2 \sum (\Delta x_1)^2 + \left(\frac{\partial Q}{\partial x_2}\right)^2 \sum (\Delta x_2)^2 + 2 \frac{\partial Q}{\partial x_1} \frac{\partial Q}{\partial x_2} \sum (\Delta x_1 \ \Delta x_2)$$

4. Since x_1 and x_2 are independent quantities.

Dividing by the number of observations, n, and with the third term on the right-hand side of the equation tending to zero, we find

$$\sigma_Q^2 = \left(\frac{\partial Q}{\partial x_1}\right)^2 \sigma_{x1}^2 + \left(\frac{\partial Q}{\partial x_2}\right)^2 \sigma_{x2}^2 \tag{14-21}$$

Equation (14-21) was derived for Q, which is a function of two variables only. For the general case when

$$Q = f(x_1, x_2, x_3, \ldots)$$

we obtain the expression for the variance of Q,

$$\sigma_Q^2 = \left(\frac{\partial Q}{\partial x_1}\right)^2 \sigma_{x1}^2 + \left(\frac{\partial Q}{\partial x_2}\right)^2 \sigma_{x2}^2 + \left(\frac{\partial Q}{\partial x_3}\right)^2 \sigma_{x3}^2 + \cdots \tag{14-22}$$

The above derivations are based on the assumption that the changes Δx_i in all the x_i (independent variables) are independent. The partial derivatives should be evaluated at the true values of the measured variables. In general, this is not possible, but evaluation at the observed values is usually good enough.

Equation (14-22) is especially useful because it can be utilized to deduce several relationships, e.g., Eq. (13-3). Let us denote by d the difference between two means $\bar{x}_1$ and $\bar{x}_2$, of sample size n_1 and n_2, respectively,

$$d = \bar{x}_1 - \bar{x}_2$$

Applying Eq. (14-22), we find the variance of the difference to be

$$\sigma_d^2 = \left(\frac{\partial d}{\partial \bar{x}_1}\right)^2 \sigma_{\bar{x}1}^2 + \left(\frac{\partial d}{\partial \bar{x}_2}\right) \sigma_{\bar{x}2}^2$$

whence

$$\sigma_d^2 = (1)^2 \sigma_{\bar{x}1}^2 + (-1)^2 \sigma_{\bar{x}2}^2 = \sigma_{\bar{x}1}^2 + \sigma_{\bar{x}2}^2$$

Using Eq. (6-9), we have $\sigma_{\bar{x}1}^2 \simeq s_{\bar{x}1}^2 = s_1^2/n_1$ and

$$\sigma_{\bar{x}2}^2 \simeq s_{\bar{x}2}^2 = \frac{s_2^2}{n_2}.$$

Thus

$$\sigma_d^2 \simeq \frac{s_1^2}{n_1} + \frac{s_2^2}{n_2} \qquad [13\text{-}3]$$

Example. A company assembles a particular machine which requires the union of male and female parts made by different manufacturers, A and B. Samples of parts were taken and their means $\bar{x}$ and standard deviations s were calculated; the results are shown in Table 14-3.

TABLE 14-3

	Female (A)	*Male (B)*
$\bar{x}$, cm	1.270	1.230
s, cm	0.040	0.045
n	50	36

Determine the 95 percent confidence limits on the tolerance of fit.

We shall assume that the data are normally distributed. From Table A-4, $z = 1.96$. Now,

$$d = 1.270 - 1.230 = 0.040 \text{ cm}$$

From Eq. (13-3),

$$\sigma_d = \left[\frac{(0.040)^2}{50} + \frac{(0.045)^2}{36}\right]^{1/2} = 0.0094 \text{ cm}$$

Thus the tolerance is $0.040 \pm (z)(\sigma_d) = 0.040 \pm (1.96)(0.0094) = 0.040 \pm 0.018$ cm. Hence at the 95 percent confidence level the maximum tolerance is 0.058 cm and the minimum tolerance is 0.022 cm. If these limits fall outside the specification limits, the company may reject the parts.

Example. The refractive index N of the glass of a prism is to be determined on the basis of several measurements of the angle of the prism, A, and the angle of minimum deviation, D. The mean values and standard deviations of the angles are A: 56°; 1° and D: 41°; 0.5°.

If the refractive index of the prism is given by

$$N = \frac{\sin \frac{1}{2}(A + D)}{\sin \frac{1}{2}A}$$

determine which angle contributes more to the error in N?

From Eq. (14-21) we can write

$$\sigma_N^2 = \left(\frac{\partial N}{\partial A}\right)^2 \sigma_A^2 + \left(\frac{\partial N}{\partial D}\right)^2 \sigma_D^2$$

Now,

$$\frac{\partial N}{\partial A} = \frac{[\frac{1}{2}\cos \frac{1}{2}(A + D)\sin \frac{1}{2}A - \frac{1}{2}\sin \frac{1}{2}(A + D)\cos \frac{1}{2}A]}{\sin^2 \frac{1}{2}A}$$

$$= -\frac{\sin \frac{1}{2}D}{2\sin^2 \frac{1}{2}A} = -\frac{(\sin 20.5°)}{2\sin^2 28°}$$

$$= -0.794468$$

and

$$\frac{\partial N}{\partial D} = \frac{\frac{1}{2}\cos\frac{1}{2}(A+D)}{\sin\frac{1}{2}A} = \frac{\frac{1}{2}(\cos 48.5^\circ)}{\sin 28^\circ}$$

$$= 0.705708$$

The standard deviations are $\sigma_A = 1^\circ = 0.017453$ radians and $\sigma_D = 0.5^\circ = 0.008726$ radians.

Substituting in the expression for σ_N^2, we obtain

$$\sigma_N^2 = (0.794468)^2(0.017453)^2 + (0.705708)^2(0.008726)^2$$

$$= 192.26 \times 10^{-6} + 37.92 \times 10^{-6}$$

Thus the angle of the prism contributes more to the error in the refractive index than the angle of minimum deviation.

CONFIDENCE LIMITS FOR VARIANCE

By analogy to the confidence limits of the mean, the confidence limits of variance give the limits within which the true population variance lies with a specified probability.

If s^2 is the variance calculated from a sample with v degrees of freedom, then the limits are given by

$$\frac{v}{\chi^2}s^2 \tag{14-23}$$

where χ^2 has a value corresponding to v degrees of freedom and to the 5 percent level of significance for the lower limit, and to the 95 percent level of significance for the upper limit.

Thus we have a 5 percent probability of σ^2 falling below the lower limit and a 5 percent probability of σ^2 falling above the upper limit. The confidence limits thus contain the true value of σ^2 with a 90 percent probability. Similarly, if we take χ^2 at the 1 percent level of significance, the confidence limits contain the true value of σ^2 with a 98 percent probability.

Example. Tests on 10 concrete tension specimens yielded an estimate of the population variance of 40,000 (standard deviation of 200 kN/m^2). Find the 90 percent confidence limits.

For $v = 9$, we find from Table A-7 that $\chi^2 = 16.919$ at the 5 percent level of significance and $\chi^2 = 3.325$ at the 95 percent level of significance. Thus the lower limit of σ^2 is

$$\frac{9}{16.919} \times 40{,}000 = 21{,}300$$

and the upper limit is

$$\frac{9}{3.325} \times 40{,}000 = 108{,}500$$

Hence we conclude, with a 90 percent probability of being correct, that

$$21{,}300 \leq \sigma^2 \leq 108{,}500$$

or

$$146 \leq \sigma \leq 329$$

In the case of large samples, say $n > 30$, we can take advantage of Eq. (14-8) for the standard deviation of the standard deviation; this is $\sigma_s = \sigma/\sqrt{2n}$.

For instance, if in the previous example we tested 50 specimens, the standard deviation remaining at 200 kN/m^2, we find the standard deviation of the standard deviation:

$$\sigma_s = \frac{200}{\sqrt{2 \times 50}} = 20 \text{ kN/m}^2$$

Since the standard deviation is not normally distributed, we have to use the t distribution. At the 5 percent level of significance and $\nu = 49$, $t \simeq 2.010$ (from Table A-8). Thus the confidence limits are

$$200 \pm 2.010 \times 20$$

i.e.,

$$160 \leq \sigma \leq 240$$

It is apparent how an increase in sample size decreases the width of the confidence interval at a given level of significance.

COMPARISON OF STANDARD DEVIATIONS OF LARGE SAMPLES

In the case of very large samples (say, $n > 120$) we can test the significance of a difference between two *standard deviations* by the normal distribution test of Chapter 13 instead of by the F test. This will not seem strange if we are quite clear about the fact that we treat the standard deviation as the variate.

The procedure is as follows: We compute a pooled estimate of the assumed common variance s_c^2. This is similar to the value given by

Eq. (13-8), except that Bessel's correction can be ignored because the samples are large. Thus

$$s_c^2 = \frac{n_1 s_1^2 + n_2 s_2^2}{n_1 + n_2} \tag{14-24}$$

where s_1^2, and s_2^2 are variances of the two samples, and n_1 and n_2 are the respective sample sizes.

We now use Eq. (14-8) to write the variance of the distribution of sample standard deviations s_c as

$$\frac{s_c^2}{2n_1} \quad \text{and} \quad \frac{s_c^2}{2n_2}$$

for the two distributions, respectively.

Since the variance of a difference is equal to the sum of variances [Eq. (14-14)], the standard deviation of the difference s_d is given by

$$s_d = s_c \sqrt{\frac{1}{2n_1} + \frac{1}{2n_2}} \tag{14-25}$$

We can now apply the normal distribution test of Eq. (13-4), i.e., compare the difference of the observed standard deviations s_1 and s_2 with s_d:

$$z = \frac{|s_1 - s_2|}{s_d}$$

The probability of encountering z at least this large is given in Table A-4.

Example. Shear test specimens of soil were obtained on two sites, as follows:

Site	*Number of specimens*	*Standard deviation, kN/m²*
1	120	300
2	150	150

Is there a significant difference between the variabilities on the two sites, i.e., are the two standard deviations significantly different? Use $\alpha = 1$ percent.

We have to apply a two-sided test, since we are checking the null hypothesis $\sigma_1^2 = \sigma_2^2$ against the alternative hypothesis $\sigma_1^2 \neq \sigma_2^2$. Thus

$$s_c^2 = \frac{120 \times 300^2 + 150 \times 150^2}{120 + 150}$$

$$= 52{,}500$$

Now

$$s_d = \sqrt{52{,}500\left(\frac{1}{2 \times 120} + \frac{1}{2 \times 150}\right)} = 19.8 \text{ kN/m}^2$$

Hence

$$z = \frac{300 - 150}{19.8} = 7.58$$

The probability of obtaining such a value of z by chance is extremely low, and we conclude that the difference between the standard deviations is highly significant.

Solved Problems

14-1. The water hardness of two samples taken from separate outlets in a power plant was checked. The coded results (parts per million) are shown in Table 14-4. Determine whether the variance of water hardness from location 1 is greater than from location 2. Use $\alpha = 5$ percent.

TABLE 14-4

Location 1	*Location 2*
$\sum x_1 = 504$	$\sum x_2 = 868$
$\sum x_1^2 = 29{,}101$	$\sum x_2^2 = 54{,}201$
$n_1 = 9$	$n_2 = 14$

SOLUTION

Estimate of variance:

$$s_1^2 = \frac{\sum x_1^2 - [(\sum x_1)^2/n_1]}{n_1 - 1} = \frac{29{,}101 - [(504)^2/9]}{8} = 109.6$$

Estimate of variance:

$$s_2^2 = \frac{\sum x_2^2 - [(\sum x_2)^2/n_2]}{n_2 - 1} = \frac{54{,}201 - [(868)^2/14]}{13} = 29.6$$

$$F = \frac{\text{greater variance}}{\text{smaller variance}} = \frac{109.6}{29.6} = 3.71$$

Using Table A-10 for $v_1 = n_1 - 1 = 9 - 1 = 8$, and $v_2 = n_2 - 1 = 14 - 1 = 13$, $F = 2.77$ at the 5 percent level of significance. Since the calculated $F > 2.77$, the difference in the water hardness of the two samples is significant at the 5 percent level.

14-2. A car manufacturing company carried out a series of tests on the rate of crack growth in tires under four different conditions. The summary of the results is shown in Table 14-5.

TABLE 14-5

Condition k	*Sample size n*	*Variance of the rate of crack growth* σ_i^2
1	20	0.0349
2	25	0.0875
3	20	0.0652
4	15	0.0445

We want to determine whether the variability of the measurements under the four conditions is the same, i.e., whether the variances are homogeneous. Use $\alpha = 10$ percent. (See Table 14-6.)

SOLUTION

TABLE 14-6

k	$v_i = n_i - 1$	s_i^2	*Sum of squares* $v_i s_i^2$	$\log s_i^2$	$v_i \log s_i^2$	$\frac{1}{v_i}$
1	19	0.0349	0.663	0.542825 − 2	10.313675 − 38	0.05263
2	24	0.0875	2.100	0.942008 − 2	22.608192 − 48	0.04167
3	19	0.0652	1.239	0.814248 − 2	15.470712 − 38	0.05263
4	14	0.0445	0.623	0.648360 − 2	9.077040 − 28	0.07143
$\sum$	76		4.625	2.947441 − 8	57.469619 − 152 = −94.530381	0.21836

$$\bar{s}^2 = \frac{\sum v_i s_i^2}{\sum v_i} = \frac{4.625}{76} = 0.060855$$

$$\log \bar{s}^2 = 0.784296 - 2 = -1.215704$$

From Eq. (14-2),

$$\begin{aligned}\chi^2 &= 2.3026[76 \times (-1.215704) - (-94.530381)] \\ &= 2.3026 \times 2.136877 \\ &= 4.920\end{aligned}$$

From Eq. (14-5), the correction factor is

$$C = 1 + \frac{1}{3(k-1)}\left[\sum\left(\frac{1}{v_i}\right) - \frac{1}{\sum v_i}\right]$$

$$= 1 + \frac{1}{3(4-1)}(0.21836 - 0.01316)$$

$$= 1.0228$$

Thus

$$\chi_c^2 = \frac{\chi^2}{C} = \frac{4.920}{1.0228} = 4.810$$

with $(k-1) = 4 - 1 = 3$ degrees of freedom.

Table A-7 gives $\chi^2 = 4.642$ at the 20 percent level of significance, and $\chi^2 = 6.251$ at the 10 percent level. We conclude, therefore, that there is no significant difference in the variances of the different tests. We would, however, be wise to apply Bartlett's test again when more test results are available.

(NOTE: The correction C is not necessary, since the uncorrected χ^2 indicates that the difference is not significant; the computation of χ_c^2 is given solely as an illustration of the method of applying the correction.)

14-3. The following data were calculated from tensile tests on two samples of glass fibers; we can assume that the observations within each sample are normally distributed:

Sample size $n_1 = 11$:

$$\sum_1^{11} x_1 = 33{,}000 \text{ kN/m}^2 \qquad \sum_1^{11} x_1^2 = 100.6 \times 10^6$$

Sample size $n_2 = 6$:

$$\sum_1^{6} x_2 = 16{,}800 \text{ kN/m}^2 \qquad \sum_1^{6} x_2^2 = 48.84 \times 10^6$$

a. Test the hypothesis that $s_1 = s_2$ against the alternative $s_1 \neq s_2$ at the 10 percent level of significance.
b. Assuming that the sample sizes are $n_1 = n_2 = 144$ and that the standard deviations of the two samples are $s_1 = 300$ kN/m^2 and $s_2 = 320$ kN/m^2, apply an appropriate test to the null hypothesis that $s_1 = s_2$.

SOLUTION

a. $$s_1 = \sqrt{\frac{\sum x_1^2 - [(\sum x_1)^2/n_1]}{n_1 - 1}} = 10^3 \times \sqrt{\frac{100.6 - [(33)^2/11]}{10}} = 400 \text{ kN/m}^2$$

$$s_2 = 10^3 \times \sqrt{\frac{48.84 - [(16.8)^2/6]}{5}} = 600 \text{ kN/m}^2$$

Hence

$$F = \frac{s_2^2}{s_1^2} = \left[\frac{600}{400}\right]^2 = 2.25$$

We enter Table A-10 with $\nu_1 = 11 - 1 = 10$, and $\nu_2 = 6 - 1 = 5$, as ν has to correspond to the larger variance. Then $F = 4.74$ at the 5 percent level of significance. As this is a two-sided test, we double this probability. Thus, since the calculated $F < 4.74$, the difference in the sample variances is not significant at the 10 percent level.

b. The samples can be considered large ($n > 30$); therefore, we can use the normal distribution test. The pooled estimate of the variance is

$$s_c^2 = \frac{300^2 + 320^2}{2} = 9.62 \times 10^4$$

The standard deviation of the difference is

$$s_d = s_c\sqrt{\frac{1}{n}} = 100 \times \sqrt{\frac{9.62}{144}} = 25.85 \text{ kN/m}^2$$

Hence

$$z = \frac{|s_1 - s_2|}{s_d} = \frac{|300 - 320|}{25.85} = 0.774$$

Table A-4 gives the probability of obtaining at least this value of z by chance as about $0.5 - 0.28 = 0.22$. The difference cannot, therefore, be deemed significant.

14-4. A simply supported girder is loaded by a uniformly distributed load w per unit length. The center deflection Δ is given by

$$\Delta = \frac{5}{384}\frac{wl^4}{EI}$$

where the length l is measured as 40 ± 0.4 m, w as 2 ± 0.02 kN/m, and the flexural rigidity of the girder, EI, is known precisely. Calculate the resulting fractional standard deviation (or relative error) in Δ; i.e., determine the quantity (σ_Δ/Δ).

SOLUTION

From Eq. (14-22) we have

$$\sigma_\Delta^2 = \left(\frac{\partial\Delta}{\partial w}\right)^2 \sigma_w^2 + \left(\frac{\partial\Delta}{\partial l}\right)^2 \sigma_l^2$$

No other terms are required, since the quantity EI does not vary. To simplify the computation, let us write

$$\Delta = \frac{5}{384}\frac{wl^4}{EI} = Cwl^4$$

Therefore,

$$\sigma_\Delta^2 = (Cl^4)^2(0.02)^2 + (4Cwl^3)^2(0.4)^2$$

Dividing by Δ^2 yields

$$\left(\frac{\sigma_\Delta}{\Delta}\right)^2 = \left(\frac{0.02}{w}\right)^2 + 16\left(\frac{0.4}{l}\right)^2$$

$$= \left(\frac{0.02}{2}\right)^2 + 16\left(\frac{0.4}{40}\right)^2 = 0.0017$$

Therefore, the fractional standard deviation in the center deflection $\sigma_\Delta/\Delta = 0.041$ or 4.1 percent, which is considerably larger than the error in either l or w. It can be observed that the error in the length l has a greater influence on the precision of the deflection than the load intensity w.

14-5. The heat transfer coefficient U for a system of two fluids separated by a partition of negligible thermal resistance is

$$U = \frac{h_1 h_2}{h_1 + h_2}$$

where h_1 and h_2 are the individual coefficients of the two fluids. The mean value of h_1 is 20 W/(m^2·°C) with a percent standard deviation (σ_{h_1}/h_1) of 5 percent; the mean value of h_2 is 30 W/(m^2·°C) with (σ_{h_2}/h_2) of 2 percent. Calculate the percentage standard error in U.

SOLUTION

From Eq. (14-22) we find

$$\sigma_U^2 = \left(\frac{\partial U}{\partial h_1}\right)^2 \sigma_{h_1}^2 + \left(\frac{\partial U}{\partial h_2}\right)^2 \sigma_{h2}^2$$

whence

$$\sigma_U^2 = \left[\frac{h_2(h_1 + h_2) - h_1 h_2}{(h_1 + h_2)^2}\right]^2 \sigma_{h_1}^2 + \left[\frac{h_1(h_1 + h_2) - h_1 h_2}{(h_1 + h_2)^2}\right]^2 \sigma_{h2}^2$$

$$= \left[\frac{h_2^2}{(h_1 + h_2)^2}\right]^2 \sigma_{h_1}^2 + \left[\frac{h_1^2}{(h_1 + h_2)^2}\right]^2 \sigma_{h2}^2$$

Dividing by U^2 yields

$$\left(\frac{\sigma_U}{U}\right)^2 = \left[\frac{h_2}{(h_1 + h_2)}\right]^2 \left(\frac{\sigma_{h_1}}{h_1}\right)^2 + \left[\frac{h_1}{(h_1 + h_2)}\right]^2 \left(\frac{\sigma_{h2}}{h_2}\right)^2$$

Therefore,

$$\left(\frac{\sigma_U}{U}\right)^2 = \left[\frac{30}{(20 + 30)}\right]^2 (0.05)^2 + \left[\frac{20}{(20 + 30)}\right]^2 (0.02)^2$$

$$= \frac{9}{25}\left(\frac{25}{10^4}\right) + \frac{4}{25}\left(\frac{4}{10^4}\right)$$

whence

$$\left(\frac{\sigma_U}{U}\right)^2 = \frac{241}{(25)(10^4)}$$

Hence the percentage standard error in the over-all heat transfer coefficient is

$$\frac{\sigma_U}{U} = \frac{15.3}{(5)(10^2)} = 0.0306, \text{ or } 3.1 \text{ percent}$$

Problems

14-1. Two different precision instruments were compared, 20 measurements being taken with instrument A and 30 with instrument B. The errors of instrument A had a variance of 15; those of B had a variance of 10. Assuming that the population of errors is normally distributed, test the hypothesis that

a. $\sigma_A^2 = \sigma_B^2$.
b. $\sigma_A^2 = 1.75\sigma_B^2$.
c. $\sigma_A^2 \ngtr \sigma_B^2$.

Use $\alpha = 10$ percent.

14-2. In problem 13-2 verify the assumption of equal variances of the two machines, A and B. Use a 1 percent level of significance.

14-3. From numerous tests (which can be considered infinite) on a certain type of light bulb, it was found that the variance in burning time was 9000 hours. A sample of 25 new-type light bulbs is found to have a variance of 13,000. Determine at the 5 percent level of significance whether or not the two variances are different.

14-4. The variability of six planimeters is to be tested. Six observations are taken on each planimeter, and the sample variances are computed. The coded results are as follows: 6.5, 9.4, 8.7, 12.4, 10.5, and 7.7. Determine at the 1 percent level of significance whether all the planimeters have the same variance.

14-5. Verify the assumption of equal variances in content of A from the two sources 1 and 2 in Problem 13-6, using a 5 percent level of significance.

14-6. The variance of error of tests in five laboratories is given in Table 14-7. Establish whether there are any significant differences among them. Use $\alpha = 1$ percent.

TABLE 14-7

Laboratory	*Sample size*	*Estimate of variance*
A	91	267
B	92	388
C	89	552
D	90	860
E	90	480

14-7. Twenty-six shipments of cement were obtained from each of five plants. From each shipment 15 test specimens were made, their average strengths being given in Table 14-8. A control supply of cement was also obtained, and on every occasion when tests were made on the shipped-in cement, similar tests were made on the control cement (which, of course, did not vary).

TABLE 14-8

Sample number	Compressive strength, MN/m^2, for cement source number					
	Control	1	2	3	4	5
1	18.28	20.40	13.55	13.09	18.68	11.02
2	17.80	19.66	14.87	13.04	20.72	14.78
3	16.44	19.00	—	14.00	17.84	11.93
4	17.75	15.82	—	16.69	18.48	14.21
5	16.51	15.14	14.89	13.15	19.02	11.71
6	16.77	16.92	14.12	14.47	18.43	14.35
7	16.42	17.33	15.54	12.11	19.92	13.27
8	16.33	18.73	14.78	15.27	17.50	11.51
9	16.01	16.31	13.76	16.02	19.19	10.97
10	16.82	19.15	15.90	14.62	20.04	13.56
11	17.09	18.18	16.44	13.27	18.59	14.93
12	17.36	17.31	16.07	14.17	18.80	14.29
13	17.48	20.93	14.21	15.95	17.76	15.27
14	16.90	23.19	13.65	12.69	17.24	10.82
15	17.43	23.02	12.88	17.41	18.38	13.78
16	17.11	19.43	13.52	12.02	18.22	13.95
17	17.08	22.05	13.53	14.35	17.52	12.00
18	16.84	19.92	12.78	13.81	16.95	13.51
19	17.13	19.99	14.64	13.17	14.69	13.57
20	17.00	—	—	13.97	16.15	14.03
21	16.86	—	12.77	14.42	18.03	13.67
22	17.82	20.98	11.96	14.13	19.40	14.20
23	16.33	18.76	13.63	14.25	19.82	14.30
24	17.06	21.41	12.55	15.65	17.60	14.04
25	16.27	16.89	11.58	14.54	12.42	13.92
26	15.72	19.91	10.73	12.92	11.00	—

Source: S. Walker and D. L. Bloem, "Tests of Uniformity of Mortar Strengths of Cement Samples from Same Source," National Ready Mixed Concrete Assn., Washington D.C., Jan. 1957. The original data was in psi units.

Comment on the suggestion that the shipment-to-shipment variability is no greater than the variability within the control cement. (HINT: The standard deviation of control represents the testing error. Is the standard deviation for shipments from a given plant significantly greater?) Use $\alpha = 1$ percent.

14-8. For the data of Problem 13-11 test the homogeneity of standard deviations. Use $\alpha = 5$ percent.

14-9. The standard deviation of the strength of beams of four different sizes was obtained using three series of tests, each test consisting of two specimens. (See Table 14-9.) Are the standard deviations homogeneous? Use $\alpha = 1$ percent.

TABLE 14-9

Beam size (in 2.54 cm)	Number of tests			Standard deviation, kN/m^2		
	Series A	Series B	Series C	Series A	Series B	Series C
6	17	10	6	503	248	131
9	24	8	2	490	138	76
12	16	10	4	538	324	83
18	8	3	2	386	228	221

Source: A. M. Neville, "Some Aspects of the Strength of Concrete," Part I, *Civil Engineering* (London), vol. 54, Oct. 1959, p. 1156. The original data was in psi units.

14-10. The radius r of a cylinder is given as 6.1 ± 0.1 cm and the length l as 15.6 ± 0.2 cm. Find the volume of the cylinder and the associated standard error.

14-11. The value of the gravitation constant g is to be determined from $g = 2s/t^2$ by observing a body fall, where s is the distance traveled in time t. The experimenter has a choice of setting s at 1.22 m with t about $\frac{1}{2}$ s or of setting s at 4.88 m with t about 1 s. Assume that the standard deviation of s is 0.0006 m and that the standard deviation of recording time is 0.05 s. Calculate the variance of g for the two choices of s and t. Which value of s is better in the sense of giving a smaller variance of the estimated value of g? Given

$$\frac{\partial g}{\partial s} = \frac{2}{t^2}, \quad \frac{\partial g}{\partial t} = -\frac{4s}{t^3}$$

14-12. The volume of a sphere is $V = \pi D^3/6$, where D is the diameter. What is the standard error in the volume, σ_V if D is measured and found to be 10.00 ± 0.16 cm? (NOTE: The standard error in D, σ_D, is 0.16 cm, as shown.)

14-13. The deflection δ of an eccentrically loaded hinged-end column is given as

$$\delta = e\left\{\sec\left[\left(\frac{P}{EI}\right)^{1/2}\frac{l}{2}\right] - 1\right\}$$

If for a model column the load P is 10 ± 0.5 kN, the eccentricity e of 0.5 ± 0.05 mm, the moment of inertia $I = 400{,}000\ \text{mm}^4$, the modulus of elasticity $E = 100 \pm 4\ \text{kN/mm}^2$, and the length $l = 250$ mm, calculate the standard error in the deflection δ.

14-14. The energies of the various quantum states of a hydrogen atom are given by

$$E_n = -\frac{1}{2}\frac{me^4}{n^2\kappa^2} \qquad \text{with } n = 1, 2, 3, \ldots$$

If the mass of the electron m is known with a fractional standard deviation $\sigma_m/m = 0.1$ percent, the electron charge e with $\sigma_e/e = 0.2$ percent, and the constant κ with $\sigma_\kappa/\kappa = 0.1$ percent, what is the percent standard error in E_n, σ_{E_n}/E_n, for the state $n = 1$? For the $n = 2$ state?

Determine which of the measurements (m, e, or κ) should be increased in accuracy for best improvement in the precision of E_n.

Chapter 15

Method of Least Squares and Regression

In elementary work we often establish numerical relations by determining the values of the variables at a number of points equal to the total number of variables. For example, if a linear relation $y = a + bx$ is postulated, two pairs of values (x_1, y_1) and (x_2, y_2) determine the constants in the equation. This is satisfactory, provided that the observed quantities are free from error.

In practice, error enters all our observations, and if we take further observations, say (x_3, y_3), we may obtain a point that does not fit exactly on the straight line through the original two points. This also applies, of course, to curves involving powers of x and y. Statistical methods help us to fit the "best" line to a given set of data, instead of simply drawing a line "by eye."

Our main interest is in studying the association between two variables, rather than estimating one variable from the other.

METHOD OF LEAST SQUARES

The principle underlying the fitting of the "best" line is that of least squares; this states that if y is a linear function of an independent variable x, the most probable position of a line $y = a + bx$ is such that the sum of squares of deviations of all points (x_i, y_i) from the line is a minimum; the deviations are measured *in the direction of the y-axis*. It should be stressed that the underlying assumption is that x is either free from error (being assigned) or subject to negligible error only, while y is the observed or measured quantity, subject to errors that have to be "eliminated" by the method of least squares. The observed y is thus a random value from the population of values of y corresponding to a given x. Such a situation exists in controlled experiments, where we are interested in finding a mean value of $\bar{y}_i$ for each given value of x_i.

Suppose that our observations consist of n pairs of values:

$$\begin{cases} x_1, x_2, \ldots, x_n \\ y_1, y_2, \ldots, y_n \end{cases}$$

and imagine that the various pairs plot as points shown in Fig. 15-1. Assume further that from the physical nature of the relation between y and x we

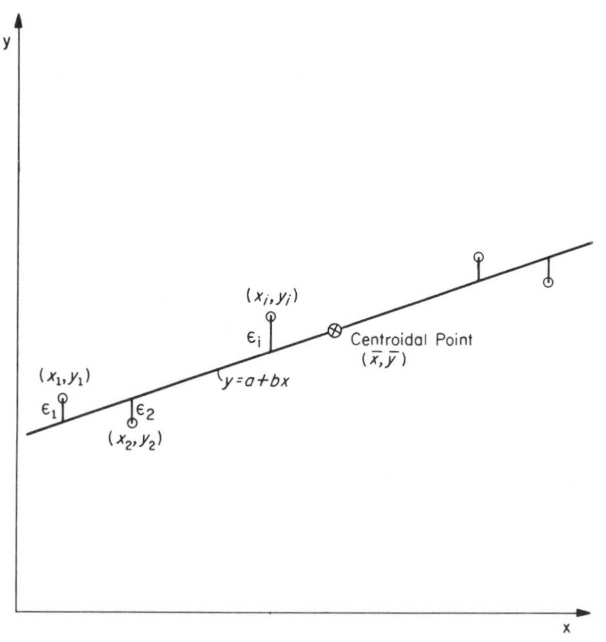

Fig. 15-1. Regression line.

know that the relation is linear, or alternatively expect or suspect it to be linear. Therefore, we postulate the relation as

$$y = a + bx \tag{15-1}$$

Our problem is to find the values of a and b for the line of "best fit."

For a point i on this line:

$$y_i - (a + bx_i) = 0 \tag{15-2}$$

but if there is error in the measurement, there will be a residual ϵ_i such that

$$y_i - (a + bx_i) = \epsilon_i \tag{15-3}$$

With n observations we have n equations:

$$\begin{aligned} y_1 - (a + bx_1) &= \epsilon_1 \\ y_2 - (a + bx_2) &= \epsilon_2 \\ \vdots \qquad \vdots \quad &\quad \vdots \\ y_n - (a + bx_n) &= \epsilon_n \end{aligned}$$

Using the summation notation, we can write the sum of squares of residuals as

$$P = \sum \epsilon_i^2 \tag{15-4}$$

or

$$P = \sum [y_i - (a + bx_i)]^2 \tag{15-5}$$

the summation extending from $i = 1$ to $i = n$.

As stated before, we have to satisfy the condition that the sum of squares of residuals is a minimum; i.e., P is a minimum. This occurs when

$$\left.\begin{aligned} \frac{\partial P}{\partial a} &= 0 \\ \frac{\partial P}{\partial b} &= 0 \end{aligned}\right\} \tag{15-6}$$

or

$$\sum [y_i - (a + bx_i)] = 0 \tag{15-7}$$

and

$$\sum x_i[y_i - (a + bx_i)] = 0 \tag{15-8}$$

Omitting the subscripts, we can write Eq. (15-7) as

$$\sum y - \sum a - b \sum x = 0$$

Since a is a constant, we have

$$\sum y = na + b \sum x \qquad (15\text{-}9)$$

or

$$\frac{\sum y}{n} = a + b\frac{\sum x}{n}$$

Thus

$$\bar{y} = a + b\bar{x} \qquad (15\text{-}10)$$

Equation (15-10) states that the line passes through the point $(\bar{x}, \bar{y})$, i.e., through the point whose coordinates are the appropriate means of all observations; we can call this point the centroidal point of all observations. From the fact that the point $(\bar{x}, \bar{y})$ lies on the line, it follows that Eq. (15-1) can be written also as

$$y - \bar{y} = b(x - \bar{x}) \qquad (15\text{-}1a)$$

Returning now to Eq. (15-8), we have

$$\sum xy = a \sum x + b \sum x^2 \qquad (15\text{-}11)$$

Equations (15-9) and (15-11) are called the *normal equations.*

REGRESSION LINE

Solving the normal equations (15-9) and (15-11), we obtain

$$a = \frac{\sum x^2 \sum y - \sum x \sum xy}{n \sum x^2 - (\sum x)^2} \qquad (15\text{-}12)$$

and[1]

$$b = \frac{n \sum xy - \sum x \sum y}{n \sum x^2 - (\sum x)^2} \qquad (15\text{-}13)$$

Hence the equation to the line of best fit can be written as

$$y = \frac{\sum x^2 \sum y - \sum x \sum xy}{n \sum x^2 - (\sum x)^2} + \frac{n \sum xy - \sum x \sum y}{n \sum x^2 - (\sum x)^2} x \qquad (15\text{-}14)$$

1. Note that

$$n \sum xy - \sum x \sum y = n \sum (x - \bar{x})(y - \bar{y})$$

since

$$n \sum (x - \bar{x})(y - \bar{y}) = n(\sum xy - \bar{x} \sum y + \bar{x} \sum y - \bar{y} \sum x)$$

In practice, it is more convenient to compute a and b separately [using Eqs. (15-12) and (15-13)], and to use the numerical values of a and b directly in writing $y = a + bx$.

The line given by Eq. (15-14) is called the line of *regression of y on x*. In deriving the line, we assumed that x is the assigned variable (i.e., sensibly free from error) and that y is the observed quantity.

If, however, the properties of the variables are reversed, i.e., if y is the assigned variable and x is the observed quantity, we find the constants in the equation to the line

$$x = a' + b'y \tag{15-15}$$

by minimizing the sum of squares of the x *residuals*. The equation to the line, known as the line of *regression of x on y*, is

$$x = \frac{\sum y^2 \sum x - \sum y \sum xy}{n \sum y^2 - (\sum y)^2} + \frac{n \sum xy - \sum x \sum y}{n \sum y^2 - (\sum y)^2}\, y \tag{15-16}$$

In general,

$$a \neq -\frac{a'}{b'}$$

and

$$b \neq \frac{1}{b'}$$

but both lines intersect at $(\bar{x}, \bar{y})$. An example of the two regression lines is shown in Fig. 15-2. We may note here that it is possible to calculate regression when both variables are subject to error.[2]

The computation of a and b is best performed by arranging the values of x, y, x^2, and xy in a tabular form. Tables of squares or logarithmic tables are, of course, useful. If a calculating machine is available, the values of $\sum x^2$ and $\sum xy$ can be found by one series of operations: x and y are set at the two extremes of the calculator keyboard; a multiplication by x then yields x^2 and xy, respectively, at the two ends of the carriage. The operation is repeated for all values of (x_i, y_i), the sum being carried so that at the end $\sum x^2$ and $\sum xy$ are obtained.

The computation of a and b may be rather laborious and may involve large numbers. The effort can be reduced by taking advantage of the fact that $(\bar{x}, \bar{y})$ is a point on the line. We can, therefore, transform the axes of coordinates to a new origin $(\bar{x}, \bar{y})$. The new coordinates (X, Y) are then:

$$\left.\begin{aligned} X &= x - \bar{x} \\ Y &= y - \bar{y} \end{aligned}\right\} \tag{15-17}$$

2. See A. Hald, *Statistical Theory with Engineering Applications*. New York: (John Wiley & Sons, Inc., 1952).

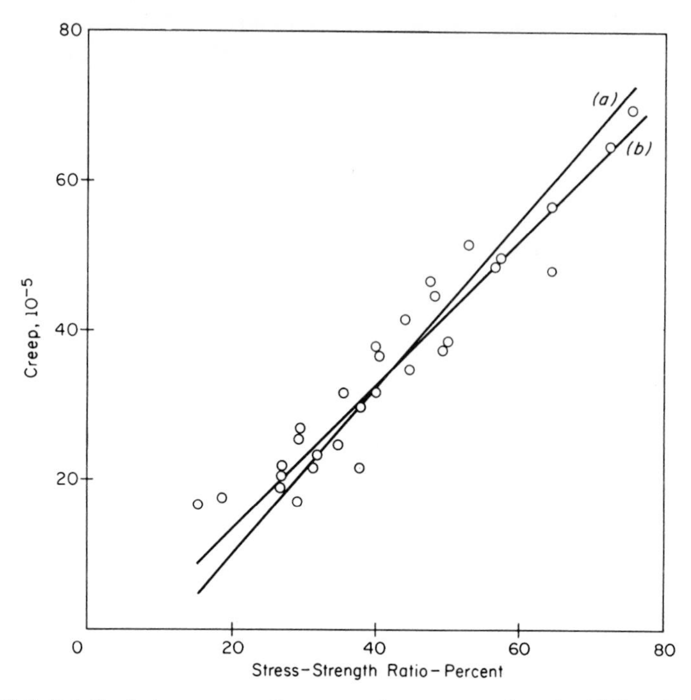

Fig. 15-2. Relation between creep of mortar and stress-strength ratio: (a) line of regression of stress-strength ratio upon creep; and (b) line of regression of creep upon stress-strength ratio.

Since the origin of coordinates (X, Y) is at the centroidal point, it follows that

$$\sum X = \sum Y = 0$$

Therefore, from Eq. (15-12),

$$a = 0$$

and from Eq. (15-13),

$$b = \frac{\sum XY}{\sum X^2} \tag{15-18}$$

This is equivalent to writing Eq. (15-13) in the form:

$$b = \frac{\sum (x - \bar{x})(y - \bar{y})}{\sum (x - \bar{x})^2} \tag{15-19}$$

which is of more theoretical interest.

Using Eq. (15-18), we find that the equation to the line of regression of y on x (or Y on X) becomes

$$Y = \frac{\sum XY}{\sum X^2} X \tag{15-20}$$

The use of (X, Y), of course, requires computing $(x - \bar{x})$ and $(y - \bar{y})$ for all the observations. This may be tedious if $\bar{x}$ or $\bar{y}$ involves several decimal places, and the computation of products and squares may in consequence be more laborious than operation on (x, y) directly when the latter are integers. As an example, computation by both methods is given here.

Example. To determine the relation between the normal stress and the shear resistance of soil, a shear-box experiment was performed, giving the following results:

Normal stress x, kN/m^2	11	13	15	17	19	21
Shear resistance y, kN/m^2	15.2	17.7	19.3	21.5	23.9	25.4

In the test the value of x is assigned, and y is the derived quantity. The relation between the two is of the form

$$y = a + bx$$

where a is the cohesion of soil, $b = \tan \phi$, and ϕ is the angle of friction.

Tabular arrangement is most convenient (see Table 15-1).

TABLE 15-1

x	y	x^2	xy	$X = x - \bar{x}$	$Y = y - \bar{y}$	XY	X^2
11	15.2	121	167.2	−5	−5.3	26.5	25
13	17.7	169	230.1	−3	−2.8	8.4	9
15	19.3	225	289.5	−1	−1.2	1.2	1
17	21.5	289	365.5	1	1.0	1.0	1
19	23.9	361	454.1	3	3.4	10.2	9
21	25.4	441	533.4	5	4.9	24.5	25
$\sum = 96$	123.0	1606	2039.8	$8 - 8 = 0$	$9.3 - 9.3 = 0$	71.8	70

$$\bar{x} = \frac{\sum x}{n} = \frac{96}{6} = 16.0 \text{ kN/m}^2 \qquad \bar{y} = \frac{\sum y}{n} = \frac{123.0}{6} = 20.5 \text{ kN/m}^2$$

From Eq. (15-12),

$$a = \frac{\sum x^2 \sum y - \sum x \sum xy}{n \sum x^2 - (\sum x)^2}$$

$$= \frac{1606 \times 123 - 96 \times 2039.8}{6 \times 1606 - 96 \times 96}$$

$$= 4.089$$

From Eq. (15-13),

$$b = \frac{n \sum xy - \sum x \sum y}{n \sum x^2 - (\sum x)^2}$$

$$= \frac{6 \times 2039.8 - 96 \times 123}{6 \times 1606 - 96 \times 96}$$

$$= 1.026$$

Hence the relation between the normal stress x and the shear resistance y is given by

$$y = 4.089 + 1.026x$$

We can solve the same problem using coordinates (X, Y) referred to the centroidal point; the appropriate values are computed in the right-hand part of the preceding table. From Eq. (15-18),

$$b = \frac{\sum XY}{\sum X^2} = \frac{71.8}{70} = 1.026$$

Substituting in Eq. (15-20), we obtain

$$Y = 1.026X$$

Transforming to the original coordinates yields

$$y - 20.5 = 1.026(x - 16.0)$$

whence

$$y = 4.089 + 1.026x \qquad \text{(as before)}$$

The line of regression of y on x, as well as the experimental points, is shown in Fig. 15-3.

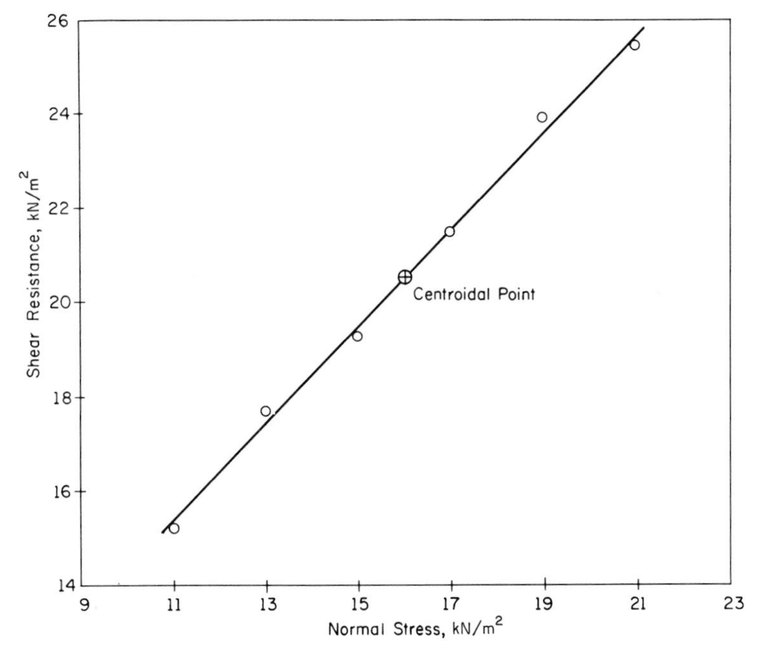

Fig. 15-3. Regression line of shear resistance on normal stress for the data of the example on p. 261.

LIMITATIONS OF METHOD

It may be worth stating explicitly that the method of least squares is applicable only when the observed values of y_i correspond to assigned (or error-free) values of x_i; furthermore, the error in y (expressed as variance of y) must be independent of the level of x. (Of course, x and y may be reversed.)

For inferences and estimates to be made about regression (but not for the method of least squares), it is also necessary that the values of y_i corresponding to a given x_i be normally distributed, with the mean of the distribution satisfying the regression equation. Furthermore, the variance of the values of y for a given value of x should be independent of the magnitude of x. In many practical problems this is not the case,[3] and transformation is then necessary; the usual transformations are by taking logarithms, square roots, etc. The transformation stabilizes the variance of y and makes the distribution closer to normal. This procedure is discussed in Chapters 10 and 13.

3. For example, the scatter of values of temperature determined by the pyrometer is smaller the higher the temperature, since the source of light is brighter. Contrariwise, the scatter of values of strength of concrete is greater the higher the mean strength.

CONFIDENCE LIMITS OF REGRESSION ESTIMATES

When we write the equation to the regression line

$$y = a + bx \qquad [15\text{-}1]$$

y should, strictly speaking, be written as $\hat{y}$, to denote that it is an estimate. To reduce the complexity of symbols, the circumflex will be omitted in what follows, except where required for clarity.

An estimate of y is given by Eq. (15-1) for any value of x, not necessarily one at which y was observed but within the range of values for which Eq. (15-14) was established. The question arises: What confidence can we have in our estimate of the "true" equation to the line?

The variance of an estimate enables us to form confidence limits of the estimate. In a manner similar to the variance of a sample, we shall consider variance about a regression line; in this case the deviations are reckoned from the line instead of from the mean. Thus the variance of y, estimated by the regression line, is the sum of squares of deviations divided by the number of degrees of freedom ν available for calculating the regression line; i.e.,

$$s_{y|x}^2 = \frac{\sum \epsilon_i^2}{\nu} \qquad (15\text{-}21)$$

where ϵ_i is defined by Eq. (15-3).

Two constraints determine the regression line: the centroidal point $(\bar{x}, \bar{y})$ and either slope b or intercept a, or simply b and a. Since these parameters have been estimated from the sample of size n (number of observations x_i, y_i), there will be two linear restrictions or constraints on the values of ϵ_i as defined by Eq. (15-3). Thus, from the definition of degrees of freedom in Chapter 3, the number of degrees of freedom ν for the variance $s_{y|x}^2$ is

$$\nu = n - 2$$

Hence

$$s_{y|x}^2 = \frac{\sum \epsilon_i^2}{n - 2} \qquad (15\text{-}22)$$

The variance of the mean value of y, i.e., $\bar{y}$, is given in a manner similar to Eq. (6-7) by

$$s_{\bar{y}|x}^2 = \frac{s_{y|x}^2}{n} \qquad (15\text{-}23)$$

We can now write the confidence limits for $\bar{y}$. As in the case of the sample mean, we find the value of t for the desired level of significance and the appropriate number of degrees of freedom (Table A-8). We can then state that the true value of $\bar{y}$ lies within the interval

$$\bar{y} \pm ts_{\bar{y}|x}$$

The probability of our being wrong is equal to the level of significance of the value of t.

Since the regression line must pass through the centroidal point, an error in the value of $\bar{y}$ leads to a constant error in y for all points on the line, the line being translated up or down without a change in slope (Fig. 15-4a).

The variance of the slope b is derived as follows. From Eq. (15-18), we have

$$b = \frac{\sum XY}{\sum X^2}$$

Thus

$$\text{variance of } b = \sigma_b^2 = \text{var}\left(\frac{\sum XY}{\sum X^2}\right)$$

From the discussion of the propagation of errors (Chapter 14) and recognizing that X has a negligible error, we find that

$$\begin{aligned}
\sigma_b^2 &= \left(\frac{1}{\sum X^2}\right)^2 \text{var}\left(\sum XY\right) \\
&= \left(\frac{1}{\sum X^2}\right)^2 \sum \text{var}\,(XY) \\
&= \left(\frac{1}{\sum X^2}\right)^2 (X_1^2\sigma_{y|x}^2 + X_2^2\sigma_{y|x}^2 + X_3^2\sigma_{y|x}^2 + \cdots) \\
&= \frac{\sum X^2}{(\sum X^2)^2}\sigma_{y|x}^2 = \frac{\sigma_{y|x}^2}{\sum X^2}
\end{aligned}$$

An estimate of σ_b^2 from the sample would be

$$s_b^2 = \frac{s_{y|x}^2}{\sum X^2}$$

or

$$s_b^2 = \frac{s_{y|x}^2}{\sum (x - \bar{x})^2} \tag{15-24}$$

The confidence band for the slope is represented by a double fan-shaped area with slopes of $b \pm ts_b$ and apex at $(\bar{x}, \bar{y})$ (Fig. 15-4b).

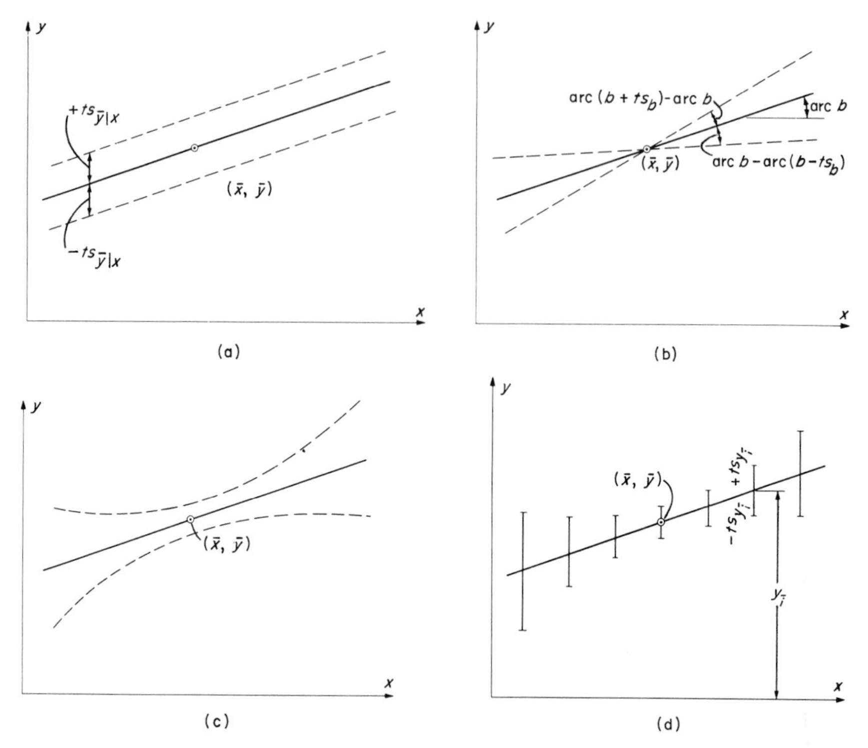

Fig. 15-4. (a) Influence of the confidence limits of $\hat{y}$ on the confidence limits of the regression line; (b) confidence band for slope; (c) approximate confidence area of the regression line; and (d) confidence limits of an estimate of y .

The confidence area of the regression can be approximated by smooth curves asymptotic to the confidence interval of the slope near the ends of the range of observations and including the confidence limits of $\bar{y}$. This is illustrated by Fig. 15-4c.

If we require a closer estimate of the confidence limits, we can calculate the limits for $\hat{y}_i$ corresponding to a specified value of x_i. To do this, we need the variance of $\hat{y}_i$, denoted further by y.

If the limits are for the *mean* estimated value for y_i, i.e., $y_{\bar{i}}$, we require the variance of this mean value. The variance is

$$s_{\bar{y}_i}^2 = s_y^2 \left[\frac{1}{n} + \frac{(x_i - \bar{x})^2}{\sum (x - \bar{x})^2} \right] \tag{15-25}$$

Equation (15-25) is derived in the same manner as Eq. (15-24). From the equation of the line, Eq. (15-1a), we have

$$y_{\bar{i}} - \bar{y} = b(x_i - \bar{x})$$

or

$$y_i = \bar{y} + bX_i$$

Hence

$$\text{variance of } y_i = \text{variance of } \bar{y} + X_i^2 \text{ variance of } b$$

i.e.,

$$\sigma_{y_i}^2 = \sigma_{\bar{y}|x}^2 + X_i^2 \sigma_b^2$$

$$= \frac{\sigma_{y|x}^2}{n} + \frac{X_i^2 \sigma_{y|x}^2}{\sum X^2}$$

or

$$\sigma_{y_i}^2 = \sigma_{y|x}^2 \left(\frac{1}{n} + \frac{X_i^2}{\sum X^2} \right)$$

An estimate of $\sigma_{y_i}^2$ from the sample would be

$$s_{y_i}^2 = s_{y|x}^2 \left(\frac{1}{n} + \frac{X_i^2}{\sum X^2} \right)$$

Putting $X_i = x_i - \bar{x}$ will yield Eq. (15-25). It is obvious that if $X_i = 0$, then y_i is at the centroid and hence Eq. (15-25) reduces to Eq. (15-23).

The confidence interval for the mean estimated value of y_i is then

$$y_i \pm t s_{y_i}$$

For a specified level of significance we can thus predict the limits within which a future *mean* estimated value of $y_{\bar{i}}$ will lie with the appropriate probability of error. The limits are wider the larger the value of $(x_i - \bar{x})$, i.e., the further x_i is removed from the centroidal point. This is shown in Fig. 15-4d.

If we are interested in predicting the confidence interval of a *single* estimated value of $\hat{y}_i$, we have to use the variance of such a single value; this variance is

$$s_{yi}^2 = s_{y|x}^2 \left[1 + \frac{1}{n} + \frac{(x_i - \bar{x})^2}{\sum (x - \bar{x})^2} \right] \qquad (15\text{-}26)$$

This variance is larger than $s_{y_i}^2$ because the variance of the single value is equal to the variance of the mean plus the variance of $\hat{y}$ estimated by the line, i.e.,

$$s_{yi}^2 = s_{y_i}^2 + s_{y|x}^2 \qquad (15\text{-}27)$$

The confidence interval for a single value is correspondingly greater, namely,

$$y \pm t s_{y_i} \tag{15-28}$$

A diagram representing this would be similar to Fig. 15-4d.

If we want to use the intercept a to define the regression line, we find the variance of a as a particular case of the variance of any mean estimated value $\hat{y}_i$. Therefore, we substitute $x_i = 0$ in Eq. (15-25), and the variance of a is given by

$$s_a^2 = s_{y|x}^2 \left[\frac{1}{n} + \frac{\bar{x}^2}{\sum (x - \bar{x})^2} \right] \tag{15-29}$$

For all the confidence intervals of this section the number of degrees of freedom used in determining t is $\nu = n - 2$, for the reasons stated on p. 264.

SIGNIFICANCE TEST FOR SLOPE

In some cases we have a theoretical value b_0 of the slope b in Eq. (15-1), and we want to determine whether there is a significant difference between b_0 and the value of b given by the regression line. This is done by means of the t test applied to $|b - b_0|$.

The standard deviation of $|b - b_0|$ is equal to the standard deviation of b, s_b, because the theoretical value of b is free from error. We find

$$t = \frac{|b - b_0|}{s_b} \tag{15-30}$$

and if the calculated t is greater than t given in Table A-8 for a required level of significance, we conclude that there is a significant difference between b and b_0. The level of significance represents the probability of our drawing an erroneous conclusion.

The number of degrees of freedom for which t is found is equal to $(n - 2)$, where n is the number of observations used in deriving the regression line.

A similar test can be applied to the intercept a. The difference between the value given by regression and a theoretical value a_0 is tested by

$$t = \frac{|a - a_0|}{s_a} \tag{15-31}$$

The number of degrees of freedom is the same as before.

Sometimes, it is useful to test whether the slope differs significantly from zero. We then find

$$t = \frac{b}{s_b} \tag{15-32}$$

and proceed as before. If b does not differ significantly from zero, y is independent of x, and the computation of the regression line is pointless.

Example. Referring to the preceding example on the relation between normal stress x and shear resistance of soil y:

1. Check whether the intercept a and slope b are significantly different from zero.
2. Find whether the slope differs significantly from a theoretically predicted slope $b = 1.000$.
3. Find the 95 percent confidence interval for the estimates of a and b.
4. Assuming that the regression line can be extrapolated to $x = 25$ kN/m^2, find the estimate of the mean value of shear resistance $\hat{y}_i$ for $x = 25$ kN/m^2 and also the 90 percent confidence interval for this estimate.
5. Find the 95 percent confidence interval for a single estimated value of the shear resistance corresponding to $x = 25$ kN/m^2.

The regression line of the preceding example gives $y = 4.089 + 1.026x$. From this we compute the estimates $\hat{y}$, and the deviations $(y_i - \hat{y}_i)$ for all assigned values of x. (See Table 15-2.)

TABLE 15-2

x	$\hat{y}$	$\epsilon_i = y_i - \hat{y}_i$	ϵ_i^2
11	15.375	−0.175	0.0306
13	17.427	+0.273	0.0745
15	19.479	−0.179	0.0320
17	21.531	−0.031	0.0010
19	23.583	+0.317	0.1005
21	25.635	−0.235	0.0552
$n = 6$			$\sum \epsilon_i^2 = 0.2938$

1. From Eq. (15-22),

$$s_{y|x} = \sqrt{\frac{\sum \epsilon_i^2}{n - 2}} = \sqrt{\frac{0.2938}{6 - 2}} = 0.271$$

From Eq. (15-24) and using the value of $\sum X^2 = \sum (x - \bar{x})^2 = 70$, we have

$$s_b = \frac{s_{y|x}}{\sqrt{\sum (x - \bar{x})^2}} = \frac{0.271}{\sqrt{70}} = 0.0324$$

From Eq. (15-29),

$$s_a = s_{y|x}\sqrt{\frac{1}{n} + \frac{\bar{x}^2}{\sum (x - \bar{x})^2}} = 0.271\sqrt{\frac{1}{6} + \frac{16^2}{70}}$$
$$= 0.5298$$

To test the significance of b, we use Eq. (15-32) and find that

$$t = \frac{b}{s_b} = \frac{1.026}{0.0324} = 31.667$$

For $v = 4$, Table A-8 shows this value of t as significant at better than 0.1 percent. For a,

$$t = \frac{a}{s_a} = \frac{4.089}{0.5298} = 7.718$$

and this is significant at the 1 percent level.

2. From Eq. (15-30),

$$t = \frac{|b - b_0|}{s_b} = \frac{1.026 - 1}{0.0324} = 0.8025$$

From Table A-8 for $v = 4$, this value of t is not significant, and we conclude that the observed value of b accords with the predicted value $b_0 = 1.000$.

3. The 95 percent confidence intervals for the estimates of a and b are $a \pm ts_a$ and $b \pm ts_b$, respectively. For $v = 4$, and a 5 percent level of significance, $t = 2.776$. Hence the required estimates are for a:

$$4.089 \pm 2.776 \times 0.5298 = (2.62,\ 5.56)\ \text{kN/m}^2$$

and for b:

$$1.026 \pm 2.776 \times 0.0324 = (0.94,\ 1.12)$$

4. From the regression equation $y = a + bx$, for $x = 25$ kN/m^2 the estimate of the mean value of y_i is

$$y_i = 4.089 + 1.026 \times 25 = 29.74\ \text{kN/m}^2$$

The 90 percent confidence interval for this estimate is

$$y_i \pm ts_{y_i}$$

where s_{y_i} is given by Eq. (15-25):

$$s_{y_i} = s_{y|x}\sqrt{\frac{1}{n} + \frac{(x_i - \bar{x})^2}{\sum (x - \bar{x})^2}}$$
$$= 0.271\sqrt{\frac{1}{6} + \frac{(25 - 16)^2}{70}}$$
$$= 0.312$$

and for $v = 4$ and a 10 percent level of significance, $t = 2.132$. The confidence interval is, therefore,

$$29.74 \pm 2.132 \times 0.312 = (29.07, 30.40) \text{ kN/m}^2$$

5. For a single estimated value of y_i, the standard deviation is given by Eq. (15-26):

$$s_{y_i} = 0.271\sqrt{1 + \frac{1}{6} + \frac{(25 - 16)^2}{70}} = 0.413$$

For the 5 percent level of significance, $t = 2.776$. Hence the confidence interval is

$$29.74 \pm 2.776 \times 0.413 = (28.59, 30.88) \text{ kN/m}^2$$

USE OF THE METHOD OF LEAST SQUARES IN SURVEYING

It may be of interest to mention that the principle of minimizing errors by the method of least squares can be used also in computing most probable values where these have to satisfy a condition different from that of linearity. For example, in surveying operations we measure angles, lengths, or levels; these may have to satisfy a condition equation of the type $\sum$ angles $= 360°$ or $\sum$ differences in level $= 0$. The procedure is illustrated by the following example.

Example. Bench marks at three locations A, B, and C were required. Precise leveling was carried from a known bench mark R, as shown in

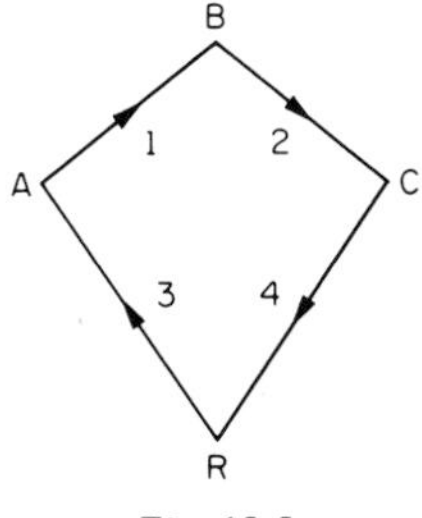

Fig. 15-5

Fig. 15-5. Due to differences in lengths of lines, number of instruments settings, etc., the observations[4] were weighted as shown in Table 15-3. Find the most probable values of the elevations of A, B, and C with respect to R.

4. The "weight" is proportional to our confidence in the observed value.

TABLE 15-3

Leg	*Observed difference in elevation, m*	*Most probable value*	*Error*	*Weight w*
R to *A*	$\alpha_1 = 94.775$	θ_1	$\epsilon_1 = \alpha_1 - \theta_1$	4
A to *B*	$\alpha_2 = -21.739$	θ_2	$\epsilon_2 = \alpha_2 - \theta_2$	2
B to *C*	$\alpha_3 = 18.631$	θ_3	$\epsilon_3 = \alpha_3 - \theta_3$	3
C to *R*	$\alpha_4 = -91.317$	θ_4	$\epsilon_4 = \alpha_4 - \theta_4$	4

We denote the most probable value of an elevation by θ. Because of the closed circuit of observations, the most probable values must satisfy the condition equation:

$$\theta_1 + \theta_2 + \theta_3 + \theta_4 = 0$$

Hence we can write

$$\epsilon_4 = \alpha_4 + \theta_1 + \theta_2 + \theta_3$$

We now apply the principle of least squares to the circuit by minimizing the *weighted* sum of squares of errors, $\sum w\epsilon^2$. Now

$$\sum w\epsilon^2 = w_1\epsilon_1^2 + w_2\epsilon_2^2 + w_3\epsilon_3^2 + w_4\epsilon_4^2 = P$$

or

$$P = w_1(\alpha_1 - \theta_1)^2 + w_2(\alpha_2 - \theta_2)^2 + w_3(\alpha_3 - \theta_3)^2 + w_4(\alpha_4 + \theta_1 + \theta_2 + \theta_3)^2$$

This has a minimum value when

$$\frac{\partial P}{\partial \theta_1} = \frac{\partial P}{\partial \theta_2} = \frac{\partial P}{\partial \theta_3} = 0$$

i.e., when

$$-2w_1(\alpha_1 - \theta_1) + 2w_4(\alpha_4 + \theta_1 + \theta_2 + \theta_3) = 0$$

$$-2w_2(\alpha_2 - \theta_2) + 2w_4(\alpha_4 + \theta_1 + \theta_2 + \theta_3) = 0$$

$$-2w_3(\alpha_3 - \theta_3) + 2w_4(\alpha_4 + \theta_1 + \theta_2 + \theta_3) = 0$$

These are the normal equations. Substituting the values of w and α and simplifying, we get

$$8\theta_1 + 4\theta_2 + 4\theta_3 = 744.368$$

$$4\theta_1 + 6\theta_2 + 4\theta_3 = 321.790$$

$$4\theta_1 + 4\theta_2 + 7\theta_3 = 421.161$$

Solving these three equations by elimination, we obtain

$$\theta_1 = +94.709, \qquad \theta_2 = -21.870, \qquad \theta_3 = +18.543, \qquad \theta_4 = -91.382$$

Hence the most probable values of elevation with respect to R are $R = 0$ m, $A = 94.709$ m, $B = 72.839$ m, and $C = 91.382$ m.

NONLINEAR RELATIONS

The method of fitting a regression line can be extended to the case where the known, expected, or suspected relation is not in the form of a straight line. The procedure is to write the equation to the curve in its general form, tabulate the deviations of y from the assumed curve, and to find the constants in the equation which satisfy the condition that the sum of the squares of deviations is a minimum.

Example. Suppose we observed the values of y for four values of x, x being the assigned variable. (For example, x can be time, and y the measured distance). Assume further that the general relation between y and x is of the form

$$y = a + bx^2$$

We can tabulate the results as in Table 15-4.

TABLE 15-4

x	0	1	2	3
Observed value y_i	0.5	6.5	21.3	48.6
Most probable value $\hat{y}$	a	$a + b$	$a + 4b$	$a + 9b$
Deviation $(y_i - \hat{y})$	$0.5 - a$	$6.5 - (a + b)$	$21.3 - (a + 4b)$	$48.6 - (a + 9b)$

Hence the sum of squares of deviations is

$$P = (0.5 - a)^2 + (6.5 - a - b)^2 + (21.3 - a - 4b)^2 + (48.6 - a - 9b)^2$$

For P to be a minimum, we have to satisfy the conditions:

$$\frac{\partial P}{\partial a} = 0 \quad \text{and} \quad \frac{\partial P}{\partial b} = 0$$

or

$$(0.5 - a) + (6.5 - a - b) + (21.3 - a - 4b) + (48.6 - a - 9b) = 0$$

and

$$(6.5 - a - b) + 4(21.3 - a - 4b) + 9(48.6 - a - 9b) = 0$$

These reduce to the normal equations:

$$4a + 14b = 76.9$$

$$14a + 98b = 529.1$$

Hence $a = 0.219$ and $b = 5.430$.

Thus the equation to the best line of the general form of $y = a + bx^2$ is

$$y = 0.219 + 5.430x^2$$

It should be stressed, however, that a better fit may be obtained with an equation of a different form, e.g., $y = a + bx + cx^2$. This can be determined only by trial and error, the procedure depending on the problem in hand, and, in many cases, of course, the search for a "better fit" may not be warranted.

RECTIFICATION

The application of the method of least squares to nonlinear relations usually requires a great deal of computational effort. However, in many cases, a nonlinear relation can be transformed to a straight-line relation, i.e., rectified. This not only simplifies the handling of the data, but also results in a graphical presentation that is more revealing as far as the assessment of scatter is concerned. Extrapolation, if this is warranted (and often it is not), is also easier, and so is the computation of various statistics, such as standard deviation or confidence limits. Clearly, the statistics calculated for rectified variables apply to them and not to the original data. Several simple cases will be illustrated.

The *exponential function* $y = ab^x$ can be rectified by log transformation, i.e., by taking logarithms of both sides of the equation:

$$\log y = \log a + x \log b$$

This will plot as a straight line if the ordinates give $\log y$ (i.e., are to a logarithmic scale), while the abscissae are to a linear scale. Log a and $\log b$ are the fitting constants of the equation. In other words, $\log y$ and x are treated as new (and linear) variables to which the principle of least squares is applied.

The *power function* $y = ax^b$ can be rectified even more simply, again by taking logarithms:

$$\log y = \log a + b \log x$$

The fitting constants are now log a and b, and the new variables log x and log y are linearly related.

The *hyperbola* $y = a + b/x$ can be rectified by treating $1/x = u$ as the new variable. Then y and u are linearly related.

If the equation is in the form

$$y = \frac{x}{a + bx}$$

we can invert it to

$$\frac{1}{y} = \frac{a}{x} + b$$

Then $1/x$ and $1/y$ are linearly related. Alternatively, we can multiply both sides of the above equation by x, thus obtaining:

$$\frac{x}{y} = a + bx$$

We then plot x/y versus x. The choice depends on the nature of the case considered.

The *polynomial function* of the form $y = a + bx + cx^2$ is concave up or down, depending on the signs of the coefficients. We differentiate both sides of the equation with respect to x:

$$\frac{dy}{dx} = b + 2cx$$

A straight-line relation is given by plotting dy/dx versus x.

If no advance information on the shape of the curve fitting the experimental data is available, trial-and-error methods may be necessary. As a first step, the data should be plotted using linear x and y coordinates; a smooth curve is then drawn, and a function likely to fit is chosen from the knowledge of the shapes of curves corresponding to simple algebraic functions.

It is important to note that when transformation is used, the deviation minimized is not in y but in the transformed variable. We should remember this when drawing conclusions from an experiment, as in some cases the difference may be significant. If we have reason to believe that, from physical considerations of an experiment, it is the original and not the transformed variable that should have its deviation minimized, then the transformed variable should be weighted in inverse proportion of some function of the error of the original variable. Often, the weight is taken as proportional to $1/(\text{error})^2$.

If the fitting of the straight line is done by eye, the standard error of each point representing a mean of a set of observations can be indicated by a bar, and the curve is then drawn so that the smaller the error associated with a given point the greater the probability of the line's passing through this point. This, of course, is often done intuitively when we have reason to believe that readings at, say, low temperatures are less reliable (i.e., have a lesser weight) than at high temperatures. In Solved Problem 15-1 the difference between minimizing the deviation of the original variable and of the transformed variable is shown.

Standard computer programs for the fitting of various least square curves are available. For nonlinear functions, iteration using a computer is practical.

Solved Problems

15-1. In an experiment the volume of a gas was measured at different pressures, the temperature remaining constant. The results were as follows:

Pressure p, pascals	20	25	30	35	40
Volume v, m^3	0.31	0.22	0.18	0.15	0.13

The assumed relation is given by $p = A/v + B$, where A and B are constants (B can be considered as instrument error). Assuming that only v contains errors,

a. Find the best values of A and of B.
b. Plot the relation between p and $1/v$.

SOLUTION

By minimizing the deviation of the transformed variable:

a. The problem is to determine the best value for the constant A and is best approached by employing a change of variable:

$$v' = \frac{1}{v}$$

then $p = Av' + B$, and we have

Pressure p, pascals	20	25	30	35	40
v'	3.226	4.545	5.556	6.667	7.692

This approach implies that the error of all $1/v$ values is the same, which may or may not be the case in practice. The centroidal point of the five observations (p, v') is (30, 5.537). When the pressure is, say p_1, the estimated value for v'_1 from the equation is

$$\hat{v}'_1 = \frac{p_1 - B}{A}$$

Therefore, the deviation in v' is

$$\epsilon = \frac{p_1 - B}{A} - v'_1$$

Then

$$\sum \epsilon^2 = \left(\frac{p_1 - B}{A} - v'_1\right)^2 + \left(\frac{p_2 - B}{A} - v'_2\right)^2 + \left(\frac{p_3 - B}{A} - v'_3\right)^2 + \left(\frac{p_4 - B}{A} - v'_4\right)^2 + \left(\frac{p_5 - B}{A} - v'_5\right)^2$$

To obtain the regression line, $\sum \epsilon^2$ is to be a minimum, i.e.,

$$\frac{\partial(\sum \epsilon^2)}{\partial A} = 2\left(\frac{p_1 - B}{A} - v'_1\right)\left(\frac{-(p_1 - B)}{A^2}\right) + \cdots + 2\left(\frac{p_5 - B}{A} - v'_5\right)\left(\frac{-(p_5 - B)}{A^2}\right) = 0$$

Hence

$$A = \frac{(p_1 - B)^2 + (p_2 - B)^2 + (p_3 - B)^2 + (p_4 - B)^2 + (p_5 - B)^2}{v'_1(p_1 - B) + v'_2(p_2 - B) + v'_3(p_3 - B) + v'_4(p_4 - B) + v'_5(p_5 - B)}$$

$$= \frac{p_1^2 + p_2^2 + p_3^2 + p_4^2 + p_5^2 - 2B(p_1 + p_2 + p_3 + p_4 + p_5) + 5B^2}{v'_1 p_1 + v'_2 p_2 + v'_3 p_3 + v'_4 p_4 + v'_5 p_5 - B(v'_1 + v'_2 + v'_3 + v'_4 + v'_5)}$$

By substituting, we find that

$$A = \frac{4750 - 300B + 5B^2}{885.85 - 27.686B}$$

Now, since the regression line must pass through the centroid (30, 5.537), we can write

$$5.537 = \frac{30 - B}{A}$$

Eliminating A, we obtain

$$\frac{30 - B}{5.537} = \frac{4750 - 300B + 5B^2}{885.85 - 27.686B}$$

whence

$$55.3B = 274.7$$

$$B = \frac{274.7}{55.3} = 4.967$$

and

$$A = \frac{30 - 4.967}{5.537} = \frac{25.033}{5.537} = 4.521$$

Thus

$$p = 4.967 + 4.521v'$$

Hence

$$p = 4.967 + 4.521\left(\frac{1}{v}\right)$$

is the equation of the "best" line.

b. Figure 15-6 shows the plot of p versus $1/v$.

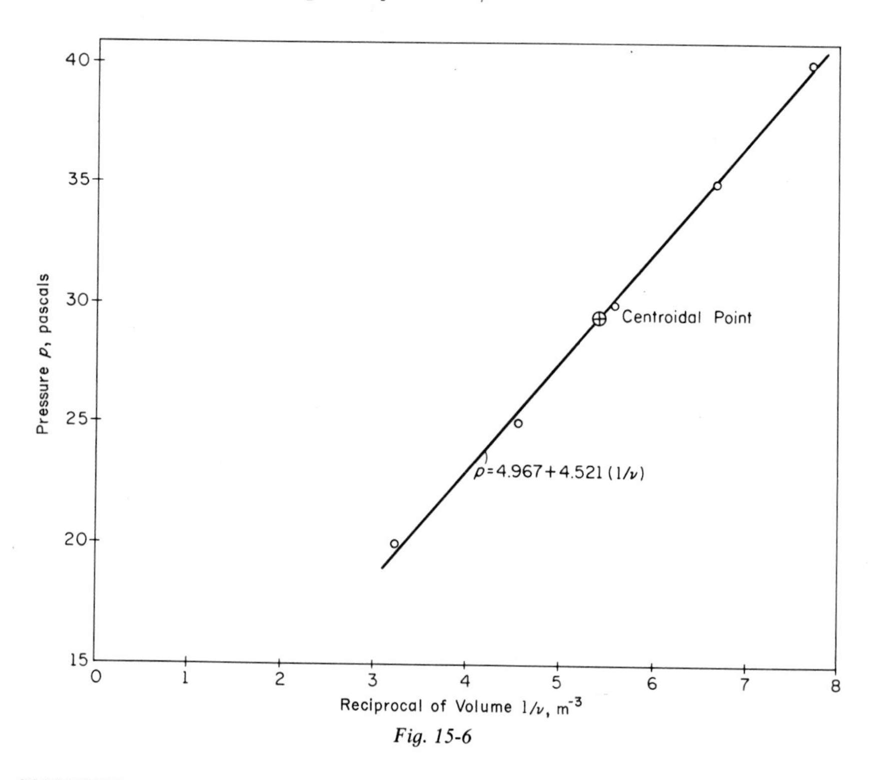

Fig. 15-6

SOLUTION

By minimizing the deviations of the original variable: In general, assuming that all values of v have equal precision does not mean that the error of all the values $1/v$ is the same. In fact, the values of $v' = 1/v$ will not be of equal precision, and weighting considerations will have to be applied.

Equation (14-22) states that

$$\sigma_Q^2 = \left(\frac{\partial Q}{\partial x_1}\right)^2 \sigma_{x_1}^2 + \left(\frac{\partial Q}{\partial x_2}\right)^2 \sigma_{x_2}^2 + \cdots$$

With $v' = 1/v$, this gives

$$\sigma_{v'}^2 = \left(-\frac{1}{v^2}\right)^2 \sigma_v^2$$

whence

$$\sigma_{v'} = \frac{\sigma_v}{v^2}$$

Adopting the well-known criterion that the weight of a value is proportional to $(\text{error})^{-2}$, we can say that the weight of each value of v' is proportional to $(\sigma_v/v^2)^{-2}$. Since it is understood that σ_v is constant, it follows that the weight of each value of v' is proportional to v^4. (See Table 15-5.)

TABLE 15-5

Pressure p, pascals	20	25	30	35	40
v'	3.226	4.545	5.556	6.667	7.692
Relative weight of v', $w(= v^4 \times 10^4)$	92.35	23.43	10.50	5.06	2.86

The weighted centroidal point is found from

$$\bar{p} = \frac{\sum_i w_i p_i}{\sum_i w_i} = 22.647$$

and

$$\bar{v}' = \frac{\sum_i w_i v_i'}{\sum_i w_i} = 3.863$$

The method of least squares requires that the sum of weighted squares of the errors is a minimum. Thus we have to differentiate the equation

$$P = \sum_i w_i \epsilon_i^2 = \sum_i w_i \left[\frac{p_i - B}{A} - v_i'\right]^2$$

which gives

$$\frac{\partial P}{\partial A} = -2 \sum_i w_i \left[\frac{p_i - B}{A^2}\right]\left[\frac{p_i - B}{A} - v_i'\right] = 0$$

whence

$$A = \frac{\sum_i w_i(p_i - B)^2}{\sum_i w_i v_i'(p_i - B)} = \frac{71{,}808.25 - 6078.50B + 134.20B^2}{12{,}431.49 - 518.48B} \qquad (15\text{-}33)$$

Since the regression line must pass through the centroidal point ($\bar{p}$, $\bar{v}' = 22.647$, 3.863), it follows that

$$\bar{v}' = \frac{\bar{p} - B}{A}$$

Substituting for $\bar{p}$ and $\bar{v}'$, we obtain

$$3.863 = \frac{22.647 - B}{A} \tag{15-34}$$

From Eqs. (15-33) and (15-34) we find that $A = 4.314$ and $B = 5.979$. Hence the equation of the "best" line on the basis of minimum deviation of the original variable is

$$p = 5.979 + 4.314\left(\frac{1}{v}\right)$$

It can be seen that minimizing the deviation of the original variable leads to a different result from that obtained by minimizing the transformed variable: Which of the two is more appropriate depends on the physical basis to the data.

15-2. The following table gives experimental data obtained from measuring plate current versus plate voltage from a triode with grid voltage held at a certain constant level.

Plate voltage $\times \frac{1}{10} = x$, *volts*	1	2	3	4	5
Plate current $\times \frac{1}{10} = y$, *mA*	1.7	5.2	8.9	14.2	19.9

It is assumed that y only is subject to error. Find a polynomial of the form $y = Ax + Bx^2$ to fit best the data given.

SOLUTION

The values of y for $x = 1, 2, \ldots, 5$ from the formula are, respectively,

$$A + B, \qquad 2A + 4B, \qquad 3A + 9B, \qquad 4A + 16B, \qquad 5A + 25B$$

The deviations ϵ of the experimental values of y from the above are, respectively,

$$A + B - 1.7, \qquad 2A + 4B - 5.2, \qquad 3A + 9B - 8.9,$$
$$4A + 16B - 14.2, \qquad 5A + 25B - 19.9$$

Therefore,

$$\sum \epsilon^2 = (A + B - 1.7)^2 + (2A + 4B - 5.2)^2 + (3A + 9B - 8.9)^2$$
$$+ (4A + 16B - 14.2)^2 + (5A + 25B - 19.9)^2$$

Making $\sum \epsilon^2$ a minimum with respect to A and B, we obtain

$$\frac{\partial(\sum \epsilon^2)}{\partial A} = 2[(A + B - 1.7) + 2(2A + 4B - 5.2) + 3(3A + 9B - 8.9)$$
$$+ 4(4A + 16B - 14.2) + 5(5A + 25B - 19.9)] = 0$$

and

$$\frac{\partial(\sum \epsilon^2)}{\partial B} = 2[(A + B - 1.7) + 4(2A + 4B - 5.2) + 9(3A + 9B - 8.9)$$

$$+ 16(4A + 16B - 14.2) + 25(5A + 25B - 19.9)] = 0$$

From these two equations, we get the normal equations:

$$55A + 225B = 195.1$$

$$225A + 979B = 827.3$$

Solving for A and B, we find that $A = 1.509$ and $B = 0.498$. Therefore, the "best" second-degree polynomial is of the form:

$$y = 1.509x + 0.498x^2$$

It is now possible to calculate the "true" current y for the given voltages x from the above formula.

15-3. A round of angles α_1, α_2, α_3, and α_4 was observed at a station R. The sum of the four angles was less than 360° by 4″. The angles were measured several times. The average readings and their weights are shown in Table 15-6.

TABLE 15-6

	Weight, w
$\alpha_1 = 75°44'30''$	2
$\alpha_2 = 89°15'40''$	3
$\alpha_3 = 110°27'35''$	2
$\alpha_4 = 84°32'11''$	4

Find the most probable values of the angles θ_1, θ_2, θ_3, and θ_4.

SOLUTION

Let ϵ_1, ϵ_2, ϵ_3, and ϵ_4 be the errors in the four angles, respectively. Then

$$\epsilon_1 = \alpha_1 - \theta_1, \qquad \epsilon_2 = \alpha_2 - \theta_2, \qquad \epsilon_3 = \alpha_3 - \theta_3, \qquad \epsilon_4 = \alpha_4 - \theta_4$$

The sum of the squares of the weighted errors is

$$\sum w\epsilon^2 = w_1\epsilon_1^2 + w_2\epsilon_2^2 + w_3\epsilon_3^2 + w_4\epsilon_4^2$$

$$= w_1(\alpha_1 - \theta_1)^2 + w_2(\alpha_2 - \theta_2)^2 + w_3(\alpha_3 - \theta_3)^2 + w_4(\alpha_4 - \theta_4)^2$$

Since

$$\theta_4 = 360° - (\theta_1 + \theta_2 + \theta_3)$$

we have only three unknowns.

By the principle of least squares, $\sum \epsilon^2$ is a minimum when the differential coefficients with respect to θ_1, θ_2, and θ_3 are zero. Thus

$$\frac{\partial(\sum \epsilon^2)}{\partial \theta_1} = \frac{\partial(\sum \epsilon^2)}{\partial \theta_2} = \frac{\partial(\sum \epsilon^2)}{\partial \theta_3} = 0$$

or

$$\frac{\partial(\sum w\epsilon^2)}{\partial \theta_1} = -2w_1(\alpha_1 - \theta_1) + 2w_4(\alpha_4 - 360° + \theta_1 + \theta_2 + \theta_3) = 0$$

$$\frac{\partial(\sum w\epsilon^2)}{\partial \theta_2} = -2w_2(\alpha_2 - \theta_2) + 2w_4(\alpha_4 - 360° + \theta_1 + \theta_2 + \theta_3) = 0$$

$$\frac{\partial(\sum w\epsilon^2)}{\partial \theta_2} = -2w_3(\alpha_3 - \theta_3) + 2w_4(\alpha_4 - 360° + \theta_1 + \theta_2 + \theta_3) = 0$$

After substituting the numerical values of w_1, w_2, w_3, w_4, α_1, α_2, α_3, and α_4, the preceding equations reduce to the following normal equations:

$$6\theta_1 + 4\theta_2 + 4\theta_3 = 1253.337778$$

$$4\theta_1 + 7\theta_2 + 4\theta_3 = 1369.637777$$

$$4\theta_1 + 4\theta_2 + 6\theta_3 = 1322.773888$$

Solving these equations by elimination, we obtain

$$\theta_1 = 75°44'31.26''$$

$$\theta_2 = 89°15'40.84''$$

$$\theta_3 = 110°27'36.27''$$

and

$$\theta_4 = 360° - (\theta_1 + \theta_2 + \theta_3) = 84°32'11.63''$$

15-4. To determine the effects of rising and falling stages on a particular stage discharge curve, the following measurements for falling and rising stages were taken at a constant value of rate of stage change (Table 15-7). The instrument used for the falling stage was new and, therefore, assumed subject to error, while the values for the rising stage are believed to be sensibly free from error.

TABLE 15-7

Measured discharge for falling stage in m^3/s, *y*	350	770	1240	1640	1980	2430	3000	3430	4020	4370
Measured discharge for rising stage in m^3/s, *x*	400	800	1300	1680	2010	2500	3050	3510	4050	4450

a. Determine the regression line of discharge for the falling stage upon discharge for the rising stage.
b. Test the hypothesis that the slope of the regression line is 1.00.
c. If the true slope = 1.01, what is the probability of accepting the hypothesis in (b)?
d. Estimate the discharge for the falling stage when the measured discharge for rising stage is 2010 m^3/s.
e. Find the 95 percent confidence limits for a single value of discharge for falling stage corresponding to a rising stage discharge of 2010 m^3/s.

SOLUTION

a. The constants in the regression line $y = a + bx$ will be found using Table 15-8.

TABLE 15-8

x	y	$x^2 \times 10^{-4}$	$xy \times 10^{-4}$	$X = x - \bar{x}$	$Y = y - \bar{y}$	$XY \times 10^{-4}$	$X^2 \times 10^{-4}$	$Y^2 \times 10^{-4}$
400	350	16.00	14.00	−1975	−1973	389.668	390.062	389.273
800	770	64.00	61.60	−1575	−1553	244.598	248.063	241.181
1300	1240	169.00	161.20	−1075	−1083	116.423	115.562	117.289
1680	1640	282.24	275.52	−695	−683	47.468	48.303	46.649
2010	1980	404.01	397.98	−365	−343	12.519	13.322	11.765
2500	2430	625.00	607.50	125	107	1.338	1.563	1.145
3050	3000	930.25	915.00	675	677	45.698	45.562	45.833
3510	3430	1232.01	1203.93	1135	1107	125.644	128.823	122.545
4050	4020	1640.25	1628.10	1675	1697	284.247	280.562	287.981
4450	4370	1980.25	1944.65	2075	2047	424.752	430.563	419.021
$\sum =$ 23,750	23,230	7343.01	7209.48	5685 − 5685 = 0	5635 − 5635 = 0	1692.355	1702.385	1682.682

$$\bar{x} = \frac{\sum x}{n} = \frac{23{,}750}{10} = 2375 \text{ m}^3/\text{s}$$

$$\bar{y} = \frac{\sum y}{n} = \frac{23{,}230}{10} = 2323 \text{ m}^3/\text{s}$$

From Eqs. (15-12) and (15-13),

$$a = \frac{\sum x^2 \sum y - \sum x \sum xy}{n \sum x^2 - (\sum x)^2}$$

$$= \frac{7343.01 \times 23{,}230 - 23{,}750 \times 7209.48}{10 \times 7343.01 - (237.5)^2}$$

$$= -38.007$$

and

$$b = \frac{n \sum xy - \sum x \sum y}{n \sum x^2 - (\sum x)^2}$$

$$= \frac{10 \times 7209.48 - 237.5 \times 232.3}{10 \times 7343.01 - (237.5)^2}$$

$$= 0.994$$

Therefore,

$$y = -38.007 + 0.994x \text{ [with respect to origin (0, 0)]}$$

Using the XY method

$$B = \frac{\sum XY}{\sum X^2} = \frac{1692.355}{1702.385} = 0.994$$

or

$$Y = 0.994X \text{ [with respect to centroid (2375, 2323) as origin]}$$

b. The sum of the squares of the y residuals $\sum \epsilon^2$ is obtained from Table 15-9.

TABLE 15-9

x	$\hat{y}$	$\epsilon = \hat{y} - y_{\text{observed}}$	ϵ^2
400	359.593	9.593	92.026
800	757.193	−12.807	164.019
1300	1254.193	14.193	201.441
1680	1631.913	−8.087	65.340
2010	1959.933	−20.067	402.684
2500	2446.993	16.993	288.762
3050	2993.693	−6.307	39.778
3510	3450.933	20.933	438.190
4050	3987.693	−32.307	1043.742
4450	4385.293	15.293	233.876
Total			$\sum \epsilon^2 = 2969.858$

$$s_{y|x} = \sqrt{\frac{\sum \epsilon^2}{n-2}} = \sqrt{\frac{2969.858}{8}} = 19.267$$

To test the hypothesis that $b_0 = 1.00$ calculate:

$$s_b = s_{y|x}\sqrt{\frac{1}{\sum X^2}} = \frac{19.267}{100}\sqrt{\frac{1}{1702.385}}$$

$$= 0.00467$$

Then

$$t = \frac{|b - b_0|}{s_b} = \frac{1.00 - 0.994}{0.00467} = 1.285$$

From statistical tables, for $v = n - 2 = 10 - 2 = 8$, $t = 1.108$ at a probability level of 30 percent and 1.397 at 20 percent. The difference in slope is, therefore, not significant and the null hypothesis can be considered valid.

c. Calculate

$$t = \frac{1.01 - 1.00}{s_b} = \frac{0.01}{0.00467} = 2.141$$

For $v = 8$, Table A-8 gives $t = 2.306$ at the 5 percent level of significance. Therefore, the probability of correctly accepting the hypothesis $b_0 = 1.00$ is a little more than 5 percent.

d. From the equation of the regression line for $x = 2010$ m^3/s

$$y = 1959.933 \text{ m}^3/\text{s}$$

e. The 95 percent confidence limits are

$$1959.933 \pm ts_{y_i}$$

where t is the value at the 5 percent level of significance for 8 degrees of freedom, and

$$s_{y_i} = s_{y|x}\sqrt{1 + \frac{1}{n} + \frac{(x_i - \bar{x})^2}{\sum X^2}}$$

Thus the limits are

$$1959.933 \pm 2.306 \times 19.267\sqrt{1 + \frac{1}{10} + \frac{(2010 - 2375)^2}{17{,}023{,}850}}$$

$$= 1959.933 \pm 46.784$$

$$= (1913.1, 2006.7) \text{ m}^3/\text{s}$$

15-5. In order to establish bench marks, precise leveling was carried out between four stations A, B, C, and D, as shown in Fig. 15-7. Because of various factors, the results were weighted as indicated in Table 15-10.

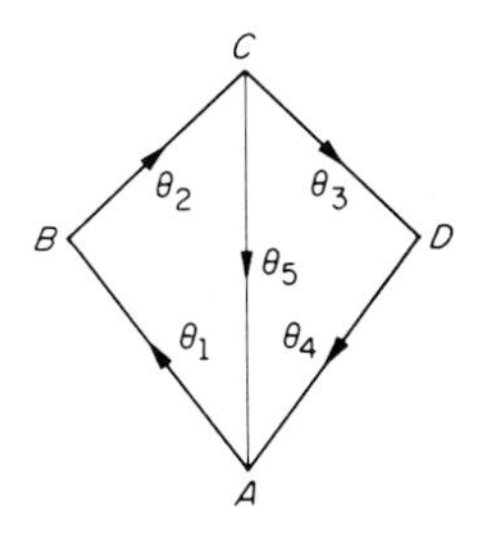

Fig. 15-7

TABLE 15-10

Leg	*Observed difference in elevation* α, *m*	*Weight* w
A to B	4.912 (rise)	2
B to C	2.638 (rise)	1
C to D	−6.382 (fall)	2
D to A	−1.075 (fall)	3
C to A	−7.450 (fall)	2

Using the method of least squares, find the most probable values of the differences in level between the various points.

SOLUTION

Let the most probable values of the differences in levels be $\theta_1, \theta_2, \theta_3, \theta_4$, and θ_5 for the legs numbered in Fig. 15-7. Considering the two closed circuits $ABCA$ and $ACDA$, with a common link AC, we have

$$\theta_1 + \theta_2 + \theta_5 = 0$$

and (15-35)

$$\theta_3 + \theta_4 - \theta_5 = 0$$

If $\epsilon_1, \epsilon_2, \epsilon_3, \epsilon_4$, and ϵ_5 are the errors in levels, then $\epsilon_1 = \alpha_1 - \theta_1, \epsilon_2 = \alpha_2 - \theta_2$, etc. These errors are not all independent, since they are related by means of Eq. (15-35). Thus only three errors are independent. Let us assume that ϵ_1 and ϵ_3 are dependent errors on ϵ_2, ϵ_4, and ϵ_5. Then, using Eq. (15-35), we have

$$\epsilon_1 = \alpha_1 + \theta_2 + \theta_5$$

and (15-36)

$$\epsilon_3 = \alpha_3 + \theta_4 - \theta_5$$

The method of least squares requires that the sum of weighted squares of the errors is a minimum, with respect to the independent, most probable differences in levels, θ_2, θ_4, and θ_5. Thus if

$$P = w_1\epsilon_1^2 + w_2\epsilon_2^2 + w_3\epsilon_3^2 + w_4\epsilon_4^2 + w_5\epsilon_5^2$$

then

$$\frac{\partial P}{\partial \theta_2} = \frac{\partial P}{\partial \theta_4} = \frac{\partial P}{\partial \theta_5} = 0$$

Writing these in full yields

$$w_1 \epsilon_1 \frac{\partial \epsilon_1}{\partial \theta_2} + w_2 \epsilon_2 \frac{\partial \epsilon_2}{\partial \theta_2} + w_3 \epsilon_3 \frac{\partial \epsilon_3}{\partial \theta_2} + w_4 \epsilon_4 \frac{\partial \epsilon_4}{\partial \theta_2} + w_5 \epsilon_5 \frac{\partial \epsilon_5}{\partial \theta_2} = 0$$

$$w_1 \epsilon_1 \frac{\partial \epsilon_1}{\partial \theta_4} + w_2 \epsilon_2 \frac{\partial \epsilon_2}{\partial \theta_4} + w_3 \epsilon_3 \frac{\partial \epsilon_3}{\partial \theta_4} + w_4 \epsilon_4 \frac{\partial \epsilon_4}{\partial \theta_4} + w_5 \epsilon_5 \frac{\partial \epsilon_5}{\partial \theta_4} = 0$$

$$w_1 \epsilon_1 \frac{\partial \epsilon_1}{\partial \theta_5} + w_2 \epsilon_2 \frac{\partial \epsilon_2}{\partial \theta_5} + w_3 \epsilon_3 \frac{\partial \epsilon_3}{\partial \theta_5} + w_4 \epsilon_4 \frac{\partial \epsilon_4}{\partial \theta_5} + w_5 \epsilon_5 \frac{\partial \epsilon_5}{\partial \theta_5} = 0$$

Hence we obtain the following three normal equations:

$$w_1 \epsilon_1 - w_2 \epsilon_2 = 0$$

$$w_3 \epsilon_3 - w_4 \epsilon_4 = 0$$

$$w_1 \epsilon_1 - w_3 \epsilon_3 - w_5 \epsilon_5 = 0$$

Substituting the numerical values of w and expressing ϵ in terms of α and the unknowns θ_2, θ_4, and θ_5, we obtain

$$3\theta_2 + 2\theta_5 = -7.186$$

$$5\theta_4 - 2\theta_5 = +9.539$$

$$\theta_2 - \theta_4 + 3\theta_5 = -18.744$$

The solution of these three simultaneous equations yields

$$\theta_2 = 2.584 \text{ m}, \qquad \theta_4 = -1.080 \text{ m}, \qquad \theta_5 = -7.469 \text{ m}$$

and from Eq. (15-35):

$$\theta_1 = 4.885 \text{ m}, \qquad \theta_3 = -6.389 \text{ m}$$

15-6. The design of certain metallic alloys for high temperature is usually based on stress rupture curves. In many cases the deformation involved is intolerable, and to ensure a life of a given number of hours it is necessary to keep the applied stress below the value that would produce rupture in the same number of hours. An experiment on one such alloy was run at 700°C, and the data listed in Table 15-11 were obtained relating applied stress to rupture time.

TABLE 15-11

Stress f, MN/m^2	*Rupture time T, h*
80	22
70	57
60	205
50	1324

a. Estimate the constants in the relation $\log T = a + bf$.
b. Estimate the rupture time by a point estimate corresponding to a stress $f = 65\ \text{MN/m}^2$.
c. Estimate the predicted rupture time by a 95 percent confidence interval for the stress given in (b).

SOLUTION

a. Let $y = \log T$ and $x = f$. (See Table 15-12.)

TABLE 15-12

T	y	x	x^2	xy
22	1.3424	80	6400	107.392
57	1.7559	70	4900	122.913
205	2.3118	60	3600	138.708
1324	3.1219	50	2500	156.095
$\sum =$	8.5320	260	17,400	525.108

From Eqs. (15-12) and (15-13), we obtain

$$a = \frac{\sum x^2 \sum y - \sum x \sum xy}{n \sum x^2 - (\sum x)^2} = \frac{17{,}400 \times 8.5320 - 260 \times 525.108}{4 \times 17{,}400 - 260 \times 260}$$

$$= 5.96436$$

and

$$b = \frac{n \sum xy - \sum x \sum y}{n \sum x^2 - (\sum x)^2} = \frac{4 \times 525.108 - 260 \times 8.5320}{4 \times 17{,}400 - 260 \times 260}$$

$$= -0.058944$$

Hence

$$\log T = 5.96436 - 0.058944f$$

b. For $f = 65\ \text{MN/m}^2$, the above equation for T gives

$$\log T = 5.96436 - 0.058944 \times 65 = 2.1330$$

$$T = 135.8\ \text{h}$$

c. In order to predict the required confidence interval, the variance of $\log T$, as estimated from the regression line, must be first computed. (See Table 15-13.)

TABLE 15-13

f, MN/m^2	$y(=\log T)$	$\hat{y}(=\log T$ *computed*)	$\epsilon_i = y_i - \hat{y}_i$	ϵ_i^2
80	1.3424	1.24884	+0.09356	0.00875347
70	1.7559	1.83828	−0.08238	0.00678646
60	2.3118	2.42772	−0.11592	0.01343745
50	3.1219	3.01716	+0.10474	0.01097047
Total				$\sum \epsilon_i^2 = 0.03994785$

From Eq. (15-22),

$$s_{y|x}^2 = \frac{\sum \epsilon_i^2}{n-2} = \frac{0.03994785}{2} = 0.0199739$$

or

$$s_{y|x} = 0.14133$$

Since the confidence interval is required for $x = 65$, and $\bar{x} = 65$, we have from Eq. (15-26):

$$s_{y_i} = 0.14133\sqrt{1 + \tfrac{1}{4}}$$
$$= 0.15799$$

Hence the confidence interval for a single value

$$= 2.1330 \pm t \times 0.15799$$
$$= 2.1330 \pm 4.303 \times 0.15799$$

since $t = 4.303$ for $\nu = 4 - 2$ at the 5 percent probability level. Thus the limits for $\log T$ are 2.8128 and 1.4531, and for T, 649.8 and 28.4 h.

Problems

15-1. In a laboratory experiment the lateral pressure and failure load were measured, with the following (coded) results:

Pressure x	0	1	2	3	4	5	8
Failure load y	10	10.7	12.1	12.6	13.8	16.2	18.9

Obtain the regression equation of y on x, and x on y, and plot these together with the experimental data.

15-2. Tests on the fuel consumption of a vehicle traveling at different speeds yielded the following (coded) results:

Speed x	20	30	40	50	60	70	80	90
Consumption y	18.3	18.8	19.1	19.3	19.5	19.7	19.8	20.0

It is believed that the relation between the two variables is of the type $y = a + b/x$. Obtain the equation to the regression line.

15-3. An experimental determination of the relation between x and y yielded the following results:

x	4	5	6	7	8	9	10	11
y	4	6	8	13	18	23	26	31

a. Find the equation to the regression line of y on x.
b. Estimate the 99 percent confidence limits of a predicted single observation of y when $x = 8.5$.

15-4. In a study of the relation between y and x (believed to be free from error), the following data were obtained: $n = 18$; $\sum (y - \bar{y})^2 = 720$; $\sum (x - \bar{x})^2 = 144$; $\sum (x - \bar{x})(y - \bar{y}) = 288$; and $\bar{x} = 5$; $a = 10$.

a. Obtain the equation to the regression line.
b. Test the hypothesis: $b = 0$.
c. Test the hypothesis: $a = 8$.
d. Obtain the 95 percent confidence interval for b.

Use the expression:

$$(n-2)s_{y|x}^2 = \sum (y - \bar{y})^2 - \frac{[\sum (x - \bar{x})(y - \bar{y})]^2}{\sum (x - \bar{x})^2}$$

(See Appendix E.)

15-5. Obtain the equation to the regression line of the modulus of elasticity (y) on content of a certain compound (x) in a plastic (see Table 15-14).

TABLE 15-14

x	55.8	55.0	54.3	49.9	53.0	50.6	58.3	63.7
y	83.9	66.4	73.1	30.1	36.2	66.7	87.3	135.0
x	67.0	65.0	58.8	57.6	57.5	54.4	54.2	55.8
y	153.1	158.2	65.8	72.1	83.1	72.1	71.3	58.0

Hence calculate the standard deviation of the regression line $s_{y|x}$. If $x = 60$, estimate the expected modulus of elasticity and determine a 95 percent confidence range for the expected modulus at this value of x. Find a prediction interval such that the probability is 95 percent that the value of the modulus of elasticity corresponding to $x = 60$ will lie within the interval.

15-6. Leveling was carried out from a station P at a known elevation of 234.15 m above datum, the weights of the different legs being given in Table 15-15. Find the most probable values of the levels of points Q, R, and S.

TABLE 15-15

Leg	*Difference in level, m*	*Weight*
P to Q	6.32 rise	1
Q to S	5.68 rise	2
Q to R	3.15 rise	1
R to S	2.59 rise	1
S to P	12.04 fall	2

15-7. A, B, C, and D form a round of angles at a station such that they add up to 360°. The observed values are $A = 82°15'35''$; $B = 110°37'45''$; $C = 66°24'40''$; and $D = 100°42'10''$.

The angle $(A + B)$ was measured separately twice, and the average value was found to be $192°53'25''$. If each of the six measurements is of equal reliability (weight), find the most probable values of all the angles.

15-8. The number of bacteria per unit volume found in a tillage after x hours is given in the following table:

Number of hours x	0	1	2	3	4	5	6	7
Number of bacteria y	47	64	81	107	151	209	298	841

a. Estimate the constants in the relation $\log y = a + bx$.
b. Estimate the number of bacteria per unit volume by a point estimate corresponding to $x = 4.5$ h.
c. Estimate the 99 percent confidence interval for the predicted number of bacteria per unit volume for $x = 4.5$ h.
d. Plot the data on semilogarithmic graph paper.

15-9. The strengths of concrete cylinders of the same proportions, but different size, were recorded as shown in Table 15-16. Obtain an equation to the curvilinear regression line relating the strength (y) and diameter (x).

TABLE 15-16

Diameter (in 2.54 cm)	2	3	6	8	12	18	24	36
Strength (coded)	108	106	100	96	92	86	84	84

Source: A. M. Neville, "Some Aspects of the Strength of Concrete," Part II, *Civil Engineering* (London), vol. 54, (Nov. 1959), p. 1309.

15-10. The compressive strengths of concrete specimens of three types are shown in Table 15-17.

TABLE 15-17

Test number	*Strength, MN/m²*		
	15 cm cubes	*12.7 cm × 12.7 cm cylinders*	*15 cm × 30 cm cylinders*
	52.4	49.3	44.3
1	52.3	51.0	43.6
	51.9	51.4	43.4
	44.7	43.1	36.1
2	44.2	45.5	37.1
	44.4	43.4	36.1
	55.3	55.5	46.3
3	55.8	55.5	44.4
	53.0	53.8	44.5
	43.6	42.4	37.6
4	43.9	42.7	37.8
	44.3	45.2	39.0
	42.1	43.1	38.5
5	46.1	45.5	38.5
	44.2	43.4	38.3
	41.2	44.1	37.8
6	44.2	40.0	38.3
	45.2	40.7	38.7
	46.5	47.6	40.7
7	45.5	45.2	40.7
	45.2	45.2	39.2
	49.3	47.6	42.2
8	48.8	47.9	42.9
	48.8	45.9	41.7
	42.4	39.6	37.1
9	43.4	39.0	36.6
	43.6	40.3	36.3
	50.7	50.0	43.4
10	—	51.7	43.8
	—	52.4	44.1
	37.9	40.0	35.9
11	42.7	44.8	37.2
	43.1	45.5	38.0
	44.5	45.2	38.7
12	43.6	45.2	38.5
	43.6	45.5	38.5
	41.0	41.4	34.8
13	41.8	42.1	35.2
	42.5	42.7	35.6

TABLE 15-17—*Continued*

Test number	*Strength, MN/m^2*		
	15 cm cubes	*12.7 cm × 12.7 cm cylinders*	*15 cm × 30 cm cylinders*
	32.5	31.4	27.7
14	32.5	31.0	27.2
	32.4	33.8	28.5
	22.6	26.5	22.1
15	22.3	26.9	22.4
	23.2	26.9	23.8

Source: M. W. Cormack, "Note on Cubes v. Cylinders," *New Zealand Engineering*, vol. 11, no. 3, March 1956, p. 99. The original data was in psi units.

Using the means of each test, obtain equations to the regression lines of (a) 15 cm cubes and 12.7 cm cylinders, and (b) 15 cm cubes and 15 cm cylinders. Test the significance of these relations. Test also the hypotheses that the slopes are for (a) 1.00 and (b) 0.85.

15-11. Experimental measurements of y and x are given in Table 15-18. It is expected that the relation is in the form $y = ax^b$. Fit the constants of the regression line.

TABLE 15-18

x	y	x	y
667	54	3619	106
727	42	3865	98
823	34	4266	261
1086	75	4299	197
1529	103	4382	106
1941	87	5560	216
2266	53	5955	251
2515	113	6358	347
3187	137	7165	339
3218	114	7910	282

15-12. From a certain station, angles A, B, and C are observed a number of times by the method of repetition, with the results, all of equal weight:

$$3A = 172°45'50''$$

$$3A + 3B = 400°25'30''$$

$$A + B + C = 225°46'30''$$

$$2B + 2C = 336°22'10''$$

$$3C = 276°52'50''$$

$$3A + 3C = 449°38'50''$$

$$3B = 227°40'10''$$

Find to the nearest second the most probable values of A, B, and C.

15-13. In establishing levels of B, C, and D from an ordnance bench mark at A, precise leveling was carried out from A to each point B, C, and D, and in addition between B and C and between C and D with the results given in Table 15-19.

TABLE 15-19

	Approximate distance between stations, km
B 10.714 m higher than A	2.0
A 51.762 m higher than C	1.5
D 32.840 m higher than A	1.0
B 62.463 m higher than C	1.5
D 84.624 m higher than C	1.5

Assuming that measurement errors are proportional to the square root of the distance, weight the observations and find the most probable values of the heights of B, C, and D with respect to A.

Chapter 16

Correlation

We should stress the fact that just because we have fitted a straight-line relation to a number of observations, it does not mean that the physical data really follow a straight line. For example, there may be a cyclic (or any other) relation with a general rise of y with x that could be represented by a straight line. An example of this is shown in Fig. 16-1, and it is clear that although we have fitted a line satisfying the minimum value of the sum of squares of deviations, their sum is large. We can distinguish thus between the deviations of the y observations from the regression line (this representing variation about the regression) and the total variation of the y observations about their mean. The difference between the two variations, expressed in an appropriate mathematical form, gives the amount of variation accounted for by regression, and the higher this amount the better the fit.

It is clear, therefore, that the operation of fitting the best line must be followed by a test of the goodness of fit. Before doing this, however, some more detailed comments on the variation about the regression line may be of value.

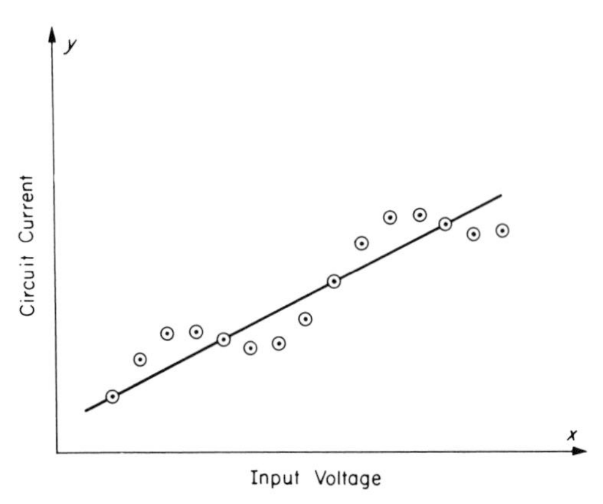

Fig. 16-1. Relation between y and x and regression line that would show poor correlation. The experimental points in the form shown would be obtained when a circuit contains resistors with two or more tunnel diodes in series.

Consider Fig. 16-2, which shows the regression line of y upon x; this line must, of course, pass through the point $(\bar{x}, \bar{y})$ (see p. 259). Let the point (x_i, y_i) represent an observed value of y_i at a given value of x_i. At this value of x_i, the regression line estimates the value of y as $\hat{y}_i$. Then the difference in ordinates $(y_i - \hat{y}_i) = (\Delta y)_1$ represents the deviation from the regression line or the variation that we wish to study.

More generally, the problem is to distinguish between that part of the variation in the dependent variable y_i which is associated with the relation between y and x and the part which is not associated. The deviation of y_i from $\bar{y}$, $(y_i - \bar{y}) = (\Delta y)_2$, can be seen from Fig. 16-2 to be equal to $(\Delta y)_1$ plus the distance $(\hat{y}_i - \bar{y}) = (\Delta y)_3$. Thus

$$(y_i - \bar{y}) = (y_i - \hat{y}_i) + (\hat{y}_i - \bar{y}) \tag{16-1}$$

or

$$(\Delta y)_2 = (\Delta y)_1 + (\Delta y)_3$$

The term $(\Delta y)_3$ depends on the slope of the regression line, i.e., is associated with the relation between y and x; therefore, this part of the deviation can be considered "explained." On the other hand, the term $(\Delta y)_1$ represents that part of the total deviation $(\Delta y)_2$ which exists over and above that explained by regression, and is therefore "unexplained."

Squaring both sides of Eq. (16-1) and summing for all observed points, we obtain

$$\sum (y_i - \bar{y})^2 = \sum (y_i - \hat{y}_i)^2 + \sum (\hat{y}_i - \bar{y})^2 \tag{16-2}$$

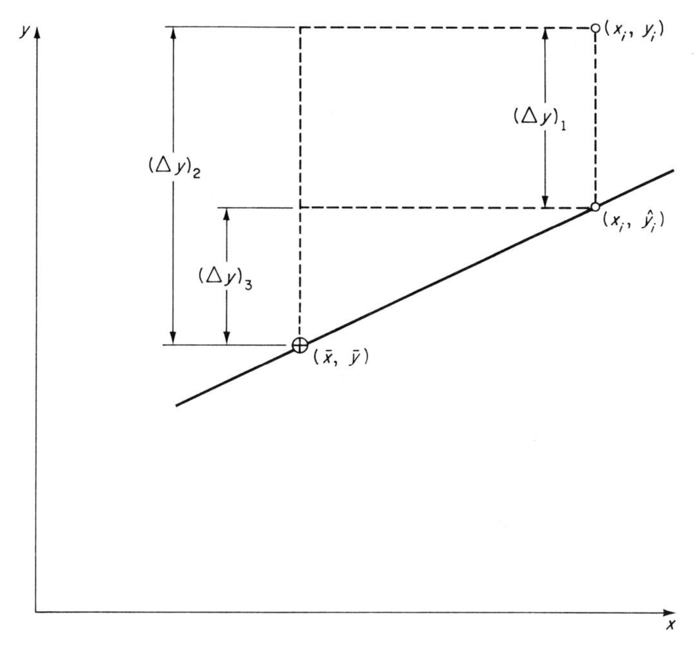

Fig. 16-2. Regression line of y upon x: "explained" and "unexplained" deviations.

We should note that the term $2 \sum (y_i - \hat{y}_i)(\hat{y}_i - \bar{y})$ does not appear, as it is equal to zero by virtue of the relation $\hat{y} = a + bx$, derived by the method of least squares.[1]

The left-hand side of Eq. (16-2) represents the total variation. Following the argument of the preceding paragraph, we can call the first term on the right-hand side the *unexplained variation* (or the residual sum of squares), and the second term the *explained variation*. Thus the total variation can be partitioned and the ratio of the explained variation to the total variation gives an indication of how well the regression line fits the observed data. This ratio is called the *coefficient of determination*, and must clearly lie between zero and unity. In the latter case, all variation has been explained and we have a perfect fit with all the points lying on the regression line. If the

1. Since $\hat{y}_i = a + bx_i$,

$$\begin{aligned}\sum (y_i - \hat{y}_i)(\hat{y}_i - \bar{y}) &= \sum (y_i - a - bx_i)(a + bx_i - \bar{y}) \\ &= a \sum (y_i - a - bx_i) + b \sum x_i(y_i - a - bx_i) \\ &\quad - \bar{y} \sum (y_i - a - bx_i)\end{aligned}$$

This is equal to zero by virtue of the normal equations (15-7) and (15-8).

coefficient is zero, the regression line does not explain anything; i.e., it is horizontal and y is not a function of x.

The coefficient of determination can also be calculated for nonlinear lines of fit as well as for linear and nonlinear multiple correlations.

We may note that the coefficient of determination is the square of the correlation coefficient considered in the next section.

CORRELATION COEFFICIENT

Referring to Eq. (15-14), we can observe that if there is no correlation between y and x, i.e., if y is independent of x, the coefficient of x (the slope b) is zero and the line plots as a horizontal line.

$$\frac{n\sum xy - \sum x \sum y}{n\sum x^2 - (\sum x)^2} = 0 \qquad (16\text{-}3)$$

If we consider now the regression of x on y, there is no correlation if x is independent of y, i.e., if the line described by Eq. (15-16) is vertical. Thus, referring the slope to a vertical axis, we obtain

$$\frac{n\sum xy - \sum x \sum y}{n\sum y^2 - (\sum y)^2} = 0 \qquad (16\text{-}4)$$

and expressing slope in the usual way (y vertical and x horizontal):

$$\frac{n\sum y^2 - (\sum y)^2}{n\sum xy - \sum x \sum y} = \infty \qquad (16\text{-}5)$$

If there is no correlation between the two variables being studied, the product of the slopes given by Eqs. (16-3) and (16-4) is zero,

$$\frac{n\sum xy - \sum x \sum y}{n\sum x^2 - (\sum x)^2} \times \frac{n\sum xy - \sum x \sum y}{n\sum y^2 - (\sum y)^2} = 0$$

Conversely, when there is a perfect correlation, i.e., all the points lie exactly on each of the two regression lines, the lines coincide; their slopes are therefore equal, namely,

$$\frac{n\sum xy - \sum x \sum y}{n\sum x^2 - (\sum x)^2} = \frac{n\sum y^2 - (\sum y)^2}{n\sum xy - \sum x \sum y}$$

or

$$\frac{n\sum xy - \sum x \sum y}{n\sum x^2 - (\sum x)^2} \times \frac{n\sum xy - \sum x \sum y}{n\sum y^2 - (\sum y)^2} = 1 \qquad (16\text{-}6)$$

Thus we find that the value of the product on the left-hand side of Eq. (16-6) gives a measure of correlation: When the value is zero, there is no correlation; when it is unity, the correlation is perfect. We call the square root of this product[2] the *correlation coefficient* and denote it by r:

$$r = \frac{n \sum xy - \sum x \sum y}{\sqrt{[n \sum x^2 - (\sum x)^2][n \sum y^2 - (\sum y)^2]}} \tag{16-7}$$

In terms of the variables X, Y referred to $(\bar{x}, \bar{y})$, r can be written as

$$r = \frac{\sum XY}{\sqrt{\sum X^2 \sum Y^2}} \tag{16-8}$$

Equation (16-8) can also be derived from the definition of the coefficient of determination (see p. 297). The correlation coefficient was given as

$$r = \sqrt{\frac{\text{explained variation}}{\text{total variation}}} = \sqrt{\frac{\sum (\hat{y}_i - \bar{y})^2}{\sum (y_i - \bar{y})^2}} \tag{16-9}$$

or, referred to the X, Y coordinate system,

$$r = \sqrt{\frac{\sum \hat{Y}^2}{\sum Y^2}} \tag{16-10}$$

Using Eqs. (15-18) and (15-20), we can write

$$\hat{Y} = bX \tag{16-11}$$

Now,

$$r^2 = \frac{\sum \hat{Y}^2}{\sum Y^2} = \frac{b^2 \sum X^2}{\sum Y^2}$$

Substituting for b from Eq. (15-18), we find that

$$r^2 = \frac{(\sum XY)^2 \sum X^2}{(\sum X^2)^2 \sum Y^2} = \frac{(\sum XY)^2}{\sum X^2 \sum Y^2}$$

Hence

$$r = \frac{\sum XY}{\sqrt{\sum X^2 \sum Y^2}}$$

which is Eq. (16-8).

We can note that r is symmetrical with respect to x and y so that the correlation coefficient of a line of regression of y on x is the same as that of regression of x on y. Correlation is, in fact, concerned only with the association between the variables and not with their dependence or independence.

The correlation coefficient r must lie in the range $0 \leq |r| \leq 1$, but in practice, because of random errors, $0 < |r| < 1$.

2. Or the geometric mean of two values.

To interpret the meaning of r calculated by Eq. (16-7) or Eq. (16-8), we use Table A-11. This gives the values of r that can be expected at a given level of significance from observations drawn by chance when there is no correlation. If the absolute value of the calculated r exceeds the tabulated value, we conclude that correlation exists, and the level of significance represents the probability of our having drawn the wrong conclusion. We may note that r is related to t by the equation $t = r\sqrt{n-2}/\sqrt{1-r^2}$ when y is a function of one independent linear variable x (see Appendix E).

Table A-11 is entered with the appropriate number of degrees of freedom and the total number of variables. In the case of regression of the type given by Eq. (15-1), the total number of variables is two and the number of degrees of freedom is $v = n - 2$, where n is the total number of observations (readings x_i, y_i).

The sign of r tells us whether y increases with an increase in x (r is positive) or whether y decreases with an increase in x (r is negative). Figure 16-3 shows diagrammatically some of the possible correlations.

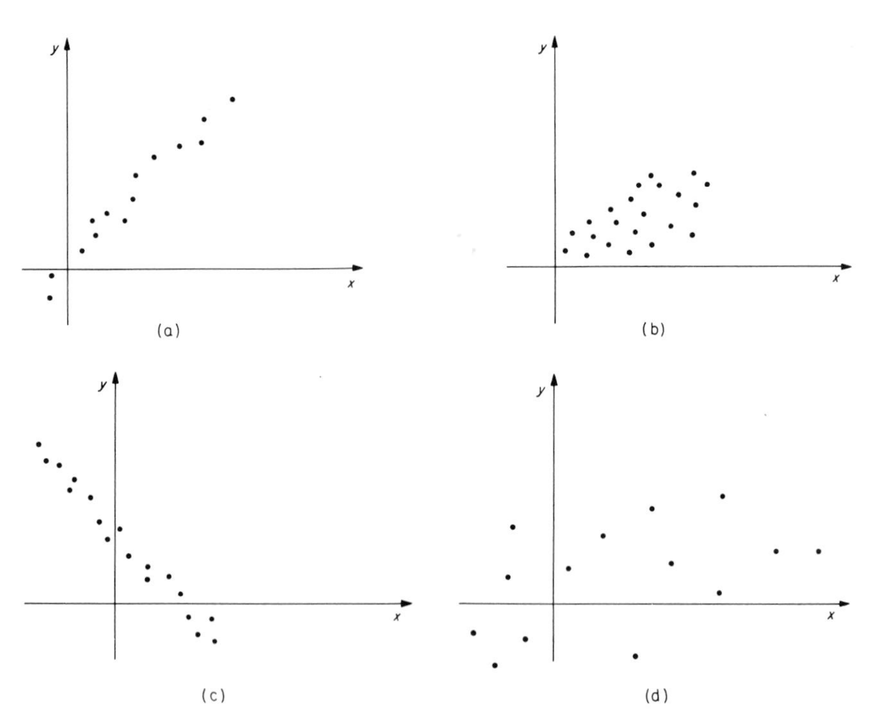

Fig. 16-3. Correlations: (a) high positive; (b) low positive; (c) high negative; and (d) none (the variables are independent).

It can be shown that the square of the correlation coefficient is equal to the ratio of the sum of squares of deviations accounted for by regression to the total sum of squares of deviations about the mean. The correlation coefficient is thus an estimate of association between the variables and is valid only when the observations are randomly drawn.

The level of significance of correlation given by the correlation coefficient is the same as the level of significance of the slope of the regression line given by the t test of Eq. (15-32). (See Appendix E.)

Example. Find the correlation coefficient r for the first example in Chapter 15 on the relation between normal stress and shear resistance of a soil.

To calculate r given by Eq. (16-8), we require $\sum XY$, $\sum X^2$ and $\sum Y^2$. (See Table 16-1.)

TABLE 16-1

X	Y	XY	X^2	Y^2
−5	−5.3	26.5	25	28.09
−3	−2.8	8.4	9	7.84
−1	−1.2	1.2	1	1.44
+1	+1.0	1.0	1	1.00
+3	+3.4	10.2	9	11.56
+5	+4.9	24.5	25	24.01
$\sum = 0$	0	71.8	70	73.94

Hence

$$r = \frac{71.8}{\sqrt{70 \times 73.94}} = 0.998$$

For two variables and $\nu = 6 - 2 = 4$, Table A-11 gives $r = 0.917$ at the 1 percent level of significance. The value at the 0.1 percent level is 0.974. Thus there is an excellent correlation between the normal stress and shear resistance of the soil tested.

CORRELATION AND CAUSATION

We have seen how the correlation coefficient is used to find the degree of association that exists between two variables y and x. A high correlation coefficient proves the existence of a close mathematical relation between the variables, but it is important to realize that this does not necessarily imply causation. For example, if we find correlation between some measured quantity, such as speed of reaction, and time (say, between 8 A.M. and noon), it does not follow that the observed quantity is a function of time of operation. It may well be that time is not the true *independent* variable but enters the picture only insofar as it is associated with temperature, which naturally rises as the day progresses. It is thus the temperature that is the causative variable.

On the other hand, when a low correlation coefficient is obtained, the conclusion that the independent variable does not influence the quantity being measured is not necessarily correct. It is possible that other factors exist, masking or nullifying the effect of the independent variable on the dependent variable. In either case the determination of causation involves a scientific study of the subject, possibly using additional experimental data, statistics being merely one of the more powerful tools in arriving at the right answer.

Solved Problem

16-1. Find the correlation coefficient for Solved Problem 15-4.

SOLUTION

The coefficient of correlation is given by

$$r = \frac{\sum XY}{\sqrt{\sum X^2 \sum Y^2}} = \frac{1692.355}{\sqrt{1702.385 \times 1682.682}}$$

$$= 0.9999$$

This means that the correlation is virtually perfect.

Problems

16-1. Find the correlation coefficient of the regression of Problem 15-5 and determine the significance of the regression.

16-2. Is the correlation between y and x significant?

x	56	58	60	70	72	75	77	77	82	87	92	104	125
y	51	60	69	54	70	65	49	60	63	61	64	84	75

16-3. A correlation coefficient based on a sample of size 16 was calculated to be 0.38. Determine whether we can conclude that the corresponding population correlation differs from zero at a significance level of (a) 5 percent, (b) 1 percent. (Use $t = r\sqrt{n-2}/\sqrt{1-r^2}$ in Appendix E.)

16-4. Find by a trial method the minimum sample size necessary in order to conclude that a correlation coefficient of 0.38 differs significantly from zero at the 1 percent level.

16-5. The compressive strength of concretes with three different aggregate-cement ratios was measured on four test specimens for each mix (see Table 16-2). Using the mean strength of each mix, obtain the regression line of strength on aggregate-cement ratio, and the correlation coefficient. Test the significance of the slope.

TABLE 16-2

Specimen number \ *Aggregate cement ratio*	*Strength, MN/m²*		
	4	*5*	*6*
1	40.2	45.4	43.7
2	42.3	42.5	48.5
3	40.8	44.5	48.1
4	40.0	43.9	45.7

Source: P. J. F. Wright, "Statistical Methods in Concrete Research," *Magazine of Concrete Research*, vol. 5, no. 15, March 1954, p. 147. The original data was in psi units.

Chapter 17

Multiple Linear Regression

In many practical cases a variable may depend on more than one independent variable. If the independent variables vary entirely randomly, we can use simple regression as in the case of one independent variable, although this leads to some loss in the precision of our estimate. However, if the independent variables tend to vary according to some pattern, simple regression leads to misleading results, and multiple regression has to be used. The difference between the two methods lies in the fact that multiple regression establishes the effect of one independent variable with the other independent variables kept constant, while simple regression does not control the other variables.

An example of a problem involving multiple regression is afforded by the influence of the temperatures of air and of the coolant on the efficiency of an engine. Because of the influence of weather, the two temperatures are likely to be low or high at the same time, and a single correlation cannot eliminate the effect of one variable when the effect of the other is measured; it is the multiple correlation that achieves this.

REGRESSION EQUATION

Let us consider the general case of a linear relation between the mean value of the dependent variable y and independent variables $x_1, x_2, \ldots, x_k$; this can be expressed as

$$y = b_0 + b_1 x_1 + b_2 x_2 + \cdots + b_k x_k \qquad (17\text{-}1)$$

where b_0 is a constant and $b_1, b_2, \ldots, b_k$ are partial regression coefficients. Equation 17-1 represents a plane in $(k + 1)$ dimensions.

As shown in Appendix F, the plane passes through the centroid of all the observed values, i.e., Eq. (17-1) is satisfied by $(\bar{y}, \bar{x}_1, \bar{x}_2, \ldots, \bar{x}_k)$, where $\bar{y}, \bar{x}_1, \ldots, \bar{x}_k$ are means of all the values of $y, x_1, \ldots, x_k$, respectively. Hence

$$\bar{y} = b_0 + b_1 \bar{x}_1 + b_2 \bar{x}_2 + \cdots + b_k \bar{x}_k$$

or

$$b_0 = \bar{y} - b_1 \bar{x}_1 - b_2 \bar{x}_2 - \cdots - b_k \bar{x}_k$$

Substituting in Eq. (17-1), we obtain

$$y - \bar{y} = b_1(x_1 - \bar{x}_1) + b_2(x_2 - \bar{x}_2) + \cdots + b_k(x_k - \bar{x}_k) \qquad (17\text{-}2)$$

The coefficients are determined using the method of least squares; for simplicity we shall consider the case of two independent variables only, i.e.,

$$y = b_0 + b_1 x_1 + b_2 x_2$$

with n sets of observations. In each case, the residual is given as

$$\epsilon = y - (b_0 + b_1 x_1 + b_2 x_2) \qquad (17\text{-}3)$$

and the sum of squares of residuals in the n sets is

$$\sum \epsilon^2 = \sum [y - (b_0 + b_1 x_1 + b_2 x_2)]^2 \qquad (17\text{-}4)$$

Using the principle of least squares, we minimize $\sum \epsilon^2$, i.e., we satisfy the condition that the partial derivatives of $\sum \epsilon^2$ with respect to b_0, b_1, and b_2 are all zero:

$$\frac{\partial(\sum \epsilon^2)}{\partial b_0} = -2 \sum [y - (b_0 + b_1 x_1 + b_2 x_2)] = 0$$

$$\frac{\partial(\sum \epsilon^2)}{\partial b_1} = -2 \sum x_1[y - (b_0 + b_1 x_1 + b_2 x_2)] = 0$$

$$\frac{\partial(\sum \epsilon^2)}{\partial b_2} = -2 \sum x_2[y - (b_0 + b_1 x_1 + b_2 x_2)] = 0$$

Hence the normal equations can be written:

$$\left.\begin{aligned} \sum y &= nb_0 + b_1 \sum x_1 + b_2 \sum x_2 \\ \sum x_1 y &= b_0 \sum x_1 + b_1 \sum x_1^2 + b_2 \sum x_1 x_2 \\ \sum x_2 y &= b_0 \sum x_2 + b_1 \sum x_1 x_2 + b_2 \sum x_2^2 \end{aligned}\right\} \tag{17-5}$$

The solution of this system of three simultaneous equations gives the values of b_0, b_1, and b_2.

SHORTER COMPUTATION

We can simplify the process of finding the equation to the regression plane by choosing the centroid as origin. Let

$$\begin{aligned} Y &= y - \bar{y} \\ X_1 &= x_1 - \bar{x}_1 \\ \vdots \quad & \quad \vdots \quad \quad \vdots \\ X_k &= x_k - \bar{x}_k \end{aligned}$$

Thus Eq. (17-2) becomes

$$Y = b_1 X_1 + b_2 X_2 + \cdots + b_k X_k$$

and the residual

$$\epsilon = Y - (b_1 X_1 + b_2 X_2 + \cdots + b_k X_k)$$

Considering two independent variables X_1 and X_2 only and taking partial derivatives of $\sum \epsilon^2$ with respect to b_1 and b_2, we obtain

$$\begin{aligned} \sum X_1 Y &= b_1 \sum X_1^2 + b_2 \sum X_1 X_2 \\ \sum X_2 Y &= b_1 \sum X_1 X_2 + b_2 \sum X_2^2 \end{aligned} \tag{17-6}$$

The solution of the system of these two equations gives the values of the regression coefficients, and hence the equation to the regression plane.

Example. From an experimental study on the stabilization of a highly plastic clay, molding water content for optimum density was found to be linearly dependent on the percentages of lime and pozzolan mixed with the clay. The results in Table 17-1 were obtained. Fit an equation of the form $y = b_0 + b_1 x_1 + b_2 x_2$ to the data in Table 17-1.

TABLE 17-1

Water content in percent y	*Percent of lime* x_1	*Percent of pozzolan* x_2
27.5	2.0	18.0
28.0	3.5	16.5
28.8	4.5	10.5
29.1	2.5	2.5
30.0	8.5	9.0
31.0	10.5	4.5
32.0	13.5	1.5

The equation can readily be obtained by means of Table 17-2. Using Eq. (17-6), we find that

$$118.21170b_1 - 110.54078b_2 = 40.99563$$

and

$$-110.54078b_1 + 258.23270b_2 = -53.20647$$

Solving for b_1 and b_2 by elimination, we get

$$b_1 = 0.257004, \qquad b_2 = -0.096026$$

Now,

$$\bar{y} = b_0 + b_1\bar{x}_1 + b_2\bar{x}_2$$

Substituting for b_1, b_2, $\bar{y}$, $\bar{x}_1$, and $\bar{x}_2$, we obtain

$$b_0 = 28.691$$

Thus the required equation is

$$y = 28.691 + 0.257x_1 - 0.0960x_2$$

USE OF MATRICES

As we have seen, multiple linear regression leads to a set of simultaneous equations, which have to be solved. Such simultaneous equations can be represented in the following compact matrix form:

$$AX = B$$

where A and B are matrices obtained from the experimental data, and X is the unknown vector (the regression coefficients to be determined).

In the case when only two coefficients b_1 and b_2 are involved,

$$A = \begin{bmatrix} \sum X_1^2 & \sum X_1X_2 \\ \sum X_1X_2 & \sum X_2^2 \end{bmatrix} \qquad X = \begin{bmatrix} b_1 \\ b_2 \end{bmatrix} \qquad B = \begin{bmatrix} \sum X_1Y \\ \sum X_2Y \end{bmatrix} \tag{17-7}$$

TABLE 17-2

y	x_1	x_2	$Y = y - \bar{y}$	$X_1 = x_1 - \bar{x}_1$	$X_2 = x_2 - \bar{x}_2$	YX_1	YX_2	X_1^2	X_2^2	X_1X_2
27.5	2.0	18.0	−1.985	−4.428	9.072	8.78958	−18.00792	19.60718	82.30118	−40.17081
28.0	3.5	16.5	−1.486	−2.929	7.572	4.35239	−11.25199	8.57904	57.33518	−22.17838
28.8	4.5	10.5	−0.686	−1.929	1.571	1.32329	−1.07771	3.72104	2.46804	−3.03046
29.1	2.5	2.5	−0.386	−3.928	−6.429	1.51621	2.48159	15.42918	41.33204	25.25311
30.0	8.5	9.0	0.514	2.071	0.071	1.06449	0.03649	4.28904	0.00504	0.14704
31.0	10.5	4.5	1.514	4.071	−4.429	6.16349	−6.70551	16.57304	19.61604	−18.03046
32.0	13.5	1.5	2.515	7.072	−7.428	17.78608	−18.68142	50.01318	55.17518	−52.53082
$\sum$ = 206.4	45.0	62.5	0	0	0	40.99563	−53.20647	118.21170	258.23270	−110.54078

$$\bar{y} = \frac{\sum y}{n} = \frac{206.4}{7} = 29.486$$

$$\bar{x}_1 = \frac{\sum x_1}{n} = \frac{45.0}{7} = 6.429$$

$$\bar{x}_2 = \frac{\sum x_2}{n} = \frac{62.5}{7} = 8.929$$

Writing the matrix equation in full, we have

$$b_1 \sum X_1^2 + b_2 \sum X_1 X_2 = \sum X_1 Y$$
$$b_1 \sum X_1 X_2 + b_2 \sum X_2^2 = \sum X_2 Y \qquad [17\text{-}6]$$

The matrices A and B are written in a similar manner for more than two unknowns.

In many instances the inverse of matrix A, denoted by A^{-1}, is required. The inverse of a matrix is defined by the relation:

$$A^{-1}A = \text{unit matrix}$$

Writing this in full, and if

$$A^{-1} = \begin{bmatrix} e_{11} & e_{12} \\ e_{21} & e_{22} \end{bmatrix} \qquad (17\text{-}8)$$

then we obtain

$$\begin{bmatrix} e_{11} & e_{12} \\ e_{21} & e_{22} \end{bmatrix} \begin{bmatrix} \sum X_1^2 & \sum X_1 X_2 \\ \sum X_1 X_2 & \sum X_2^2 \end{bmatrix} = \begin{bmatrix} 1 & 0 \\ 0 & 1 \end{bmatrix} \qquad (17\text{-}9)$$

The simultaneous equations are

$$\left.\begin{aligned} e_{11} \sum X_1^2 + e_{12} \sum X_1 X_2 &= 1 \\ e_{11} \sum X_1 X_2 + e_{12} \sum X_2^2 &= 0 \\ e_{21} \sum X_1^2 + e_{22} \sum X_1 X_2 &= 0 \\ e_{21} \sum X_1 X_2 + e_{22} \sum X_2^2 &= 1 \end{aligned}\right\} \qquad (17\text{-}10)$$

The solution of Eq. (17-10) gives

$$\left.\begin{aligned} e_{11} &= \frac{\sum X_2^2}{\sum X_1^2 \sum X_2^2 - (\sum X_1 X_2)^2} \\ e_{22} &= \frac{\sum X_1^2}{\sum X_1^2 \sum X_2^2 - (\sum X_1 X_2)^2} \\ e_{12} = e_{21} &= \frac{-\sum X_1 X_2}{\sum X_1^2 \sum X_2^2 - (\sum X_1 X_2)^2} \end{aligned}\right\} \qquad (17\text{-}11)$$

provided that the determinant of $A \neq 0$, i.e., the denominator in Eq. (17-11) is different from zero. Otherwise, values for e_{11}, e_{22}, e_{12}, and e_{21} will be infinite.

There are two main uses of finding A^{-1} [Eq. (17-8)]. The first is when we want to check whether there is an association between the independent variables that have been used previously and a new dependent variable, e.g., between the water-cement ratio of concrete, the specific gravity of aggregate, and the density of the resulting concrete. If there is an association, then the matrix A is unchanged and so is A^{-1}. Thus the unknown vector X can easily be found in the following manner.

Multiply both sides of $AX = B$ by A^{-1}. Thus

$$A^{-1}AX = A^{-1}B$$

But $A^{-1}A =$ unit matrix. Therefore,

$$X = A^{-1}B$$

Then

$$\begin{bmatrix} b_1 \\ b_2 \end{bmatrix} = \begin{bmatrix} e_{11} & e_{12} \\ e_{21} & e_{22} \end{bmatrix} \begin{bmatrix} \sum X_1 Y \\ \sum X_2 Y \end{bmatrix} \tag{17-12}$$

or

$$\left.\begin{aligned} b_1 &= e_{11} \sum X_1 Y + e_{12} \sum X_2 Y \\ b_2 &= e_{21} \sum X_1 Y + e_{22} \sum X_2 Y \end{aligned}\right\} \tag{17-13}$$

The second main use of the inverse matrix A^{-1} is in calculating the standard error of the partial regression coefficients, and hence the confidence intervals for such coefficients.

CONFIDENCE LIMITS OF A PARTIAL REGRESSION COEFFICIENT

To obtain the standard deviation of each partial regression coefficient, we need the residual variance of the dependent variable y, denoted by $s^2_{y|x}$. We can write

$$s^2_{y|x} = \frac{\sum \epsilon_i^2}{n - (k + 1)} \tag{17-14}$$

where $\epsilon_i =$ deviation of observed value of y from the value given by the regression plane, i.e.,

$$\epsilon_i = y - \hat{y}$$

where $\hat{y}$ is the estimate of y from the regression plane, n the number of observations of y, and k the number of independent variables on which y depends.

The denominator term $n - (k + 1)$ is, in fact, the number of degrees of freedom to compute $s^2_{y|x}$, since $(k + 1)$ parameters $(b_0, b_1, \ldots, b_k)$ were estimated from the data to yield the estimate $\hat{y}$. Hence there are $(k + 1)$ constraints in computing ϵ_i in Eq. (17-14).

For two independent variables, x_1 and x_2, the residual variance becomes, after shifting the origin to the centroid ($\bar{y}$, $\bar{x}_1$, $\bar{x}_2$),

$$s_{y|x}^2 = \frac{\sum \epsilon_i^2}{n-(k+1)} = \frac{\sum (Y-\hat{Y})^2}{n-(2+1)} = \frac{\sum [Y-(b_1X_1+b_2X_2)]^2}{n-3}$$

$\hat{Y}$ being the estimated value given by the regression equation. Expanding the term in the brackets, we find that

$$s_{y|x}^2 = \frac{\sum Y^2 - 2b_1 \sum YX_1 - 2b_2 \sum YX_2 + b_1^2 \sum X_1^2 + b_2^2 \sum X_2^2 + 2b_1b_2 \sum X_1X_2}{n-3}$$

Multiplying the first of Eq. (17-6) by b_1 and the second by b_2, and substituting the values of $b_1^2 \sum X_1^2$ and $b_2^2 \sum X_2^2$ in the above expression, we obtain

$$s_{y|x}^2 = \frac{\sum Y^2 - b_1 \sum YX_1 - b_2 \sum YX_2}{n-3} \tag{17-15}$$

The standard deviation of a partial regression coefficient is estimated from the sample as

$$s_{b_j} = s_{y|x}\sqrt{e_{jj}} \tag{17-16}$$

where in the general case $j = 1, 2, \ldots, k$ and e_{jj} is the corresponding diagonal element of A^{-1}. Thus, for the two regression coefficients b_1 and b_2, we have

$$s_{b_1} = s_{y|x}\sqrt{e_{11}}$$

and (17-17)

$$s_{b_2} = s_{y|x}\sqrt{e_{22}}$$

To obtain the standard deviation of the constant b_0 in Eq. (17-1), but for simplicity taking the terms up to b_2x_2 only, we write

$$b_0 = \bar{y} - b_1\bar{x}_1 - b_2\bar{x}_2$$

Hence the variance of b_0 is

$$\sigma_{b_0}^2 = \sigma_{\bar{y}|x}^2 + \bar{x}_1^2\sigma_{b_1}^2 + \bar{x}_2^2\sigma_{b_2}^2 + 2\bar{x}_1\bar{x}_2 \text{ cov } (b_1b_2)$$

where cov (b_1b_2) is the covariance of x_1 and x_2, and the procedure is based on the argument of Eq. (14-12).

Thus the standard deviation of the constant b_0 is estimated from the sample by

$$s_{b_0} = s_{y|x}\left(\frac{1}{n} + \bar{x}_1^2e_{11} + \bar{x}_2^2e_{22} + 2\bar{x}_1\bar{x}_2e_{12}\right)^{1/2} \tag{17-18}$$

We can now test the significance of the regression coefficients. This is important, since we may have assumed independent variables that do not significantly influence y. The significance of b_1 is tested by

$$t = \frac{b_1}{s_{b_1}} \tag{17-19}$$

Similar tests are applied to the other coefficients. The number of degrees of freedom is $n - 3$, as 3 constraints were used in fixing the plane, i.e., in determining the values of $\bar{y}$, b_1, and b_2 (or b_0, b_1, and b_2).

If a regression coefficient is found not to be statistically significant, we have to revise our equation. The independent variable that does not significantly influence the dependent variable is deleted, and new regression coefficients are computed.

If b_1 is significant, its confidence interval is given by $b_1 \pm t s_{b_1}$.

If we want to establish whether a regression coefficient b_j differs significantly from a value (e.g., a theoretical value) b_j^0, we apply the t test:

$$t = \frac{|b_j - b_j^0|}{s_b} \tag{17-20}$$

The null hypothesis is rejected at the stipulated level of significance if t exceeds the critical value given in Table A-8, with $v = n - (k + 1)$ degrees of freedom.

SIGNIFICANCE OF MULTIPLE REGRESSION AS A WHOLE

Sometimes, the assumed regression equation may prove to be statistically not significant. Whether this is so is determined by a comparison of the variance contributed by the regression and the error variance $s_{y|x}^2$, using the F test.

For three coefficients, b_0, b_1, and b_2 ($k = 2$) the sum of squares of deviations in y accounted for by regression is

$$\sum c^2 = b_1 \sum YX_1 + b_2 \sum YX_2$$

Therefore, we compute F as

$$F = \frac{(\sum c^2)/k}{s_{y|x}^2} \tag{17-21}$$

with the number of degrees of freedom $v_1 = k = 2$ for the numerator, since there are only two parameters, and $v_2 = n - (k + 1) = n - 3$ for the denominator.

If the computed F is greater than the value tabulated in Table A-10 for the given level of significance, then the hypothesis that all the true partial regression coefficients are equal to zero is rejected.

MULTIPLE-CORRELATION COEFFICIENT

The square of the population multiple-correlation coefficient is defined as the fraction of the total variance of y which is contributed by its regression upon the variables $x_1, x_2, \ldots$. This coefficient may be estimated from the square of the sample multiple-correlation coefficient:

$$r^2 = \frac{\sum c^2}{\sum Y^2} \tag{17-22}$$

where r is the multiple-correlation coefficient. As in the case of simple linear regression, a value of zero gives no correlation between y and the variables $x_1, x_2, \ldots,$ whereas a value of 1 means that all the sample points lie exactly on the regression plane (in the case of three independent variables).

To test the significance of r, we use Table A-11, with the total number of variables of $(k + 1)$, and the number of degrees of freedom equal to $v = n - (k + 1)$. We reject the null hypothesis that the population multiple correlation coefficient is zero if $|r|$ exceeds the tabulated value at the specified level of significance. When the hypothesis is rejected, we say that the regression of y on the variables $x_1, x_2, \ldots$ accounts for a significant amount of variation in y.

The degree of association between any two variables can be checked in a manner similar to that described in Chapter 16. Thus

$$r_{yx1} = \frac{\sum X_1 Y}{\sqrt{\sum X_1^2 \sum Y^2}}, \qquad r_{yx2} = \frac{\sum X_2 Y}{\sqrt{\sum X_2^2 \sum Y^2}}$$

$$r_{x1x2} = \frac{\sum X_1 X_2}{\sqrt{\sum X_1^2 \sum X_2^2}}, \qquad \text{etc.} \tag{17-23}$$

Solved Problem

17-1. An experiment was conducted at the University of Saskatchewan to determine the relation between the thermal conductivity of sandy textured soils and their moisture content and dry density. The data in Table 17-3 were collected in the field by means of a thermal conductivity probe and a Uhland core sampler (ρ denotes the dry density of soil in kg/m^3).

TABLE 17-3

Sample number	*Thermal conductivity,* $W \cdot m/m^2 \cdot °C$ *K*	*Moisture content, by volume,* cm^3/m	$e^{0.45\rho}$
1	1.14	96.8	10.80
2	0.97	45.2	14.20
3	1.36	46.8	17.90
4	0.75	70.4	10.80
5	0.68	75.8	8.30
6	1.67	190.3	11.13
7	0.98	157.0	6.48
8	1.14	144.1	6.00
9	0.72	62.9	12.10
10	1.40	255.9	6.87
11	1.85	195.2	11.20
12	1.93	95.2	12.80
13	1.11	150.0	9.65
14	1.66	149.5	12.80

a. Fit an equation of the form:

$$K = a_0 + a_1 M + a_2 e^{0.45\rho}$$

b. Use the F test to check whether or not this form of an equation is statistically significant. Use $\alpha = 1$ percent.

c. Use the t test to check the significance of the partial regression coefficients at the 1 percent level.

d. Calculate the multiple-correlation coefficient r and test its significance. Use $\alpha = 1$ percent.

SOLUTION

a. The computation is shown in Table 17-4.

Equations (17-6) become (a being used instead of b):

$$52357.08a_1 - 1519.01a_2 = 169.62$$

$$-1519.01a_1 + 136.66a_2 = 4.39$$

Solving for a_1 and a_2 by elimination, we get $a_1 = 0.00616$ and $a_2 = 0.101$. Now

$$\bar{y} = a_0 + a_1\bar{x}_1 + a_2\bar{x}_2$$

or

$$1.24 = a_0 + 0.00616 \times 123.92 + 0.101 \times 10.79$$

Hence $a_0 = -0.608$.

TABLE 17-4

Sample Number	$y \equiv K$	$x_1 \equiv M$	$x_2 \equiv e^{0.45\rho}$	$Y = y - \bar{y}$	$X_1 = x_1 - \bar{x}_1$	$X_2 = x_2 - \bar{x}_2$	YX_1	YX_2	X_1^2	X_2^2	X_1X_2	Y^2
1	1.14	96.77	10.80	−0.10	−27.15	0.01	2.72	0.00	737.12	0.00	−0.27	0.01
2	0.97	45.16	14.20	−0.27	−78.76	3.41	21.27	−0.92	6203.14	11.63	−268.57	0.07
3	1.36	46.77	17.90	0.12	−77.15	7.11	−9.26	0.85	5952.12	50.55	−548.54	0.01
4	0.75	70.43	10.80	−0.49	−53.49	0.01	26.21	0.00	2861.18	0.00	−0.53	0.24
5	0.68	75.81	8.3	−0.56	−48.11	−2.49	26.94	1.39	2314.57	6.20	119.79	0.31
6	1.67	190.32	11.13	0.43	66.40	0.34	28.55	0.15	4408.96	0.12	22.58	0.18
7	0.98	156.99	6.48	−0.26	33.07	−4.31	−8.60	1.12	1093.62	18.58	−142.53	0.07
8	1.14	144.09	6.00	−0.10	20.17	−4.79	−2.02	0.48	406.83	22.94	−96.61	0.01
9	0.72	62.90	12.10	−0.52	−61.02	1.31	31.73	−0.68	3723.44	1.72	−79.94	0.27
10	1.40	255.91	6.87	0.16	131.99	−3.92	21.12	−0.63	17421.36	15.37	−517.40	0.03
11	1.85	195.16	11.20	0.61	71.24	0.41	43.46	0.25	5075.14	0.17	29.21	0.37
12	1.93	95.16	12.80	0.69	−28.76	2.01	−19.84	1.39	827.14	4.04	−57.81	0.48
13	1.11	150.00	9.65	−0.13	26.08	−1.14	−3.39	0.15	680.17	1.30	−29.73	0.02
14	1.66	149.46	12.80	0.42	25.54	2.01	10.73	0.84	652.29	4.04	51.34	0.18
$\Sigma =$	17.36	1734.93	151.03	0	$\simeq 0$	$\simeq 0$	169.62	4.39	52357.08	136.66	−1519.01	2.25

$$\bar{y} = \frac{17.36}{14} = 1.24$$

$$\bar{x}_1 = \frac{1734.93}{14} = 123.92$$

$$\bar{x}_2 = \frac{151.03}{14} = 10.79$$

Therefore, we can write

$$y = -0.608 + 0.00616x_1 + 0.101x_2$$

or

$$K = -0.608 + 0.00616M + 0.101e^{0.45\rho}$$

b. Compute

$$\begin{aligned}\sum c^2 &= a_1 \sum YX_1 + a_2 \sum YX_2 \\ &= 0.00616 \times 169.62 + 0.101 \times 4.39 \\ &= 1.48\end{aligned}$$

From Eq. (17-15)

$$s_{y|x}^2 = \frac{2.25 - 1.48}{14 - 3} = 0.069$$

Hence

$$s_{y|x} = 0.26\ W \cdot m/m^2 \cdot {}^\circ C.$$

From Eq. (17-21)

$$F = \frac{\sum c^2/k}{s_{y|x}^2} = \frac{1.48/2}{0.069} = 10.7$$

For $\nu_1 = 2$, $\nu_2 = n - 3 = 14 - 3 = 11$, Table A-10 gives $F = 7.20$ at the 1 percent level of significance. Since the calculated $F = 10.7$ is greater than 7.20, we reject the hypothesis that the regression is not significant. The thermal conductivity depends, therefore, on moisture content and dry density.

c. Now

$$\begin{aligned}\sum X_1^2 \sum X_2^2 - \left(\sum X_1 X_2\right)^2 &= 52{,}357.08 \times 136.66 - (-1519.01)^2 \\ &= 4{,}847{,}727.2\end{aligned}$$

Using Eq. (17-11), we obtain

$$e_{11} = \frac{136.66}{4{,}847{,}727.2} = 0.0000282$$

$$e_{22} = \frac{52{,}357.08}{4{,}847{,}727.2} = 0.0108$$

Then from Eq. (17-17),

$$s_{a_1} = 0.26 \times \sqrt{0.0000282} = 0.0014$$

$$s_{a_2} = 0.26 \times \sqrt{0.0108} = 0.027$$

From Eq. (17-19), for a_1,

$$t_{a_1} = \frac{|a_1|}{s_{a_1}} = \frac{0.00616}{0.0014} = 4.40$$

From Table A-8, for $v = n - (k + 1) = 14 - (2 + 1) = 11$, $t = 3.106$ at the 1 percent level of significance and 4.437 at the 0.1 percent level. Since $t_{a_1} > 3.106$, the coefficient a_1 is significant at the 1 percent level and, indeed, at nearly the 0.1 percent level. Also for a_2,

$$t_{a_2} = \frac{|a_2|}{s_{a_2}} = \frac{0.101}{0.027} = 3.67$$

Again a_2 is significant at the 1 percent level since $t_{a_2} > 3.106$.

d. From Eq. (17-22), the square of the multiple correlation coefficient is

$$r^2 = \frac{\sum c^2}{\sum Y^2} = \frac{1.48}{2.25} = 0.660$$

Hence

$$r = 0.812$$

From Table A-11, with $k + 1 = 3$ variables and $v = n - (k + 1) = 14 - (2 + 1) = 11$, $r = 0.753$ at the 1 percent level of significance. Since the calculated value is greater than the tabulated one, we conclude that the regression of y on the x variables accounts for a significant amount of variation in y.

Problems

17-1. It is believed that the extent of a certain reaction (y) depends on the temperature of the ingredient $A(x_1)$, temperature of ingredient $B(x_2)$, and rate of flow (x_3), the relation being of the form $y = a + b_1 x_1 + b_2 x_2 + b_3 x_3$.

The test results are as shown in Table 17-5.

a. Determine the constants of the hyperplane in four-dimensional space.

b. Use the F test to check the significance of this form of regression equation. Use $\alpha = 5$ percent.

TABLE 17-5

x_1	x_2	x_3	y
11	58	11	126
32	21	13	92
15	22	28	107
26	55	27	120
9	41	21	103
31	18	20	84
12	56	20	113
29	40	27	110
13	57	30	104
10	21	12	83
33	40	19	85
31	58	29	104

c. Use the t test to check the significance of the partial regression coefficients b_1, b_2, and b_3 at the 5 percent level.
d. Compute the multiple correlation coefficient r and test its significance. Use $\alpha = 5$ percent.
e. Find the linear correlation coefficients for y and x_1, x_2, and x_3, respectively.

17-2. Grains used to propel rockets are made by extrusion through a die under pressure. The grain diameter y is dependent not only on the die shape, but also on the powder temperature x_1, the die temperature x_2, and the rate of extrusion x_3. Previous experiments have indicated that there is a relation of the form

$$y = a + b_1 x_1 + b_2 x_2 + b_3 x_3$$

An experiment was conducted on a particular type of grain with the coded results shown in Table 17-6.

TABLE 17-6

x_1, °C	x_2, °C	x_3, *cm/min*	y
21	41	12	81
35	29	15	92
31	30	24	105
20	35	21	101
25	31	19	97
37	47	13	93
30	45	16	85
34	31	25	87
29	34	22	102
22	37	9	94
27	28	8	86
33	39	14	84
30	33	17	109
28	38	23	110
23	36	18	103

a. Find the partial regression coefficients a, b_1, b_2, and b_3.
b. Compute the residual variance of the grain diameter y.
c. Check on the significance of the regression as a whole by means of an F test. Use $\alpha = 5$ percent.
d. Use the t test to check the significance of the partial regression coefficients b_1, b_2, and b_3 at the 1 percent level of significance.
e. Calculate the multiple correlation coefficient r and test its significance. Use $\alpha = 5$ percent.

Chapter 18

Analysis of Variance

In Chapter 14 we showed that if a process of manufacture or a system of testing involves a number of independent factors each of which contributes to the scatter of results, and therefore to variance, then the variance for the whole system is equal to the sum of the component variances of the individual factors. It is important to remember that it is variance and not standard deviation that is additive.

METHOD OF ANALYSIS

This property of variance is the basis of a numerical technique, known as *analysis of variance*, which enables us to compute the variance of the component factors and to assess the relative importance of the various components. For example, if we take k samples of a product, each sample consisting of n items, the analysis of variance enables us to split the variance of all kn

items into variance *between samples* (due to variation in the process, say, from day to day) and variance *within samples* (which represents the inherent variation, or the experimental error). Each variance is calculated as the sum of squares of deviations divided by the appropriate number of degrees of freedom, and the variances are compared by the F test. The procedure is best illustrated by an example.

Example. Four different air injection systems ($k = 4$) are used, and we want to test whether there is a significant difference between them. We choose $n = 5$ items of each system and measure the efficiency of the injection in each item. The results can be tabulated as shown in Table 18-1. The table gives three estimates of variance. The first is the overall variance, which is based on the total sum of squares of deviations for all $kn = 4 \times 5$ observations and thus represents the variance of all individuals considered as forming a single sample.

TABLE 18-1

Sample (system)	*A*	*B*	*C*	*D*	*All systems*
Efficiency x	35	21	35	21	112
	24	31	27	17	99
	46	17	39	21	123
	30	37	20	23	110
	40	29	29	28	126
$\sum x$	175	135	150	110	570
$\bar{x}$	35	27	30	22	28.5
$\sum x^2$	6,417	3,901	4,716	2,484	17,518
$\frac{(\sum x)^2}{n}$	6,125	3,645	4,500	2,420	16,245
Sum of squares of deviations[a] = $\sum x^2 - \frac{(\sum x)^2}{n}$	292	256	216	64	1,273

[a] Since $\sum (x - \bar{x})^2 = \sum x^2 - [(\sum x)^2/n]$ (see Chapter 4).

The right-hand column of the table gives this sum of squares as 1273 and the number of degrees of freedom is $(4 \times 5) - 1 = 19$. Hence the mean square is $1273/19 = 67.0$.

The second estimate of population variance is obtained from the sum of squares of deviations within the samples and is the sum of values that are obtained in calculating the variance for each group separately, namely,

$$292 + 256 + 216 + 64 = 828$$

The number of degrees of freedom is the sum of the numbers for each sample, i.e., $4 \times (5 - 1) = 16$. Hence the mean square is $\frac{828}{16} = 51.75$.

The third and last estimate is obtained from the sample mean. Their deviations from the mean of means, $\bar{\bar{x}} = 28.5$, are $35 - 28.5$, $27 - 28.5$, $30 - 28.5$, and $22 - 28.5$; i.e., 6.5, -1.5, 1.5, -6.5. The sum of squares of deviations is thus $6.5^2 + 1.5^2 + 1.5^2 + 6.5^2 = 89$. The number of degrees of freedom is one less than the number of samples: 3. Thus the mean square is $\frac{89}{3} = 29.67$. This is an estimate of variance of the mean $s_{\bar{x}}^2$ of $n = 5$ items. This variance is related to the sample variance s^2 by the equation:

$$s_{\bar{x}} = \frac{s}{\sqrt{n}} \tag{18-1}$$

so that $s^2 = 29.67 \times 5 = 148.35$. The estimate is based on 3 degrees of freedom so that the sum of squares is $148.35 \times 3 = 445.05$.

These results are summarized in Table 18-2.

TABLE 18-2

Source of variance	*Sum of squares*	*Degrees of freedom*	*Mean square*
Within samples (systems)	828	16	51.75
Between samples (systems)	445	3	148.35
Total	1273	19	67.0

The row of totals shows that both the total sum of squares and the total number of degrees of freedom have been separated into two parts corresponding to the factors in the variation of the data.

The last column gives the *mean squares*, which are ratios of the sum of squares (column 2) to the appropriate number of degrees of freedom (column 3).

USUAL METHOD OF COMPUTATION

The method just outlined explains the analysis of variance but is longer than necessary for routine use. A shorter way of computing the mean squares is to omit the calculations for the individual observations so that only the values of $\sum x$ need be found in the table of the preceding example. We proceed as follows:

a. Find the sum of all observations:

$$\sum x = 175 + 135 + 150 + 110 = 570$$

b. Find the term:[1]

$$\frac{(\sum x)^2}{kn} = \frac{570^2}{4 \times 5} = 16{,}245$$

c. Find the sum of squares:

$$\sum x^2 = 35^2 + 24^2 + \cdots + 21^2 + 31^2 + \cdots + 23^2 + 28^2 = 17{,}518$$

d. Hence obtain the total sum of squares of deviations:

$$\sum x^2 - \frac{(\sum x)^2}{kn} = 17{,}518 - 16{,}245 = 1273$$

e. Compute the sum of squares for sample means:

$$\frac{\sum (\sum x)^2}{n} - \frac{(\sum x)^2}{kn} = \frac{175^2 + 135^2 + 150^2 + 110^2}{5} - 16{,}245 = 445$$

We can arrange these results as in Table 18-3, the numbers of degrees of freedom being as before. The values for "within samples" are obtained by subtraction, and the mean squares are calculated by dividing the appropriate sum of squares by the number of degrees of freedom.

TABLE 18-3

Source of variance	*Sum of squares*	*Degrees of freedom*	*Mean square*
Between samples	445	3	148.35
Within samples	828	16	51.75
Total	1273	19	—

Such a computation is quicker, but it does not offer a check on arithmetic.

We may note that it is usual to arrange the table for the analysis of variance in such a way that the "Totals" appear in the bottom line, i.e., the subtraction is made "upward."

TEST ON HOMOGENEITY OF VARIANCES

Having obtained the two values of mean squares, we test their homogeneity by means of the F test. Both these values of mean squares are estimates of the variance of the population, so that if the F test indicates no significant

1. Known as "correction due to the mean."

difference, we would conclude that the different air injection systems do not introduce a variation in excess of the variation between individual tests for the same injection system. This is so in our case, since

$$F = \frac{148.35}{51.75} = 2.86$$

with 3 degrees of freedom for the numerator and 16 degrees of freedom for the denominator. Table A-10 gives $F = 3.24$ at the 5 percent level of significance. We conclude, therefore, that the different injection systems do not differ in their efficiency.

We should note that the test we apply is one-sided[2] as we want to answer the question: Is the variance between samples significantly greater than the variance within samples? It cannot be the other way round, as the scatter in any arrangement cannot be less than the random variation between individuals.

It is, of course, possible in a particular case for the variance within samples to be greater than the variance between samples, but the difference cannot be significant.

SIMPLIFYING THE COMPUTATIONS

In the analysis of variance we are primarily concerned with comparing variances so that reducing the data in a constant proportion does not affect our conclusions. We can, therefore, simplify the data and achieve a considerable saving in computation by the subtraction of a constant number or a division by a constant number, or both. If the actual values of variance are required, the division has to be taken into account. The use of this simplification, known as coding, is illustrated in one of the examples in Chapter 3.

MULTIFACTOR ANALYSIS

The analysis of variance can be extended to cases in which a number of factors affect the observations. As a simple case of three components, we have, say, k different cements, each of which is tested once by n operators. We want to analyze the variance into the components: "between operators," "between cements," and "residual." The variance between operators is that which would be obtained if there were no variation between the individual cements and no inherent variation in the method of test. Similarly, the variance between cements is that which would be obtained if there were no

2. The F test is a one-sided test.

variation between the operators and no inherent variation in the method of test. The residual includes the error variance (which is similar to the variance within samples in the preceding example) and also the effect of interaction between the variables. The latter is the influence of the variation in one variable on another, for example, if the effect of different operators varies with the type of cement. (We can imagine that operator A tends to read "high" when values are high but reads "low" when values are low.) If we want to estimate the error variance free from interactions, it would be necessary to repeat the tests—an operation known as *replication*. This is considered in Chapter 21.

It should be realized that the present chapter is no more than an introduction to the analysis of variance, one of the most powerful methods of statistical analysis.[3]

Solved Problems

18-1. The influence of angle of dip of strata in a certain area on the form of the drainage basins was investigated. A ridge was divided into segments named "low dip," "medium-low dip," etc. In each segment, the length of stream channels on the ridge flanks was taken from the dip slope. The data on the lengths of streams in kilometers are summarized in Table 18-4. Does the steepness of the dip influence the length of the stream, i.e., do the means of all samples belong to the same population? Use $\alpha = 1$ percent.

TABLE 18-4

Low dip	*Medium-low dip*	*Medium dip*	*Steep dip*
$\bar{x} = 0.261$	$\bar{x} = 0.296$	$\bar{x} = 0.312$	$\bar{x} = 0.135$
$s = 0.21$	$s = 0.17$	$s = 0.19$	$s = 0.08$
$n = 44$	$n = 44$	$n = 44$	$n = 44$

SOLUTION

We set up a null hypothesis of no significant difference between the four means. Calculate

$$F = \frac{\text{variance from sample means}}{\text{average variance within the samples}}$$

Now,

$$\bar{\bar{x}} = \frac{0.261 + 0.296 + 0.312 + 0.135}{4} = 0.251$$

3. It may be noted that the t test is the simplest case of the analysis of variance.

Thus

$$s_{\bar{x}}^2 = \frac{(0.261 - 0.251)^2 + (0.296 - 0.251)^2 + (0.312 - 0.251)^2 + (0.135 - 0.251)^2}{4 - 1}$$

$$= 0.00643$$

But

$$s^2 = ns_{\bar{x}}^2 = 44 \times 0.00643$$
$$= 0.2831$$

The estimate of variance from the individual measurements within the samples is

$$= \frac{n_1 s_1^2 + n_2 s_2^2 + n_3 s_3^2 + n_4 s_4^2}{n_1 + n_2 + n_3 + n_4 - 4}$$

$$= \frac{44[(0.21)^2 + (0.17)^2 + (0.19)^2 + (0.08)^2]}{4 \times 44 - 4}$$

$$= 0.0295$$

Therefore,

$$F = \frac{0.2831}{0.0295} = 9.5816$$

for $v_1 = 3$ and $v_2 = 4 \times 44 - 4 = 172$, Table A-10 indicates that the probability for such a value of F occurring by chance is less than 1 percent. Therefore, the difference in means is very significant, and it is concluded that the steepness of the dip influences the length of the stream.

18-2. An experiment was carried out to measure the strain sensitivity of Stresscoat for different curing temperatures, which were applied with no sensible error. We wish to determine at the 1 percent level of significance whether the observed differences in the mean values of the strain sensitivity for the different temperatures have been influenced by random sampling errors. The results shown in Tables 18-5 and 18-6 were obtained.

TABLE 18-5

	Curing temperature, °C				
	80	92	105	118	132
Strain sensitivity ($\times 10^{-5}$)	83	75	56	50	49
	70	62	59	52	35
	78	70	48	38	48
	71	81	54	42	47
	73	71	61	53	41
	81	79	58	41	38

Note: Number of columns, $k = 5$; sample size in each column $n = 6$; and total number of observations $kn = 30$.

TABLE 18-6

	Curing temperature, °C					*Totals for all temperatures*
	80	92	105	118	132	
$\sum x$	456	438	336	276	258	1,764
$\bar{x}$	76	73	56	46	43	$\bar{\bar{x}} = 58.8$
$\sum x^2$	34,804	32,212	18,922	12,902	11,264	110,104
$(\sum x)^2/n$	34,656	31,974	18,816	12,696	11,094	103,723.2
$\sum x^2 - \frac{(\sum x)^2}{n}$	148	238	106	206	170	6,380.8

The mean square is

$$\frac{6380.8}{kn - 1} = \frac{6380.8}{29} = 220.03$$

The total for the sum of squares of the samples is

$$148 + 238 + 106 + 206 + 170 = 868$$

with the number of degrees of freedom $v = k(n - 1) = 5(6 - 1) = 25$. Therefore,

$$\text{mean square} = \frac{868}{25} = 34.72$$

The mean square deviation of sample means from the mean of means is

$$s_{\bar{x}}^2 = \frac{(76 - 58.8)^2 + (73 - 58.8)^2 + (56 - 58.8)^2 + (46 - 58.8)^2 + (43 - 58.8)^2}{5 - 1}$$

$$= \frac{918.80}{4} = 229.7$$

Hence

$$s^2 = s_{\bar{x}}^2 \times n = 229.7 \times 6 = 1378.2$$

Summarizing, we obtain the results shown in Table 18-7.

TABLE 18-7

Source of variance	*Sum of squares*	*Degrees of freedom*	*Mean square*
Within samples	868	25	34.72
Between samples	5512.8	4	1378.2
Total	6380.8	29	220.03

The variance ratio is

$$F = \frac{1378.2}{34.72} = 39.69$$

For the degrees of freedom $\nu_1 = 4$, $\nu_2 = 25$, Table A-10 gives at the 1 percent level of significance $F = 4.18$. Since the calculated $F \gg 4.18$, it can be confidently concluded that the curing temperature influences the coating strain sensitivity; i.e., the observed difference in the means is not due to random errors.

Using the simplified and more usual method of computation, we obtain the following.

1. The sum of all observations:

$$\sum x = 456 + 438 + 336 + 276 + 258$$
$$= 1764$$

2. The correction for the mean:

$$\frac{(\sum x)^2}{kn} = \frac{(1764)^2}{30} = 103{,}723.2$$

3. The total sum of squares:

$$\sum x^2 - \frac{(\sum x)^2}{kn} = 83^2 + 70^2 + \cdots + 38^2 - 103{,}723.2 = 6380.8$$

4. The sum of squares for sample means:

$$\frac{\sum (\sum x)^2}{n} - \frac{(\sum x)^2}{kn} = \frac{456^2 + 438^2 + 336^2 + 276^2 + 258^2}{6} - 103{,}723.2$$
$$= 5512.8$$

The results are shown in Table 18-8.

TABLE 18-8

Source of variance	*Sum of squares*	*Degrees of freedom*	*Mean square*
Between samples	5512.8	4	1378.2
Within samples	868.0	25	34.72
Total	6380.8	29	—

Then continue as before.

18-3. The 24-h water absorption (in percent of dry weight) of samples of concrete taken from five different types of precast concrete curbs made with different aggregates is shown in Table 18-9.

TABLE 18-9

Curb type	*A*	*B*	*C*	*D*	*E*	*All*
Absorption x, percent	6.7	5.1	4.4	6.7	6.5	
	5.8	4.7	4.9	7.2	5.8	
	5.8	5.1	4.6	6.8	4.7	
	5.5	5.2	4.5	6.3	5.9	
$\sum x$	23.8	20.1	18.4	27.0	22.9	112.2

Source: P. J. F. Wright, "Statistical Methods in Concrete Research," *Magazine of Concrete Research*, vol. 5, no. 15, March 1954.

It is required to determine whether there is a significant difference in the water absorption values of the curbs of different types. Use $\alpha = 1$ percent.

SOLUTION

a. The sum of all observations,

$$\sum x = 112.2$$

b. Correction due to the mean,

$$C = \frac{(\sum x)^2}{kn} = \frac{(112.2)^2}{20} = 629.44$$

where k is the number of types of curbs and n the sample size.

c. The total sum of squares

$$\begin{aligned} &= \sum x^2 - C \\ &= 6.7^2 + 5.8^2 + \cdots + 5.9^2 - C \\ &= 643.80 - 629.44 \\ &= 14.36 \end{aligned}$$

d. The sum of squares for sample means

$$\begin{aligned} &= \frac{\sum (\sum x)^2}{n} - C \\ &= \frac{23.8^2 + 20.1^2 + \cdots + 22.9^2}{4} - C \\ &= 640.60 - 629.44 \\ &= 11.16 \end{aligned}$$

(See Table 18-10.)

TABLE 18-10

Source of variance	Sum of squares	Degrees of freedom	Mean square
Between samples	11.16	4	2.79
Within samples	3.20	15	0.213
Total	14.36	19	—

$$\text{variance ratio, } F = \frac{2.79}{0.213} = 13.1$$

For $v_1 = 4$ and $v_2 = 15$, Table A-10 gives $F = 4.89$ at the 1 percent level of significance; therefore, we conclude that the different types of curbs differ significantly in their absorption values.

18-4. Creep, after a given period of time, was determined on mortar specimens made with 11 cements. It is suggested that creep is a function of the chemical composition of cement in the form:

$$y = a_1x_1 + a_2x_2 + a_3x_3 + a_4x_4 + a_5x_5 \tag{18-2}$$

where y = creep, and $x_1, \ldots, x_5$ are the compounds of which the cement consists, expressed as a percentage of weight. Thus the coefficients $a_1, \ldots, a_5$ represent the contribution to creep of 1 percent of the appropriate compound. This type of relationship implies that the same amount of any compound has the same effect, regardless of the total quantity of this or any other compound present. This also means that there is no interaction between the various compounds—an assumption not in disagreement with our general knowledge of the properties of cement. From physical considerations of the problem, the constant a_0 in the equation relating y to $x_1, x_2, \ldots, x_5$ is zero; i.e., the line passes through the origin. If this constraint is not imposed,

TABLE 18-11

Cement number	x_1	x_2	x_3	x_4	x_5	y	$\sum x$
1	42.26	25.31	15.65	6.92	0.75	52	90.89
2	36.63	30.05	15.32	9.10	0.83	50	91.93
3	26.18	28.13	14.97	16.13	0.75	35	86.16
4	37.30	27.29	9.53	10.61	0.65	37	85.68
5	36.62	28.42	14.69	11.14	0.78	49	91.65
6	32.64	28.06	13.88	11.58	0.72	45	86.88
7	34.65	33.28	12.52	5.47	0.56	32	86.48
8	35.11	30.53	10.76	12.10	0.74	47	89.24
9	24.50	36.38	12.14	12.56	0.33	32	85.91
10	41.07	32.04	3.73	13.27	0.54	48	90.65
11	29.09	42.06	5.30	11.30	0.45	42	88.20

the sum of squares of residuals is minimized about the mean (see p. 305), which from the purely statistical viewpoint is correct, but we must *remember* that the choice of the function should be governed by the physical nature of the problem.

On the basis of the values y and x given in Table 18-11, we want to determine the coefficients in Eq. 18-2 and to establish their significance. Use $\alpha = 1$ percent.

SOLUTION

In the matrix for the least square analysis, the variables are entered in the expected order of importance (x_1, x_5, x_2, x_3, and x_4); this order does not, of course, affect the values of the regression coefficients.

For the sum of squares of residuals to be a minimum, the partial differential coefficients of the sum with respect to the coefficients $a_1, \ldots, a_5$ must all be zero. Hence

$$\sum yx_1 = a_1 \sum x_1^2 + a_2 \sum x_1 x_2 + a_3 \sum x_1 x_3 + a_4 \sum x_1 x_4 + a_5 \sum x_1 x_5$$

$$\vdots$$

$$\sum yx_5 = a_1 \sum x_5 x_1 + a_2 \sum x_5 x_2 + a_3 \sum x_5 x_3 + a_4 \sum x_5 x_4 + a_5 \sum x_5^2$$

This can be written in matrix form as:

$$\begin{bmatrix} \sum x_1^2 & \sum x_1 x_5 & \sum x_1 x_2 & \sum x_1 x_3 & \sum x_1 x_4 \\ \sum x_5 x_1 & \sum x_5^2 & \sum x_5 x_2 & \sum x_5 x_3 & \sum x_5 x_4 \\ \sum x_2 x_1 & \sum x_2 x_5 & \sum x_2^2 & \sum x_2 x_3 & \sum x_2 x_4 \\ \sum x_3 x_1 & \sum x_3 x_5 & \sum x_3 x_2 & \sum x_3^2 & \sum x_3 x_4 \\ \sum x_4 x_1 & \sum x_4 x_5 & \sum x_4 x_2 & \sum x_4 x_3 & \sum x_4^2 \end{bmatrix} \begin{bmatrix} a_1 \\ a_5 \\ a_2 \\ a_3 \\ a_4 \end{bmatrix} = \begin{bmatrix} \sum x_1 y \\ \sum x_5 y \\ \sum x_2 y \\ \sum x_3 y \\ \sum x_4 y \end{bmatrix}$$

or $AX = Y$, where A is the known matrix on the left, X is the unknown vector, and Y is the known vector on the right. The numerical form of matrix A is shown in Table 18-12.

TABLE 18-12

x_1	x_5	x_2	x_3	x_4	y
13,177.29	246.78	11,537.07	4,377.31	4,025.53	16,324.69
246.78	4.83	214.59	86.71	77.18	309.50
11,537.07	214.59	10,833.43	3,873.86	3,742.25	14,436.63
4,377.31	86.71	3,873.81	1,664.91	1,375.21	5,497.24
4,025.53	77.18	3,742.25	1,375.21	1,399.51	5,096.13

The least squares analysis yields:

$$a_1 = 0.6339570$$

$$a_2 = 0.1794424$$

$$a_3 = 0.2073463$$

$$a_4 = -0.2681848$$

$$a_5 = 24.4356348$$

Thus the expected creep of mortar for the given conditions and range of composition of cements is:

$$y = 0.63x_1 + 0.18x_2 + 0.21x_3 - 0.27x_4 + 24.44x_5$$

The statistical significance of the coefficients of Eq. 18-2 can now be tested by the analysis of variance. For each coefficient in turn, we find the sum of squares accounted for by that coefficient. Thus if $\hat{y}$ is the value of y predicted by

$$\hat{y} = a_1 x_1$$

and since

$$\bar{y} = a_1 \bar{x}$$

we have

$$\hat{y} - \bar{y} = a_1(x_1 - \bar{x})$$

The total sum of squares is $\sum (y - \bar{y})^2$. This consists of $\sum (\hat{y} - \bar{y})^2$ accounted for by regression and a residual of $\sum (y - \hat{y})^2$, i.e.,

$$\sum (y - \bar{y})^2 = \sum (\hat{y} - \bar{y})^2 + \sum (y - \hat{y})^2$$

[since $\sum (\hat{y} - \bar{y}) = 0$].

Thus the sum of squares accounted for by x_1 is

$$\sum (\hat{y} - \bar{y})^2 = \frac{[\sum xy - (\sum x \sum y/n)]^2}{\sum (x - \bar{x})^2}$$

The residual sum of squares is

$$\sum (y - \hat{y})^2 = \sum (y - \bar{y})^2 - \sum (\hat{y} - \bar{y})^2$$

The F test applied to the ratio of the mean square accounted for by the regression coefficient to the mean square of the residual gives the significance of the regression coefficient. (See Table 18-13.)

TABLE 18-13

Source of variance	*Sum of squares*	*Degrees of freedom*	*Mean square*	*Level of significance, in percent*
Total (sum of squares of y values)	552.54443	11	—	
Regression on x_1	263.94629	1	263.95	1
Residual after fitting x_1	288.59814	10	28.86	
Reduction on fitting x_5	187.97366	1	187.97	1
Residual after fitting x_1, x_5	100.62448	9	11.18	
Reduction on fitting x_2	69.30536	1	69.31	1
Residual after fitting x_1, x_5, x_2	31.31912	8	3.91	
Reduction on fitting x_3, x_4	11.17663	2	5.59	Not significant
Residual	20.14249	6	3.36	

It can be seen that the variables x_1, x_2, and x_5 are significant at the 1 percent level; once these coefficients have been fitted, the variables x_3 and x_4 do not contribute significantly to the regression.

The regression equation may, therefore, be modified to

$$y = a_1 x_1 + a_2 x_2 + a_5 x_5$$

The new regression coefficients then become

$$a_1 = 0.6562734$$

$$a_2 = 0.2367266$$

$$a_5 = 20.0406494$$

Hence

$$y = 0.66x_1 + 0.24x_2 + 20.04x_5$$

The 95 percent confidence limits for these coefficients were found to be 0.66 ± 0.798, 0.24 ± 1.350, and 20.04 ± 39.90, respectively. These show that creep cannot be reliably predicted on the basis of the data available. It seems probable, however, that the compounds x_1 and x_2 have a similar influence on creep, and the influence of the compound x_5 is not negligible, even though its percentage content is comparatively low.

From the table of original data, it can be seen that the sum of the values of x_1, x_2, x_3, x_4, and x_5 is approximately constant, and that the values of $(x_1 + x_2)$ and $(x_3 + x_4)$ are complementary; the order of magnitude of the values of x_5 is small compared with the other variables. Hence a high value of $(x_1 + x_2)$ means a low value of $(x_3 + x_4)$, and vice versa, and there is a linear relationship between these two quantities. Thus creep may be a function of either quantity, the other one being looked upon as a complementary filling.

Problems

18-1. The resistance of wire from six sources was tested by taking five samples from each coil (source). (See Table 18-14.) Is there a significant difference between the resistances of the wires from the six sources? Use $\alpha = 1$ percent.

TABLE 18-14

	Resistance of sample				
Source	*1*	*2*	*3*	*4*	*5*
A	7.9	7.3	7.2	7.5	7.7
B	9.0	8.8	8.8	8.6	8.6
C	9.0	8.3	8.7	8.5	8.7
D	9.6	9.5	9.4	9.4	9.4
E	5.5	5.8	5.7	5.5	5.8
F	8.0	8.4	8.2	8.4	7.6

18-2. A student was checking the precision of five planimeters of different makes. He conducted his experiment by measuring with each planimeter a definite area four times. The coded results are shown in Table 18-15.

TABLE 18-15

Planimeter				
1	*2*	*3*	*4*	*5*
6.6	5.6	7.5	7.1	5.2
6.0	6.2	5.2	7.0	6.3
6.5	7.1	5.4	6.5	6.8
7.6	5.3	6.8	7.4	7.1

Prepare the analysis of variance table and check whether the planimeters are homogeneous at the 5 percent level of significance.

18-3. A company manufacturing rubber seals for expansion joints suspected that the tensile strength of their product varied with different machines. The results are given in Table 18-16.

TABLE 18-16

Machine	*Strength, kN/m^2*			
A	4900	5300	5200	5100
B	5700	5300	5600	5200
C	5200	4800	5100	5000
D	4500	5000	5300	5100

Do the machines have any effect on the tensile strength of the rubber seals? Use a 1 percent level of significance.

18-4. A procedure for measuring run-off was used to determine whether or not the type of terrain has an effect on the measured run-off. The experiment was carried out four times for each type of area with the coded results shown in Table 18-17.

TABLE 18-17

Area A	*Area B*	*Area C*
40	43	35
32	47	42
38	40	36
42	41	34

Prepare the analysis of variance table. Test for area effects at the 5 percent level of significance.

18-5. For the data of Problem 13-11 use the analysis of variance to test whether there is a significant difference among the means of group *A*, *B*, and *C*. Use $\alpha = 5$ percent.

18-6. A manufacturing company of radio receivers wished to ascertain whether or not the use of transistors instead of vacuum tubes had any significant effect on the output voltage. Four receivers were selected at random and operated first with vacuum tubes and then reassembled with transistors and operated. The results are shown in Table 18-18.

TABLE 18-18

Output Voltage, Volts

	Design	
Receiver	*Vacuum tubes*	*Transistors*
1	20	18
2	22	24
3	21	23
4	23	24

Test whether the voltage difference between vacuum tube and transistor designs is significant at $\alpha = 1$ percent.

18-7. Three filling machines in a factory are suspected of malfunction in weighing. Data from a subsequent experiment is shown in Table 18-19.

TABLE 18-19

Machine	*Weight* $\times 10^{-1}$ *grams*			
I	30.27	30.24	30.25	30.20
II	30.26	30.25	30.22	30.19
III	30.23	30.20	30.23	30.16

a. Prepare the analysis of variance table.
b. Determine whether the three machines differ at the 5 percent level of significance.
c. By comparing the mean from machine I with that of machine III, test whether these two machines differ at $\alpha = 5$ percent.

Chapter 19

Distributions of Extremes

One of the important engineering problems that must be solved by a statistical approach is the estimation of the life of a structure or of manufactured components. The variable may be age, time of usage, or the number of occasions on which load is applied, etc.

FAILURE FUNCTIONS

Let us look at the case of age, i.e., the length of use prior to failure; the problem is then to determine the life distribution, which can also be called the distribution of failure times. Such a distribution, being continuous, can be described by a probability density function (see p. 69) $p(t)$, where t is time. The cumulative distribution function [see Eq. (5-17) and p. 148]

$$P(t) = \int_0^t p(t)\, dt \qquad (19\text{-}1)$$

can then be used to describe the probability of failure at any time between zero and t.

The probability that an item will function up to a time t, i.e., will fail only after time t is

$$R(t) = 1 - P(t) \tag{19-2}$$

and this is called the *reliability function.*

In practice, we are often interested not merely in the total survival or failure numbers but in the failure rate at or near a given instant t; this is defined as the ratio of probability of failure in the interval $(t, t + \Delta t)$ divided by Δt, to the probability of survival up to time t, i.e., the failure rate is

$$G(t, \Delta t) = \frac{(1/\Delta t) \int_t^{t+\Delta t} p(t)\, dt}{R(t)} \tag{19-3}$$

The limit of this function when $\Delta t \to 0$ represents the *instantaneous failure rate* or *hazard rate* is

$$Z(t) = \frac{p(t)}{R(t)} \tag{19-4}$$

This function gives the probability that an item which has survived until time t will fail immediately thereafter; therefore, it is known also as the *conditional failure rate function.* The function is of interest in actuarial work, where the question is, if a man has survived up to the age of 89, what is the probability of his surviving up to 90? A similar problem arises when we want to decide the probability of subsequent survival of mechanical components after they have been in service for a given time.

The conditional failure rate function, therefore, is of considerable value, but its correct interpretation is important. We must realize that since $R(t) < 1$,

$$Z(t) > p(t)$$

The relation between $Z(t)$ and $R(t)$ can be established directly. From Eqs. (19-1) and (19-2)

$$p(t) = -\frac{dR(t)}{dt}$$

Substituting in Eq. (19-4), we find that

$$Z(t) = -\frac{dR(t)}{R(t)}\frac{1}{dt}$$

or

$$Z(t) = -\frac{d \log_e R(t)}{dt} \tag{19-5}$$

Hence

$$R(t) = \exp\left[-\int_0^t Z(t)\,dt\right] \tag{19-6}$$

The pattern of the failure function can be illustrated simply by the "bathtub curve" shown in Fig. 19-1. This figure shows that the highest failure rate

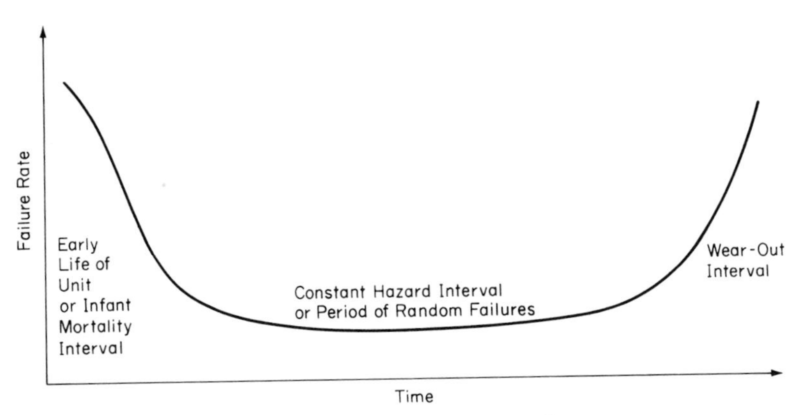

Fig. 19-1. The "bathtub curve."

occurs during the very early life of a being or a manufactured component or natural product. Then follows a long, nearly level period and finally the product enters a wear-out phase, where a drastic increase in failures occurs because of age factors.

EXPONENTIAL DISTRIBUTION

The simplest distribution describing failure times is the exponential distribution, which, it will be shown, has a constant instantaneous failure rate; i.e., the probability of failure at all ages is constant.

The exponential probability density function can be written as

$$p(t) = \frac{1}{\lambda} e^{-t/\lambda}$$

for $t \geq 0$ and $\lambda \geq 0$, where λ is the mean time to failure.

The cumulative probability of failure [Eq. (19-1)] is then

$$P(t) = \int_0^t \frac{1}{\lambda} e^{-t/\lambda}\,dt$$

or

$$P(t) = 1 - e^{-t/\lambda} \tag{19-7}$$

It can readily be observed that the probability density function is always nonnegative; its total area from $t = 0$ to $t = \infty$ is equal to 1, as can be deduced from Eq. (19-7) by putting $t = \infty$.

From Eq. (19-2)

$$R(t) = e^{-t/\lambda}$$

The failure rate is given by Eq. (19-3) as

$$G(t, \Delta t) = \frac{1}{\Delta t R(t)} \int_t^{t+\Delta t} \frac{1}{\lambda} e^{-t/\lambda}\, dt$$

i.e.,

$$G(t, \Delta t) = \frac{1}{\Delta t}(1 - e^{-\Delta t/\lambda}) \tag{19-8}$$

Equation (19-4) gives the instantaneous failure rate

$$Z(t) = \frac{1}{\lambda} \tag{19-9}$$

from which it can be seen that the instantaneous failure rate is constant and is equal to the reciprocal of the mean time to failure, λ.

The exponential distribution and its instantaneous failure rate are plotted in Fig. 19-2.

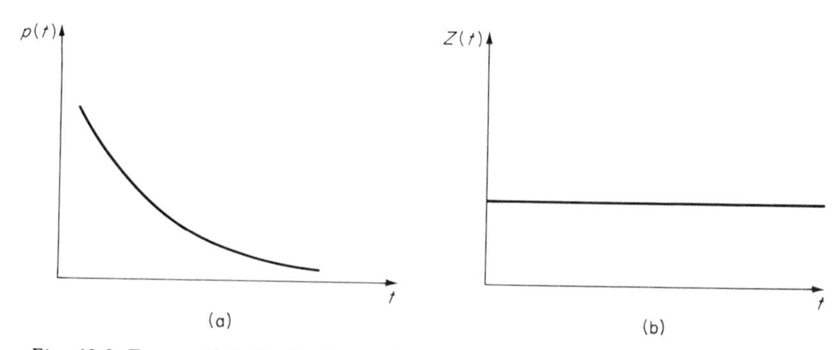

Fig. 19-2. Exponential distribution and its instantaneous failure rate: (a) Exponential distribution; (b) instantaneous failure rate.

Example. The lifetime of a particular electronic component is known to be exponentially distributed with a mean life of 1000 h. What proportion of such components will fail before 800 h?

The mean life $\lambda = 1000$ hours. Thus, from Eq. (19-7),

$$P(t) = 1 - e^{-t/1000}$$

i.e., for $t = 800$ h

$$P(800) = 1 - e^{-800/1000}$$

or

$$P(800) = 0.551$$

Therefore, we can say that approximately 55 percent of components will fail before 800 h.

The exponential distribution can be derived from the Poisson distribution (see Chapter 8) by dealing with the average time between incidents (failures, accidents, or arrivals) instead of the average number of incidents per unit time. From Eq. (8-1), the probability of no incident occurring in an interval of time of length t, i.e., for $r = 0$, is

$$P_0 = e^{-t/\lambda} \tag{19-10}$$

where t/λ is the mean frequency of incidents; i.e., $1/\lambda$ is the average number of incidents per unit time. We may note that t can be any interval of time and does not have to begin at time 0.

Let us consider the time that elapses after t until the next incident occurs, and denote this time as a random variable t_1. Then the probability that $t_1 > t$ is equal to the probability that no incidents occur in the time interval of length t. The former probability is given by the reliability function [see Eq. (19-2)] as $1 - P(t)$. Using Eq. (19-10), we can write the equality of the two probabilities as

$$1 - P(t) = e^{-t/\lambda}$$

whence

$$P(t) = 1 - e^{-t/\lambda}$$

which is Eq. (19-7).

Thus we can observe that times between incidents (failures, accidents, or other events) follow an exponential distribution. It should be mentioned that the exponential distribution, as well as the Poisson distribution, has a memoryless property, since future behavior is independent of present or past behavior.

We shall now apply the exponential distribution to predict the probability of time between arrivals of vehicles at a particular point on a highway. The determination of the length of the time interval between vehicle arrivals is important to the traffic engineer: If this length is too short, a vehicle attempting to merge or cross the traffic stream will be forced either to remain stationary or to interrupt the traffic stream. Wishing to predict the probability of a certain time between successive cars, a traffic engineer collected data of the gaps in traffic at a certain point on a freeway. From these data, he found that the mean gap length or time between successive cars is 8.33 s. What is the probability of a gap length of 9 to 11 s?

We have the mean time between arrivals

$$\lambda = 8.33 \text{ s}$$

Using Eq. (19-7), we obtain

$$P(11) - P(9) = 1 - e^{-11/8.33} - (1 - e^{-9/8.33})$$

$$= e^{-1.08} - e^{-1.32}$$

$$= 0.072$$

which is the probability of a gap length of 9 to 11 s.

WEIBULL DISTRIBUTION

Although a number of distributions may describe failure times under different conditions, we shall mention only one other because of its widespread applicability in structural failures as well as in some electronic components. This is the Weibull distribution.[1]

Here the instantaneous failure rate is given by

$$Z(t) = m\gamma t^{m-1} \tag{19-11}$$

where m is the shape parameter and γ the scale parameter.

For different values of m, different shapes of the instantaneous rate function are obtained (Fig. 19-3): For $m < 1$, the rate decreases with time; for $m = 2$, it increases linearly; for $1 < m < 2$, the rate increases at a decreasing gradient; for $m > 2$, it increases at an increasing gradient; and finally, for $m = 1$, the instantaneous rate function is constant. We note that in the case of $m = 1$, Eq. (19-11) reduces to Eq. (19-9) with $\lambda = 1/\gamma$ so that the exponential distribution can be considered a special case of the Weibull distribution, but there the connection between the two distributions ends.

Substituting in Eq. (19-6), the reliability function of the Weibull distribution can be written as

$$R(t) = \exp\left[-\gamma t^m\right] \tag{19-12}$$

1. W. Weibull, "A Statistical Distribution of Wide Applicability," *Journal of Applied Mechanics*, vol. 18, Sept. 1951, pp. 293–297.

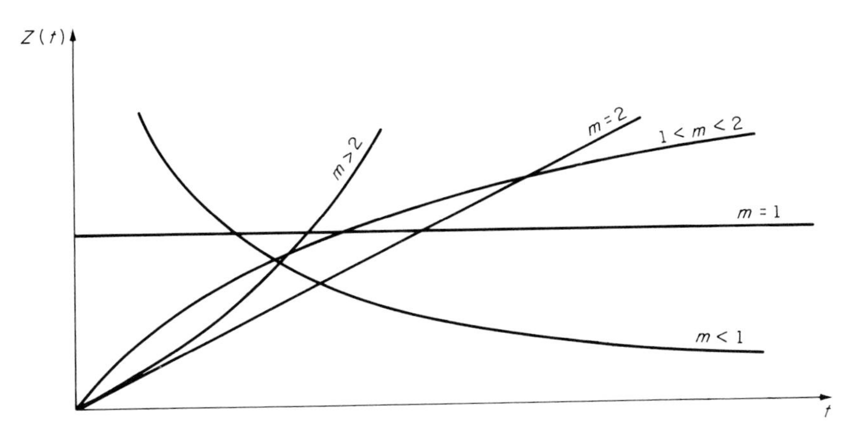

Fig. 19-3. Failure rate for Weibull distribution for different values of the shape parameter *m*.

If we substitute in Eq. (19-2) for $R(t)$ from Eq. (19-6), we can write $P(t)$ for the Weibull distribution as

$$P(t) = 1 - \exp\left[-\gamma t^m\right] \tag{19-13}$$

From Eq. (19-1) the probability density function can be expressed in terms of the cumulative distribution as

$$p(t) = \frac{dP(t)}{dt}$$

so that, for the Weibull distribution,

$$p(t) = m\gamma t^{m-1} \exp\left[-\gamma t^m\right] \tag{19-14}$$

with $t \geq 0$, $\gamma > 0$, and $m > 1$.

Thus we have obtained all the fundamental functions of the Weibull distribution, but the parameters m and γ have to be estimated from the data or, preferably, from the underlying theory of the physical behavior involved. One simple way of estimating m and γ is graphically. From Eq. (19-12),

$$\log_e R(t) = -\gamma t^m$$

so that

$$\log_e\left[-\log_e R(t)\right] = m \log_e t + \log_e \gamma \tag{19-15}$$

In an actual experiment we test N items; those that have failed are not replaced. If the ith failure occurs at time t_i, then it can be shown that an unbiased estimate of the cumulative distribution function for $t = t_i$ is

$$P(t_i) = \frac{i}{N+1} \tag{19-16}$$

Thus, from Eq. (19-2),

$$R(t_i) = \frac{N + 1 - i}{N + 1}$$

whence

$$-\log_e R(t_i) = \log_e \left(\frac{N + 1}{N + 1 - i}\right)$$

Substituting in Eq. (19-15), we obtain

$$\log_e \left[\log_e \left(\frac{N + 1}{N + 1 - i}\right)\right] = m \log_e t_i + \log_e \gamma \qquad (19\text{-}17)$$

Thus, if we plot $\log_e [(N + 1)/(N + 1 - i)]$ against t_i on log-log paper (i.e., we plot $\log_e [-\log_e R(t)]$), it follows from Eq. (19-15) that the experimental points will lie on a straight line if the distribution is of the Weibull type. The slope of the line gives m, and γ can then be found from $\log_e \gamma$ at a value of t_i corresponding to $\log_e [(N + 1)/(N + 1 - i)] = 1$.

Example. In a large production run of electronic communications equipment, a certain relay was being used for a critical application. It was decided to obtain the Weibull model for the number of actuations before failure of this relay.

TABLE 19-1

Relay number	*Number of actuations to failure* (10^5)
1	3.65
2	8.40
3	9.00
4	5.89
5	9.60
6	6.10
7	11.95
8	4.72
9	12.40
10	3.34
11	18.07
12	8.50
13	13.03
14	11.02
15	6.62
16	1.90
17	7.92
18	20.63
19	4.20
20	13.42

Twenty relays were subjected to a life test and the following numbers of actuations to failure obtained are given in Table 19-1.

SOLUTION

The data are first ranked starting with the lowest number of actuations: The number of relays tested is $N = 20$. The values of $\log_e [(N+1)/(N+1-i)]$ are obtained, where i is the rank of a particular relay failing, the least event having the rank $i = 1$. The resulting values are given in Table 19-2.

TABLE 19-2

Rank i	*Actuations to failure,* $t_i(10^5)$	$\frac{21}{21-i}$	$\log_e\left(\frac{21}{21-i}\right)$
1	1.90	1.05	0.049
2	3.34	1.11	0.104
3	3.65	1.17	0.157
4	4.20	1.24	0.215
5	4.72	1.31	0.270
6	5.89	1.40	0.336
7	6.10	1.50	0.405
8	6.62	1.62	0.482
9	7.92	1.75	0.560
10	8.40	1.91	0.647
11	8.50	2.10	0.742
12	9.00	2.33	0.846
13	9.60	2.63	0.967
14	11.02	3.00	1.099
15	11.95	3.50	1.253
16	12.40	4.20	1.435
17	13.03	5.25	1.658
18	13.42	7.00	1.946
19	18.07	10.50	2.351
20	20.63	21.00	3.045

The values of t_i and $\log_e [21/(21-i)]$ are plotted on log-log paper as shown in Fig. 19-4. Since the plotted points lie approximately on a straight line (fitted by eye), there is a good indication that the data follow a Weibull distribution. The slope of the line yields $m = 1.7$. To find γ, we require t_i corresponding to $\log_e [21/(21-i)] = 1$. From Fig. 19-4, we observe that for $\log_e [21/(21-i)] = 1$, $t = 10.2 \times 10^5$. Therefore, from Eq. (19-17),

$$m \log_e (10.2 \times 10^5) = -\log_e \gamma$$

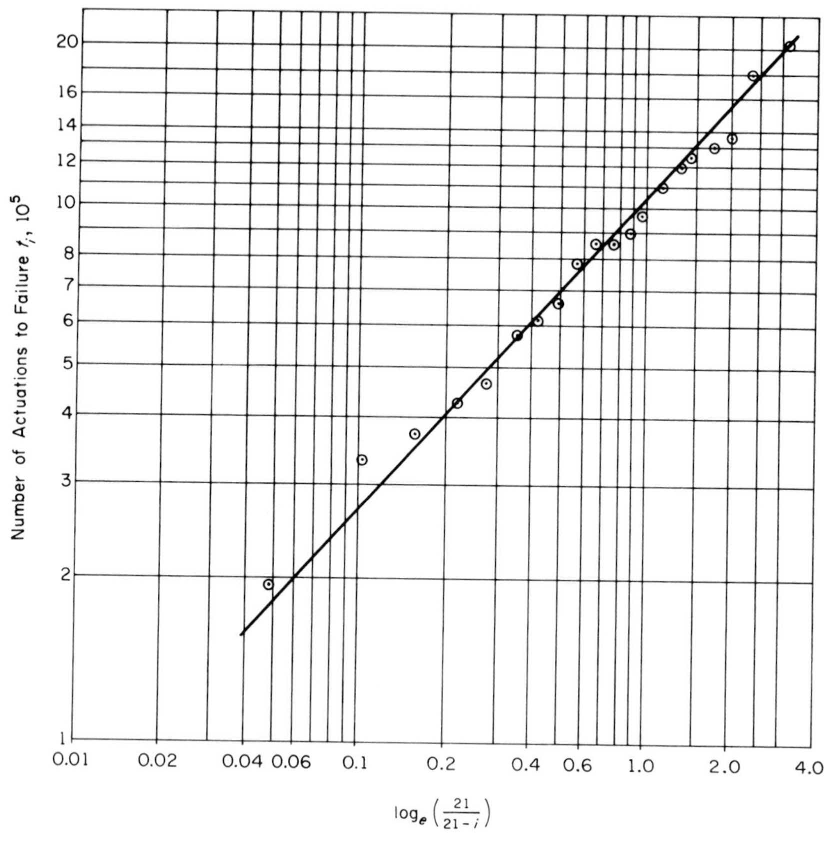

Fig. 19-4. Plot of t_i versus $\log_e [21/(21 - i)]$ on log-log scale.

whence

$$\gamma = (10.2 \times 10^5)^{-1.7}$$

$$= 0.61 \times 10^{-10}$$

With the values of the parameters m and γ determined, the probability density function $p(t)$ and the associated cumulative distribution $P(t)$ are explicitly defined. We can then estimate, e.g., the overall proportion of relays that would survive a certain number of actuations before failure.

Where the observed life times fall into a fairly narrow range, it is often necessary to include a third parameter in the Weibull distribution, the minimum life time, in order that a sufficiently sensitive measure of the parameter m may be made. If we denote this *location parameter* by L, the expression for the reliability function of the Weibull distribution [Eq. (19-12)] then becomes

$$R(t) = \exp\left[-\gamma(t - L)^m\right] \tag{19-18}$$

DISTRIBUTION OF EXTREME VALUES

The Weibull distribution was derived in a study of strength of materials, where the problem is that of finding the weakest element. This is a particular case of determining extreme values—a field of especial interest in the study of floods and in hydrology in general.

The essential problem is that of determining the distribution of smallest or largest values in samples of size n drawn at random from a certain given underlying distribution $p(x)$. The important point is that the extremes are not fixed but are new statistical variates depending on the underlying distribution and on the sample size.

In many practical engineering problems, the exact form of the distribution is not known and assumptions regarding its form have to be made. In the particular case when the underlying distribution is normal,[2] the variation in the mean and standard deviation of samples of size n in terms of the mean and standard deviation of sample of unit size is as shown in Fig. 19-5. This figure shows that the larger the specimen of a given material the lower the strength. We can also see that the distribution of strength of specimens of a given size becomes progressively more skewed (to the left) with an increase in the sample size. Since the influence of size on strength depends on the standard deviation of strength (for a given mean strength of the specimens of unit size), it follows that the size effect is smaller the greater the homogeneity of the material. Figure 19-5 can also explain why size effect virtually disappears beyond a certain size of the specimen: For instance, for each successive tenfold increase in size of the specimen it loses progressively a smaller amount of strength.[3]

In a more general case when the underlying probability density function is $p(x)$ and the cumulative distribution function is

$$P(x) = \int_{-\infty}^{x} p(x)\, dx \tag{19-19}$$

then the distribution of the smallest value in samples of size n is given by the probability density function

$$g_n(x) = np(x)[1 - P(x)]^{n-1} \tag{19-20}$$

and the corresponding cumulative distribution function is

$$G_n(x) = \int_{-\infty}^{x} g_n(x)\, dx \tag{19-21}$$

2. L. C. H. Tippett, "On the Extreme Individuals and the Range of Samples Taken from a Normal Population," *Biometrika*, vol. 17, 1925, pp. 364–387.

3. A. M. Neville, *Properties of Concrete* (London: Pitman Publishing Corporation; New York: Wiley Interscience, 1973).

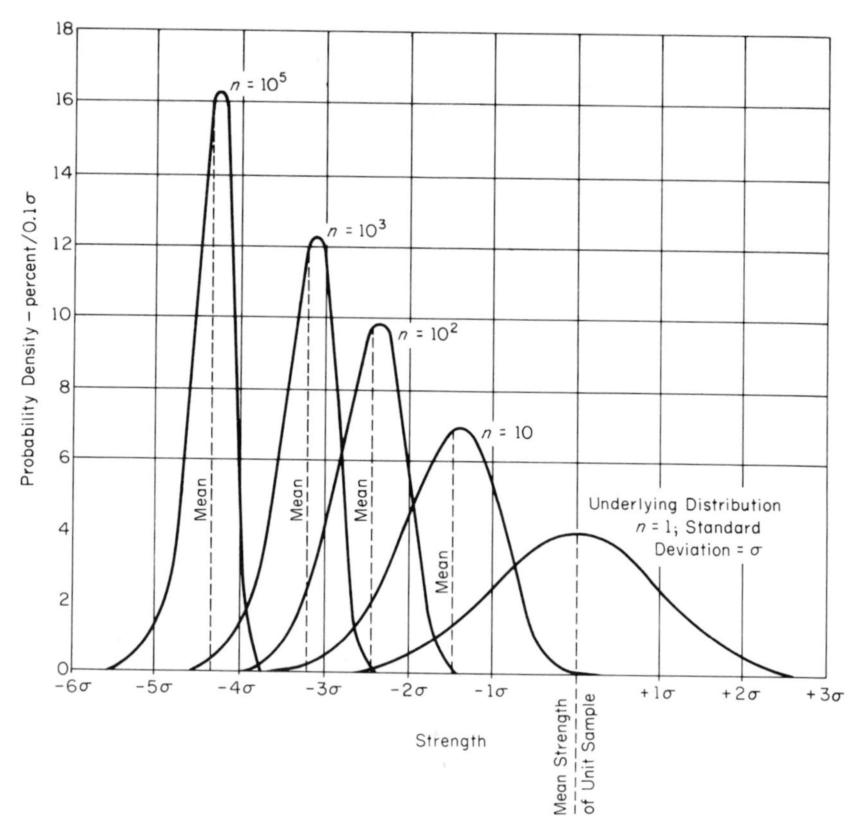

Fig. 19-5. Strength distribution in samples of size n for an underlying normal distribution. (From J. Tucker, "Statistical Theory of the Effect of Dimensions and of Method of Loading upon the Modulus of Rupture of Beams," *ASTM Proceedings*, vol. 41, 1941, pp. 1072–1088.)

or

$$G_n(x) = 1 - [1 - P(x)]^n \tag{19-21a}$$

The mode of $g_n(x)$ is the solution of the equation

$$\frac{d}{dx} g_n(x) = 0 \tag{19-22}$$

but the solution does not exist in all cases. If the underlying distribution is limited, the distribution of the smallest value decreases monotonically.

From Eq. (19-20), the solution of Eq. (19-22) can be written as

$$(n - 1)[p(\hat{x})]^2 = p'(\hat{x})[1 - P(\hat{x})] \tag{19-23}$$

where $\hat{x}$ is the mode of the smallest value and $p'(\hat{x})$ is the value of the derivative $dp(x)/dx$ at $x = \hat{x}$.

TABLE 19-3

Properties of the Normal and Weibull Distributions

$p(x)$	*Mode of smallest value $\hat{x}$ for samples of size n*	*Distribution of smallest values in samples of size n (n is large)*	*Characteristics of the distribution of the smallest value in sample of size n*
Normal [Eq. (9-16) $p(x) = \frac{1}{\sigma\sqrt{2\pi}} \exp\left[-\frac{(x-\mu)^2}{2\sigma^2}\right]$	$\hat{x} = \mu - \sigma\sqrt{2 \log_e n} + \sigma \frac{\log_e (\log_e n) + \log 4\pi}{2\sqrt{2 \log_e n}}$	$x_n = \mu - \sigma\sqrt{2 \log_e n} + \sigma \frac{\log_e (\log_e n) + \log 4\pi}{2\sqrt{2 \log_e n}} + \frac{\sigma}{\sqrt{2 \log_e n}} \log_e \xi$	Most probable value decreases as a multiple of $\sqrt{\log_e n}$. Variance decreases as n increases and is equal to $\frac{\pi^2\sigma^2}{12 \log_e n}$
Weibull [Eq. (19-14)] $p(x) = m\gamma x^{m-1} \exp[-\gamma x^m]$	$\hat{x} = \frac{1}{(\gamma n)^{1/m}}\left[1 - \frac{1}{m}\right]^{1/m}$	$x_n = \left[\frac{\xi}{n\gamma}\right]^{1/m}$	Most probable value de creases as $n^{-1/m}$. Mean value decreases as $n^{-1/m}$. Variance decreases as $n^{-2/m}$.

Source: B. Epstein, "Statistical Aspects of Fracture Problems," *Journal of Applied Physics*, vol. 19, Feb. 1948, pp. 140–147.

Equation (19-23) is used in the majority of "weakest link" problems of various sorts. (The adage that a chain is no stronger than its weakest link is a classical example of the importance of the extreme value in a distribution.) Values of $\hat{x}$ for two probability density functions $p(x)$ are given in Table 19-3.

In Table 19-3, ξ is distributed with a probability density function $h(\xi) = e^{-\xi}$ for $\xi \geq 0$. This is a simple function so that the distribution of the smallest value of x can readily be sketched or plotted.

Table 19-3 shows that the mode of the smallest value $\hat{x}$ for a sample of given size depends on the underlying distribution $p(x)$ from which the samples are drawn. The last column shows how the mode, mean, and variance of the smallest value in a sample of size n (e.g., minimum strength) vary for two common distributions: normal and Weibull.

FLOOD FREQUENCY ANALYSIS

The transformation of the variate x to its logarithmic value, log x, was discussed on p. 152. Experience has shown that the logarithms of a series of annual flood peaks, Q, are approximately normally distributed, i.e.,

$$p(Q) = \frac{1}{\sigma_y\sqrt{2\pi}} e^{-(y-\bar{y})^2/2\sigma_y} \tag{19-24}$$

Equation (19-24) is, in fact, the probability density of the log-normal distribution, where $y = \log x$, x the variate, $\bar{y}$ the mean of y, and σ_y the standard deviation of y. The probability of exceeding a flow Q—an event known as exceedance—is

$$P(Q) = \frac{1}{T_r} = 1 - \frac{1}{\sigma_y\sqrt{2\pi}} \int_{-\infty}^{y} e^{-(y-\bar{y})^2/2\sigma_y}\, dy \tag{19-25}$$

where T_r is called the return period in years and is defined as the average number of years before a certain value, i.e., a peak flow, either recurs or is again exceeded.

In practice, Eq. (19-25) is usually treated graphically by plotting the observed annual flood peaks, Q_i, versus their estimated probability of exceedance, $P(Q_i)$, on log probability paper (or log Q_i versus $P(Q_i)$ on arithmetic probability paper).

Several expressions have been suggested for estimating the probability of exceedance for an event in a annual flood series; the most common one of these is

$$P(Q_m) = \frac{1}{T_{r_m}} = \frac{m}{N+1} \tag{19-26}$$

where m is the rank of the event, Q_m, with $m = 1$ for maximum and $m = N$ for minimum event; N the length of the record in years; and $P(Q_m)$ the estimated probability of any Q's being equal or greater than Q_m. Equation (19-26) should be compared with Eq. (19-16).

Example. Flow records are available for a particular river near a proposed dam site for the period 1949–1950 to 1964–1965 (the year beginning on October 1 in each case). The data are given in Table 19-4.

TABLE 19-4

Year	*Peak flow, Q, m³/s*
1949–1950	4800
1950–1951	3000
1951–1952	1410
1952–1953	1960
1953–1954	2900
1954–1955	5100
1955–1956	5700
1956–1957	1260
1957–1958	660
1958–1959	1270
1959–1960	2400
1960–1961	1270
1961–1962	1510
1962–1963	2150
1963–1964	680
1964–1965	2550

a. Plot the annual flood frequency curve for the dam site on
   i. Arithmetic probability paper.
   ii. Logarithmic probability paper.
b. Estimate the flood peaks for 1 year in 2 years, 1 in 10, 1 in 20, 1 in 100, and 1 year in 1000 years, i.e., for return periods of 2, 10, 20, 100, and 1000 years.

SOLUTION

a. The data are first ranked, beginning with rank $m = 1$ for the maximum peak flow Q. Values of $T_r = m/(N + 1)$ and $\log_{10} Q$ are found (see Table 19-5).

It should be observed that plotting either $\log_{10} Q$ or $\log_e Q$ is equally correct, since the two are related by a constant, namely, $\log_e Q = 2.3026 \log_{10} Q$.

TABLE 19-5

Rank, m	*Peak flow* $Q(m^3/s)$	$T_r = \frac{N+1}{m}$	$\text{Log}_{10}\ Q$
1	5700	17.000	3.756
2	5100	8.500	3.708
3	4800	5.666	3.681
4	3000	4.250	3.477
5	2900	3.400	3.463
6	2550	2.833	3.406
7	2400	2.428	3.381
8	2150	2.125	3.332
9	1960	1.888	3.292
10	1510	1.700	3.179
11	1410	1.545	3.149
12	1270	1.417	3.104
13	1270	1.307	3.104
14	1260	1.214	3.101
15	680	1.133	2.832
16	660	1.063	2.819

i. The relation between the peak flow Q and the return period T_r, plotted on probability paper, is shown in Fig. 19-6. It should be noted that the return period T_r is the inverse of the probability of exceedance [Eq. (19-26)]. The curve was fitted by eye, but the nonlinear character of the relationship is readily seen.

ii. This nonlinear relationship can be rectified into a linear one if the return period T_r is plotted against $\log_{10} Q$ as shown in Fig. 19-7.

b. To estimate the peak flows for the required return periods, we use the straight-line relationship in Fig. 19-7 (fitted by eye). Thus for

$$2 \text{ years: } \log_{10} Q = 3.243, \text{ or } Q = 1750 \text{ m}^3/\text{s}$$

$$10 \text{ years: } \log_{10} Q = 3.692, \text{ or } Q = 4924 \text{ m}^3/\text{s}$$

$$20 \text{ years: } \log_{10} Q = 3.800, \text{ or } Q = 6309 \text{ m}^3/\text{s}$$

$$100 \text{ years: } \log_{10} Q = 4.000, \text{ or } Q = 10{,}000 \text{ m}^3/\text{s}$$

$$1000 \text{ years: } \log_{10} Q = 4.220, \text{ or } Q = 16{,}595 \text{ m}^3/\text{s}$$

In order to increase the reliability of the estimated frequency values,

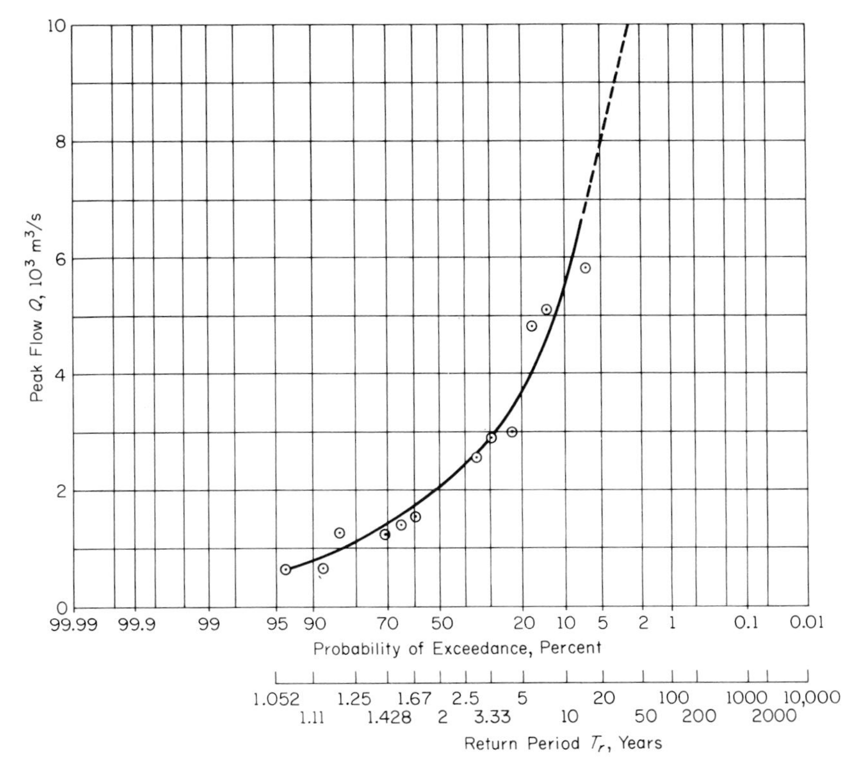

Fig. 19-6

Gumbel made use of the distribution of extreme values.[4] In this approach, we consider the extreme values $X_1, X_2, \ldots X_N$ observed, respectively, in N samples of equal size n. Now, if X is an exponentially distributed and unlimited variable, then the cumulative probability P that any of the N extremes will be less than X, as N and n approach infinity, is

$$P = e^{-e^{-y}} \tag{19-27}$$

where y is the reduced variate given by

$$y = a(X - \hat{X}) \tag{19-28}$$

$\hat{X}$ the mode of the distribution, and a the dispersion parameter. Although it is possible to determine explicit expressions for $\hat{X}$ and a, these cannot be

4. E. J. Gumbel, "Statistical Theory of Extreme Values and Some Practical Applications," National Bureau of Standards, *Applied Mathematics*, Ser. 33, Feb. 1954.

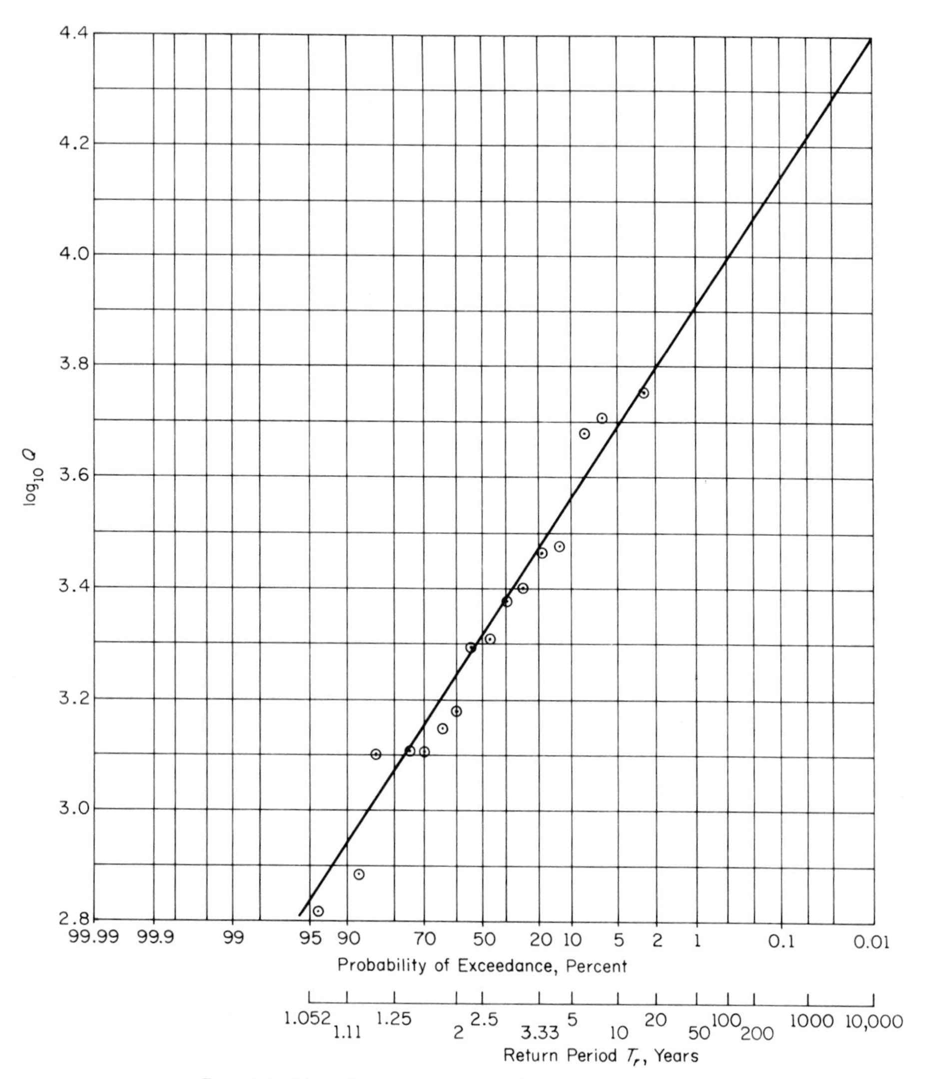

Fig. 19-7. Plot of $\log_{10} Q$ versus T_r for the data of Fig. 19-6.

strictly used to estimate return periods from *limited samples*, and Gumbel used an approach based on the method of least squares to determine these values from the annual series. By minimizing the squares of the deviations measured in a direction perpendicular to the derived straight line X versus y (on Cartesian coordinates), the following equations are found

$$\hat{X} = \bar{X} - \sigma_X \frac{\bar{y}_N}{\sigma_N} \tag{19-29}$$

and

$$a = \frac{\sigma_N}{\sigma_X} \tag{19-30}$$

The quantities of $\bar{y}_N$ and σ_N, obtained theoretically, are functions of the sample size only, the sample consisting of N largest values from the original N samples; they are given in Table 19-6.

TABLE 19-6

Expected mean $\bar{y}_N$ and expected standard deviation σ_N for different size of sample of largest values

N	$\bar{y}_N$	σ_N
15	0.51	1.02
20	0.52	1.06
30	0.54	1.11
40	0.54	1.14
50	0.55	1.16
60	0.55	1.17
70	0.55	1.19
80	0.56	1.19
90	0.56	1.20
100	0.56	1.21
150	0.56	1.23
200	0.57	1.24
500	0.57	1.26
∞	0.57	1.28

Source: E. J. Gumbel, "Statistical Theory of Extreme Values and Some Practical Applications," National Bureau of Standards, *Applied Mathematics*, Ser. 33, Feb. 1954.

Equation (19-28) indicates that a plot of X versus the reduced variate y is linear on a Cartesian coordinate system; this observation led Powell[5] to introduce the so-called *Gumbel* or *extremal probability paper*. Gumbel's distribution is such that the arithmetic average of the annual series will have a return period of 2.33 years when N is large (see Fig. 19-8). A typical Gumbel probability paper is shown in Fig. 19-8, where the data (Q versus T_r) of the example on p. 349 are plotted. (The straight line is drawn by eye.) Note the location of the mode $\hat{X}$ and the arithmetic mean of the peak flow Q.

5. R. W. Powell, "A Simple Method of Estimating Flood Frequency," *Civil Engineering*, vol. 13, 1943, pp. 105–107.

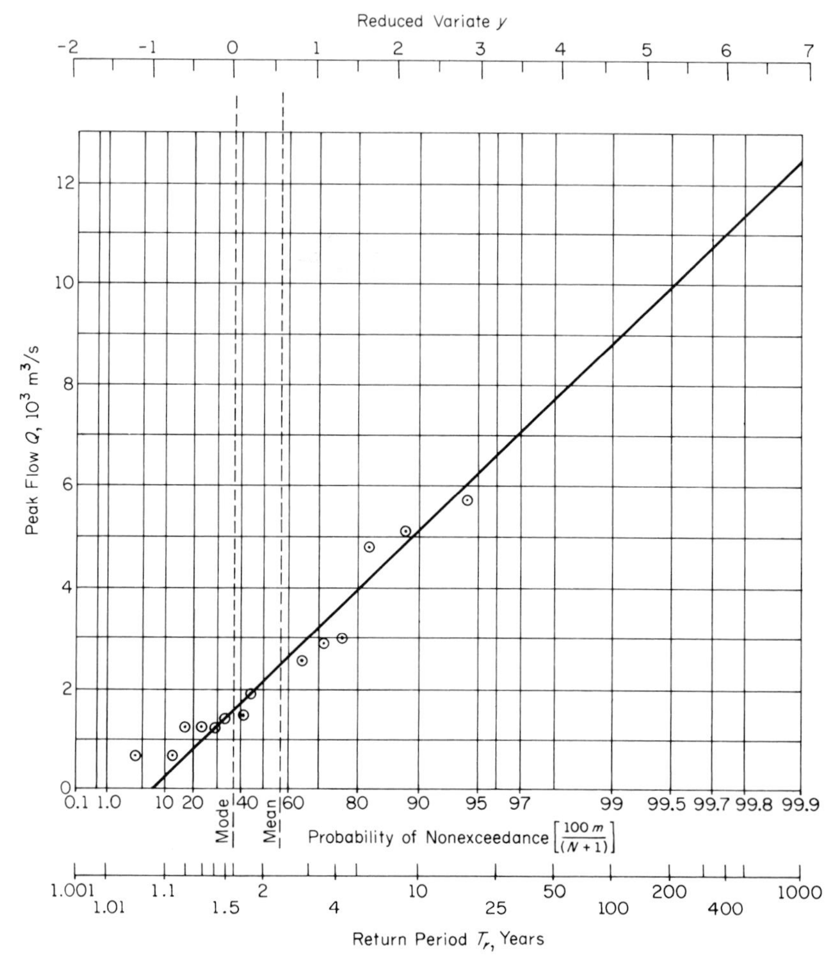

Fig. 19-8. Gumbell plot of flood frequency curve for the example on p. 349.

As a further development it has been shown that more accurate results can be predicted from a *Gumbel logarithmic plot*, i.e., by plotting $\log_{10} Q$ (instead of Q) versus the return period T_r on Gumbel probability paper. This is illustrated in the following example.

Example. Discharge data at a gauging station on a river were collected for the 40-year period from 1921 to 1960; the flow record is shown in Table 19-7.

Construct a Gumbel log plot and derive the linear equation relating the reduced variate y and X $(=\log_{10} Q)$.

TABLE 19-7

Year	*Peak discharge,* $Q(10^3\ m^3/s)$	*Year*	*Peak discharge,* $Q(10^3\ m^3/s)$
1921	46	1941	72
1922	35	1942	45
1923	47	1943	78
1924	37	1944	58
1925	22	1945	90
1926	151	1946	24
1927	21	1947	25
1928	54	1948	256
1929	34	1949	110
1930	55	1950	95
1931	56	1951	35
1932	62	1952	39
1933	109	1953	31
1934	30	1954	94
1935	56	1955	33
1936	199	1956	132
1937	66	1957	76
1938	52	1958	32
1939	148	1959	71
1940	34	1960	14

SOLUTION

The data are first ranked. The values of the return period T_r (with $N = 40$) and of $\log_{10} Q$ are then found. The results are tabulated in Table 19-8.

Now, from Eq. (19-28), the reduced variate y is given by $y = a(X - \hat{X})$. From Table 19-6, for $N = 40$, $\bar{y}_N = 0.54$ and $\sigma_N = 1.14$. Using the latter value, Eq. (19-30) gives

$$a = \frac{\sigma_N}{\sigma_X} = \frac{1.14}{0.28} = 4.07$$

Using Eq. (19-29), we obtain

$$\hat{X} = \bar{X} - \frac{\sigma_X \bar{y}_N}{\sigma_N} = 1.739 - 0.28 \times \frac{0.54}{1.14}$$

or

$$\hat{X} = 1.606$$

TABLE 19-8

Rank, m	$Q(10^3\ m^3/s)$	$T_r = \dfrac{(N+1)}{m}$	$X = \log_{10} Q$	$X - \bar{X}$	$(X - \bar{X})^2$
1	256	41.0	2.408	0.669	0.4476
2	199	20.5	2.299	0.560	0.3136
3	151	13.7	2.179	0.440	0.1936
4	148	10.3	2.170	0.431	0.1858
5	132	8.2	2.121	0.382	0.1459
6	110	6.8	2.041	0.302	0.0912
7	109	5.9	2.037	0.298	0.0888
8	95	5.1	1.978	0.239	0.0571
9	94	4.6	1.973	0.234	0.0548
10	90	4.1	1.954	0.215	0.0462
11	78	3.7	1.892	0.153	0.0234
12	76	3.4	1.881	0.142	0.0202
13	72	3.2	1.857	0.118	0.0139
14	71	2.9	1.851	0.112	0.0125
15	66	2.7	1.820	0.081	0.0066
16	62	2.6	1.792	0.053	0.0028
17	58	2.4	1.763	0.024	0.0006
18	56	2.3	1.748	0.009	0.0001
19	56	2.2	1.748	0.009	0.0001
20	55	2.1	1.740	0.001	0.0000
21	54	2.0	1.732	−0.007	0.0000
22	52	1.9	1.716	−0.023	0.0005
23	47	1.8	1.672	−0.067	0.0045
24	46	1.7	1.663	−0.076	0.0058
25	45	1.6	1.653	−0.086	0.0074
26	39	1.58	1.591	−0.148	0.0219
27	37	1.52	1.568	−0.171	0.0292
28	35	1.47	1.544	−0.195	0.0380
29	35	1.41	1.544	−0.195	0.0380
30	34	1.37	1.532	−0.207	0.0428
31	34	1.33	1.532	−0.207	0.0428
32	33	1.28	1.519	−0.220	0.0484
33	32	1.24	1.505	−0.234	0.0548
34	31	1.21	1.491	−0.248	0.0615
35	30	1.17	1.477	−0.262	0.0686
36	25	1.14	1.398	−0.341	0.1163
37	24	1.11	1.380	−0.359	0.1289
38	22	1.08	1.342	−0.397	0.1576
39	21	1.05	1.322	−0.417	0.1739
40	14	1.03	1.146	−0.593	0.3516
			$\sum$ 69.579		3.0973

$$\bar{X} = \frac{\sum X}{N} = \frac{69.579}{40} = 1.739$$

$$\sigma_X = \sqrt{\frac{\sum (X - \bar{X})^2}{N}} = \sqrt{\frac{3.0973}{40}} = 0.28$$

Thus the linear relationship is given by

$$y = 4.07(X - 1.606)$$

This line, as well as the observed values, is shown plotted in Fig. 19-9. We can readily observe the good fit afforded by this approach.

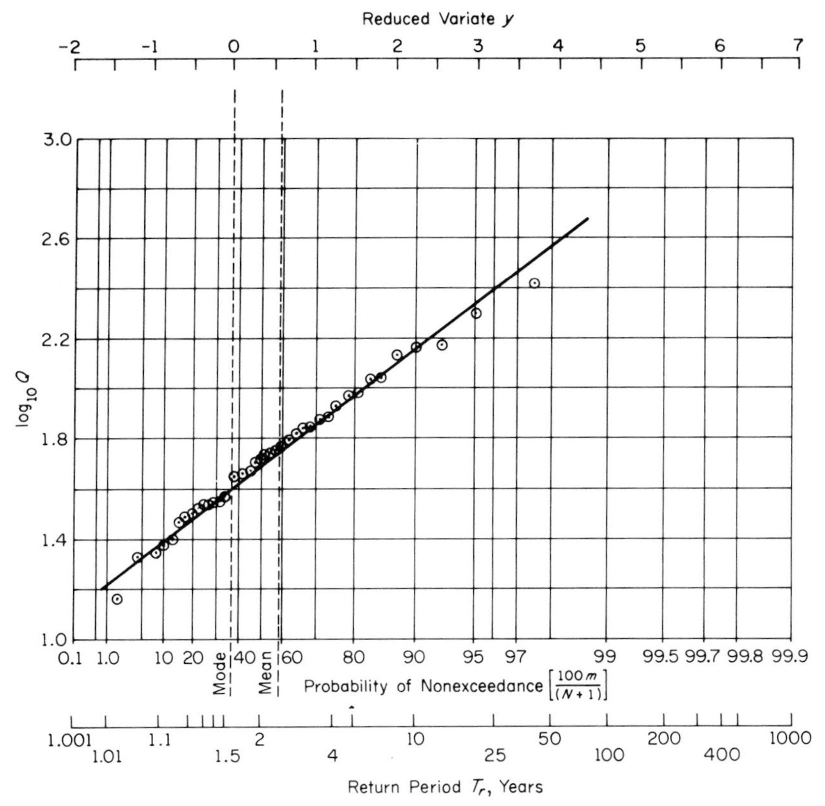

Fig. 19-9. Gumbel log plot for the data of Fig. 19-8.

WIND FREQUENCY ANALYSIS

It has been found that Eq. (19-27) can satisfactorily represent the statistical distribution of many climatological extremes, e.g., the distribution of maximum wind velocity, V. We can thus write the probability $P(V)$ that the maximum wind velocity in any one year is less than V.

$$P(V) = e^{-e^{-y}} \tag{19-31}$$

where $y = a(V - \hat{V})$, a = scale factor for the data (measuring its dispersion), and $\hat{V}$ = the mode of the data. The values of the parameters a and $\hat{V}$ are obtained from anemometer records at selected stations in any one locality. Following our definition for the return period T_r,[6] the probability that a certain velocity is *not* exceeded in any one year is $[1 - (1/T_r)]$. Thus, by taking logarithms of both sides of Eq. (19-31) and putting $P(V) = 1 - (1/T_r)$, we obtain

$$V = \hat{V} - \frac{1}{a} \log_e \left[-\log_e \left(1 - \frac{1}{T_r} \right) \right] \tag{19-32}$$

For the gradient-wind velocity $\bar{V}_G$ (which corresponds to the gradient wind, i.e., one that is moving freely under the influence of pressure gradients and is unaffected by the frictional stresses near the ground surface) with a return period T_r,

$$\bar{V}_G = \hat{V} - \frac{1}{a} \log_e \left[-\log_e \left(1 - \frac{1}{T_r} \right) \right] \tag{19-33}$$

For large values of T_r, say $T_r > 10$, Eq. (19-33) can be approximated by

$$\bar{V}_G = \hat{V} + \frac{1}{a} \log_e T_r \tag{19-34}$$

The mean wind velocity $\bar{V}_z$ at height z above the ground is related to $\bar{V}_G$ by means of a power law of the type

$$\bar{V}_z = \bar{V}_G \left(\frac{z}{z_G} \right)^{\alpha} \tag{19-35}$$

where z_G and α are functions of the ground roughness.[7] From these values of mean wind velocities, the mean wind pressures are found, and it is to resist these that the structure is designed.

The extreme value distribution has been used extensively also in interpreting fatigue tests and forecasting fatigue life of various materials.[8]

Solved Problem

19-1. The yield values for a certain grade of steel were found to be as listed in Table 19-9. Assuming that the Weibull distribution is applicable, calculate the appropriate parameters, and compare the expected and observed values by the χ^2 test.

6. For example, if we wish to design a structure to resist the once-in-50-years wind, the return period is $T_r = 50$.

7. A. Davenport, "The Application of Statistical Concepts to the Wind Loading of Structures," *Proceedings Institution of Civil Engineers*, vol. 19, Aug. 1961, pp. 449–471.

8. A. M. Freudenthal and E. J. Gumbel, "On the Statistical Interpretation of Fatigue Tests," *Proceedings Royal Society of London*, vol. 215–216, 1952–1953, p. 309.

TABLE 19-9

Class	*1*	*2*	*3*	*4*	*5*	*6*	*7*	*8*	*9*	*10*
Yield strength (kg/mm^2: 1.275)[b]	32[a]	33	34	35	36	37	38	39	40	42
Cumulative frequency	10	33	81	161	224	289	336	369	383	389

Source: W. Weibull, "A Statistical Distribution Function of Wide Applicability," *Journal of Applied Mechanics*, vol. 18, Sept. 1951, pp. 293–297.

[a] Minimum strength = 30.25 (kg/mm^2: 1.275).

[b] 1 kg/mm^2 = 9.81 N/mm^2.

Assume also that the normal distribution is applicable, and find the corresponding value of χ^2.

SOLUTION

The minimum yield strength is 30.25 (kg/mm^2 : 1.275) and this is used as an estimate of L, $\hat{L}$. This is subtracted from all the other observed yield values. The expression for $R(t_i)$ is then:

$$R(t_i) = \frac{N + 1 - i}{N + 1} = \frac{390 - i}{390}$$

We can tabulate the values as in Table 19-10.

TABLE 19-10

Class	*Yield strength* t_i *(kg/mm^2: 1.275)*	$t_i - \hat{L}$ *(kg/mm^2: 1.275)*	*Cumulative frequency* i	$\frac{390}{390 - i}$	$\log_e\left(\frac{390}{390 - i}\right)$
1	32	1.75	10	1.026	0.026
2	33	2.75	33	1.092	0.088
3	34	3.75	81	1.262	0.232
4	35	4.75	161	1.703	0.532
5	36	5.75	224	2.350	0.854
6	37	6.75	289	3.861	1.351
7	38	7.75	336	7.222	1.977
8	39	8.75	369	18.571	2.922
9	40	9.75	383	55.714	4.020
10	42	11.75	389	390.000	5.966

The values of $(t_i - \hat{L})$ and $\log_e [390/(390 - i)]$ are plotted on log-log graph paper as shown in Fig. 19-10. The slope of the straight line (fitted by eye) gives a value of 2.93 for the Weibull parameter m; the value of $(t_i - \hat{L})$ corresponding to $\log_e [390/(390 - i)] = 1$ is $6.07 \times 1.275 = 7.74$ kg/mm^2. Therefore, from Eq. (19-17),

$$m \log_e (7.74) = -\log_e \gamma$$

whence

$$\gamma = 7.74^{-2.93} = 0.0025$$

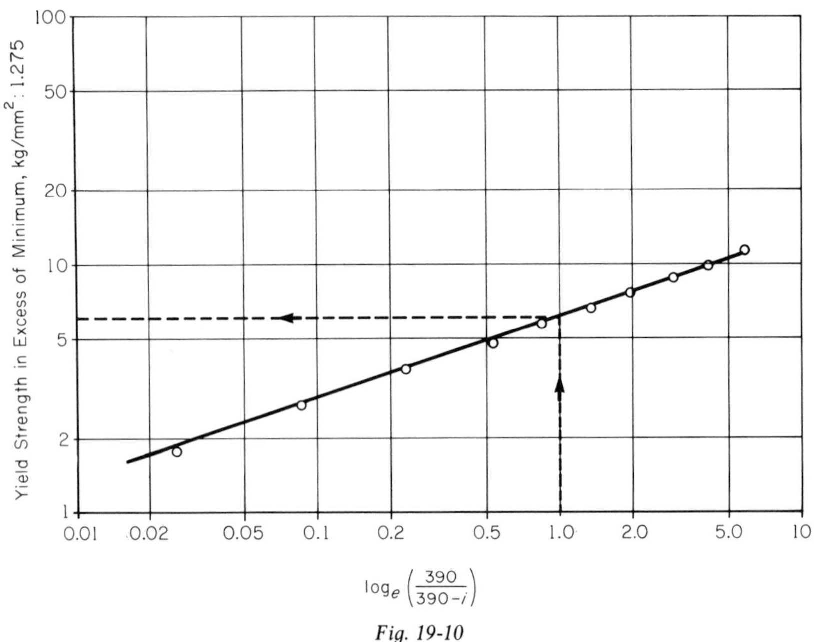

Fig. 19-10

In order to estimate the frequencies of yield as predicted by the Weibull distribution, we equate the two expressions for the reliability function $R(t)$:

$$\exp [-\gamma(t_i - \hat{L})^m] = \frac{N + 1 - i}{N + 1}$$

Therefore,

$$\exp [-0.0025(t_i - 38.57)^{2.93}] = \frac{390 - i}{390}$$

so that

$$i = 390 - 390 \exp [-0.0025(t_i - 38.57)^{2.93}]$$

where t_i is in kg/mm^2.

The estimated frequencies are calculated as in Table 19-11.

TABLE 19-11

Class	$(t_i - \hat{L})^{2.93} = T$	$0.0025 \times T$	$e^{-0.0025T}$	Estimated i
1	10.484	0.026	0.974	10
2	39.440	0.099	0.906	37
3	97.887	0.245	0.783	85
4	195.702	0.489	0.613	151
5	342.574	0.856	0.425	224
6	548.050	1.370	0.254	291
7	821.557	2.054	0.128	340
8	1172.430	2.931	0.053	369
9	1609.910	4.025	0.018	383
10	2781.311	6.953	0.001	390

To test the hypothesis that the observed yield strengths fit a Weibull distribution, we calculate χ^2 (see Table 19-12).

Since we have 9 independent classes and we must estimate three parameters from the observed data, the number of degrees of freedom is $9 - 3 = 6$. From Table A-7, a

TABLE 19-12

Class	E	O	$\lvert O - E \rvert$	$(O - E)^2/E$
1	10	10	0	0.000
2	27	23	4	0.593
3	48	48	0	0.000
4	66	80	14	2.970
5	73	63	10	1.370
6	67	65	2	0.060
7	49	47	2	0.082
8	29	33	4	0.552
9	14	14	0	0.000
10	7	6	1	0.143
				$\chi^2 = 5.770$

χ^2 value of 5.770 for 6 degrees of freedom is significant at the 50 percent level so that we have no evidence to reject the hypothesis that the Weibull distribution fits the observed data.

To fit a normal distribution to the data, we first estimate the mean and standard deviation as shown in Table 19-13.

TABLE 19-13

Class midpoint x_i	$X_i = x_0 - x_i$	X_i^2	f_i	$f_i X_i$	$f_i X_i^2$
$32 = x_0$	0	0	10	0	0
33	1	1	23	23	23
34	2	4	48	96	192
35	3	9	80	240	720
36	4	16	63	252	1008
37	5	25	65	325	1625
38	6	36	47	282	1692
39	7	49	33	231	1617
40	8	64	14	112	896
42	10	100	6	60	600
			$\sum = 389$	$\sum = 1621$	$\sum = 8373$

The mean $\bar{x}$

$$= x_0 + \frac{1621}{389} = 32 + 4.167 = 36.167$$

and the standard deviation, σ

$$= \sqrt{\frac{\sum f_i X_i^2 - [(\sum f_i X_i)^2/n]}{n - 1}}$$

$$= \sqrt{\frac{8373 - 6754.86}{388}}$$

$$= \sqrt{4.170} = 2.042$$

Ordinates of the fitted normal curve, for $n/\sigma = 190.48$, are obtained using Table A-3. (See Table 19-14.)

TABLE 19-14

$z = X/\sigma$	0	± 0.5	± 1.0	± 1.5	± 2.0	± 2.5	± 3.0	± 3.5	± 4.0
$f(z)$	0.3989	0.3521	0.2420	0.1295	0.0540	0.0175	0.0044	0.0009	0
$f(z) \times n/\sigma$	75.98	67.07	46.10	24.67	10.29	3.33	0.84	0.17	0
$X = z \times \sigma$	0	1.02	2.04	3.06	4.08	5.10	6.12	7.14	8.16
$x = \bar{x} + X$	36.17	37.19	38.21	39.23	40.25	41.27	42.29	43.31	44.33
$x = \bar{x} - X$	36.17	35.15	34.13	33.11	32.09	31.07	30.05	29.03	28.01

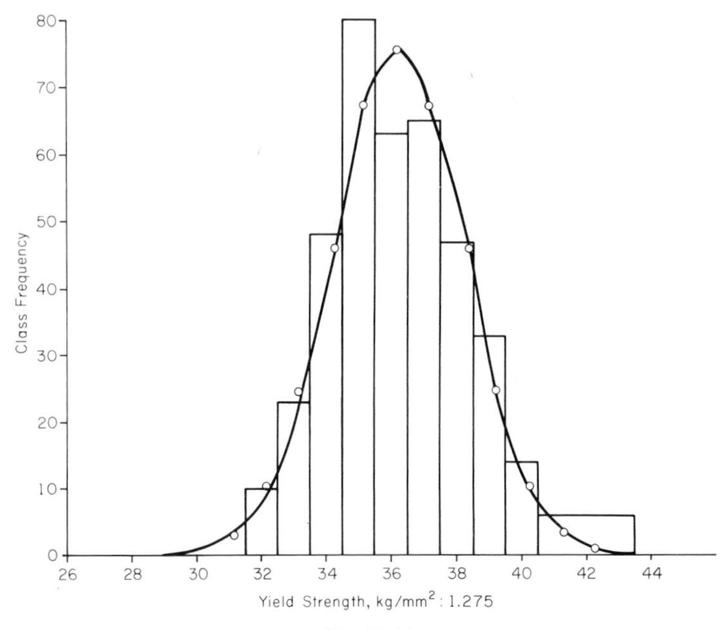

Fig. 19-11

The normal curve is shown in Fig. 19-11 and the values obtained from it are then used to obtain χ^2. (See Table 19-15.)

Since the estimated frequency of the last class is rather small, we pool the last two classes; the number of degrees of freedom is then $8 - 2 = 6$, since we estimated two

TABLE 19-15

Upper class boundary	*E*	*O*	$\lvert O - E \rvert$	$\frac{(O-E)^2}{E}$
32	8	10	2	0.500
33	19	23	4	0.842
34	41	48	7	1.195
35	66	80	14	2.970
36	76	63	13	2.224
37	69	65	4	0.232
38	53	47	6	0.679
39	29	33	4	0.552
40	12	14	2	0.333
42	2	6	4	8.000
				$\chi^2 = 17.527$

parameters from the observed data. Table A-7 gives the value of 17.527 as significant at the 1 percent level so that we conclude that the normal distribution does not satisfactorily fit the observed data.

Problems

19-1. Twenty electric motors were run to destruction and the times to failure (in hours) were as follows: 106.2, 159.6, 194.3, 215.9, 227.6, 271.6, 274.3, 297.7, 299.0, 345.5, 357.6, 373.4, 398.0, 430.5, 447.9, 485.3, 512.4, 548.9, 564.5, 608.9. Estimate the parameters for a Weibull model to fit the observed failure times using the simplified model without the location parameter [Eq. (19-12)]. What proportion of the motors will last more than two weeks (336 hours)?

19-2. Measurement of the size of fly ash revealed the distribution shown in Table 19-16. Fit a Weibull model to the observed values and compare the goodness of fit using χ^2. Also, fit a normal distribution to the data and calculate χ^2.

TABLE 19-16

Particle diameter in 20 μm, t_i	*Cumulative frequency,* i
2[a]	3
3	14
4	34
5	56
6	85
7	126
8	150
9	175
10	188
11	197
12	202
13	208
14	211

Source: W. Weibull, "A Statistical Distribution Function of Wide Applicability," *Journal of Applied Mechanics*, vol. 18, Sept. 1951, pp. 293–297; after J. M. Dalla Valle, *Micromeritics*, (New York: Pitman Publishing Corporation, 1948), p. 57, Fig. 2.

[a] Minimum diameter = 1.5 (μm : 20).

19-3. The values of annual maximum instantaneous flows of a river for the years 1942 through 1965 are listed in Table 19-17.

a. Plot the data on a graph of discharge versus return period.
b. Compute the mean and standard deviation of the extreme value distribution.
c. Estimate the return periods of a flow of 5500 m^3/s and of the average flood.
d. Estimate the magnitude of flows with return periods of 10, 15, and 25 years.

TABLE 19-17

Year	*Flow Q, m^3/s*	*Year*	*Flow Q, m^3/s*
1942	2750	1954	1150
1943	7200	1955	7150
1944	1750	1956	2000
1945	8600	1957	2600
1946	2800	1958	4670
1947	1890	1959	2350
1948	4780	1960	4300
1949	7270	1961	5900
1950	4370	1962	5300
1951	1580	1963	4300
1952	5200	1964	2800
1953	5420	1965	1800

19-4. The breakdown voltage of a capacitor is affected by the largest size of the conductivity particle present; for practical purposes it is the smallest breakdown voltage that is of interest.

The breakdown voltages listed in Table 19-18 (in rank order) have been measured on 18 capacitors. Using probability paper, obtain an expression for the proportion of capacitors having a specified breakdown voltage. (Note that because the smallest value is critical, ranking is in a decreasing order of magnitude.)

TABLE 19-18

Capacitor rank, m	*Breakdown voltage, V*
1	900
2	870
3	845
4	840
5	830
6	830
7	800
8	800
9	780
10	760
11	750
12	710
13	700
14	700
15	670
16	620
17	600
18	520

Chapter 20

Tolerance and Control Charts

This chapter gives a brief review of some of the graphical methods of presentation of data used mainly in production. Although no fundamental principles of statistics are involved, the use of charts frequently enables us rapidly to assess the behavior of a system and to take appropriate action without delay.

In general terms, the product being controlled may be judged by *attributes* or by *variables.* The former refer to a property that either is or is not possessed, e.g., a defect; thus the products are divided into two categories only. The variables refer to quantities and measurements, and it is with this type of product that we shall deal first.

SPECIFICATION AND TOLERANCE LIMITS

In design specification, limits on dimensions are usually set in the form of a nominal value and a plus or minus tolerance, e.g., 1.000 ± 0.004 cm. (The plus and minus deviations need not be equal.) Such limits ensure that the items manufactured are serviceable and can be assembled together with other parts. These are the *specification limits.* In the actual manufacture of the items a natural variation in dimensions occurs because of chance errors; from the distribution of these errors the proportion of items whose dimensions fall within *natural tolerance limits* can be calculated. An understanding of the difference between the two types of limits is of considerable importance.

A process of manufacture is said to be *stable* or *in control* if variations between individuals are caused by chance only. Under such circumstances the data obtained by sampling are consistent with the hypothesis that the observations are random values from a population.

The variability of the process can be described by the population standard deviation σ. Table A-4 shows that when the variate is normally distributed, all but 0.27 percent of observations will lie within a total range of 6σ. It follows that if the specification limits are greater than 6σ, the process will produce items with a very small proportion of defectives (Fig. 20-1a). If, on

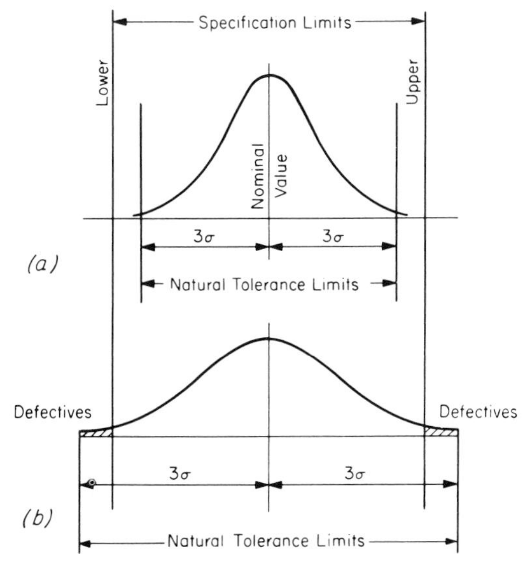

Fig. 20-1. Influence of the relation between specification and natural tolerance limits on the proportion of defectives.

the other hand, the specification limits are narrower than 6σ, then a sensible proportion of defectives will inevitably be manufactured (Fig. 20-1b). Should such a proportion of defectives not be acceptable, the only remedy lies in modifying the process, e.g., by using a more precise machine. Before attempting this, it may be wise to see whether the specification limits are not unduly restrictive and cannot be widened without ill effects.

We may also note that if the natural tolerance limits are considerably narrower than the specification limits, the process of manufacture is "too good" and, therefore, probably unnecessarily costly. In general terms, we should aim at tolerance limits approximately coinciding with the specification limits.

STABILITY OF A PROCESS

The situation discussed in the preceding section exists when both the mean and the standard deviation remain sensibly constant; we refer to such a situation as *stable*. Conversely, the process is said to be *unstable* or *not in control* when changes either in the mean or in variability take place.

In a general case, the method of manufacture is adjusted until the process becomes stable. To establish this condition, we take samples, preferably all of equal size $n \nless 4$. Not less than about 25 samples are required.

The population mean is estimated by the mean of the sample means $\bar{\bar{x}}$, and the population standard deviation may be estimated from the variance within samples, the average sample standard deviation, or the average sample range $\bar{R}$. The use of the last-mentioned estimate is most common.

WARNING AND ACTION LIMITS

Having established the parameters of the process, we can describe limits which, if exceeded, will tell us that the process is out of control or is in danger of being so.

Two pairs of control limits are used. The first pair represents warning or *inner* limits. These are usually set[1] so that there is a 2.5 percent probability of a sample mean having a value below the lower limit and a 2.5 percent probability of its having a value above the upper limit. The warning limits are thus drawn at a distance from the estimated population mean equal to $1.96 \times$ standard deviation of the mean, i.e., $\pm 1.96\sigma_{\bar{x}}$.

Action or *outer* limits are usually set[2] so that there is a probability of 0.1 percent of the mean falling above the upper limit and a probability of 0.1 percent of its falling below the lower limit. The distance of the limits from $\bar{\bar{x}}$ is then $\pm 3.09\sigma_{\bar{x}}$.

The values of $1.96\sigma_{\bar{x}}$ and $3.09_{\bar{x}}$ are obtained from Table A-4 for the normal distribution. We are justified in using this table even if the underlying population is not normally distributed, as we are dealing with sample means. Thus, once again, the Central Limit Theorem (see Chapter 10) has proved of great use.

1. Other values may be used, depending on the process.
2. A deviation from the mean of $\pm 3\sigma_{\bar{x}}$ is also used.

MEAN CHART

We have defined a process in control as one whose mean and standard deviation do not significantly change. Each of these quantities can be plotted on an appropriate chart on which both the mean value and the warning and action limits are drawn.

Let us first deal with the mean chart. The population mean is estimated from the means of samples, as explained in Chapter 6, and is given by

$$\bar{\bar{x}} = \frac{\bar{x}_1 + \bar{x}_2 + \cdots + \bar{x}_k}{k} \tag{20-1}$$

where $\bar{x}_1$, $\bar{x}_2$, ... are sample means and k is the number of samples.

To establish the warning and action limits for the mean, we require an estimate of the standard deviation of the mean. Not distinguishing between estimates and true values, we can write the standard deviation of the mean as

$$\sigma_{\bar{x}} \simeq \frac{s}{\sqrt{n}} \tag{20-2}$$

where s is the estimate of the population stnadard deviation and n the sample size. The value s can be estimated from the mean range R, using Eq. (4-11), whence

$$s = \bar{R}d \tag{20-3}$$

The values of d are given in Table A-1. Hence the warning limits for the mean (MWL) are given by

$$\bar{\bar{x}} \pm 1.96\sigma_{\bar{x}} = \bar{\bar{x}} \pm \frac{1.96s}{\sqrt{n}}$$

Thus

$$\text{MWL} = \bar{\bar{x}} \pm \frac{1.96\bar{R}d}{\sqrt{n}} \tag{20-4}$$

It is convenient to let

$$A_w = \frac{1.96d}{\sqrt{n}} \tag{20-5}$$

The values of A_w are given in Table A-12. Hence the warning limits for the mean become

$$\text{MWL} = \bar{\bar{x}} \pm A_w\bar{R} \tag{20-6}$$

By a similar argument the action limits for the mean (MAL) are

$$\bar{\bar{x}} \pm 3.09\sigma_{\bar{x}} = \bar{\bar{x}} \pm \frac{3.09s}{\sqrt{n}}$$

$$\text{MAL} = \bar{\bar{x}} \pm \frac{3.09\bar{R}d}{\sqrt{n}} \tag{20-7}$$

Let

$$A_A = \frac{3.09d}{\sqrt{n}} \tag{20-8}$$

The action limits for the mean then become

$$\text{MAL} = \bar{\bar{x}} \pm A_A\bar{R} \tag{20-9}$$

The values of A_A are given in Table A-12.

If, during the operation, points fall outside the limits, we conclude (with the appropriate probability of being wrong) that the variability between samples is significantly greater than the variability within a sample. The chart gives thus the same result as the analysis of variance for between and within samples.

Example. Set up the control limits for the mean chart for the production of steel shafts. The mean of means is $\bar{\bar{x}} = 12.000$ cm, and the mean range has been found to be $\bar{R} = 0.0093$ cm. The samples are of size $n = 5$. From Table A-12 for $n = 5$,

$$A_w = 0.377$$

$$A_A = 0.594$$

Hence

$$_{\text{lower}}^{\text{upper}}\text{MWL} = \bar{\bar{x}} \pm A_w\bar{R} = 12.000 \pm 0.377 \times 0.0093$$

$$= \left.\begin{matrix} 12.00351 \text{ cm} \\ 11.99649 \text{ cm} \end{matrix}\right\}$$

Now

$$_{\text{lower}}^{\text{upper}}\text{MAL} = \bar{\bar{x}} \pm A_A\bar{R} = 12.000 \pm 0.594 \times 0.0093$$

$$= \left.\begin{matrix} 12.00552 \text{ cm} \\ 11.99448 \text{ cm} \end{matrix}\right\}$$

RANGE CHART

As we have said earlier, the mean chart is not sufficient to determine whether the process is stable. The mean of a sample may lie within the limits, and yet individual observations may fall outside these limits. If this happened, it would mean that the standard deviation is changing. For this reason we have to keep a record of the variability of observations within each sample, and this is best done by plotting the range of each sample on a control chart, known as a range chart.[3] In doing this, we take advantage of the relation between range and standard deviation, given by Eq. (20-3).

In the range chart we mark warning and action limits in a manner similar to the limits on the mean chart. We may remember, however, that neither the range nor the standard deviation are normally distributed but have probability distributions related to the χ^2 distribution. The limits in the range chart are not symmetrically disposed about the mean range, $\bar{R}$.

We write the range warning limits (RWL):

$$ {}^{\text{upper}}_{\text{lower}}\text{RWL} = \left.\begin{matrix} D_{WU} \times \bar{R} \\ D_{WL} \times \bar{R} \end{matrix}\right\} \qquad (20\text{-}10) $$

and the range action limits (RAL):

$$ {}^{\text{upper}}_{\text{lower}}\text{RAL} = \left.\begin{matrix} D_{AU} \times \bar{R} \\ D_{AL} \times \bar{R} \end{matrix}\right\} \qquad (20\text{-}11) $$

where the values of the factors D are given in Table A-13.

The coefficients for the range chart are based on normal distribution; however, even if the distribution is not normal, the control limits, especially the action limits, are reliable for most purposes.

We may note that the control limits for range widen as the sample size increases, while the mean chart becomes narrower with an increase in sample size.

Example. Find the range control limits for the data of the preceding example.

From Table A-13 for $n = 5$,

$$ D_{WU} = 1.81 $$

$$ D_{WL} = 0.37 $$

$$ D_{AU} = 2.36 $$

$$ D_{AL} = 0.16 $$

Hence

$$ {}^{\text{upper}}_{\text{lower}}\text{RWL} = \begin{matrix} 1.81 \\ 0.37 \end{matrix} \times 0.0093 = \left.\begin{matrix} 0.016833 \text{ cm} \\ 0.003441 \text{ cm} \end{matrix}\right\} $$

3. The range chart thus tests the same hypothesis as Bartlett's test (see Chapter 14).

and

$$\substack{\text{upper}\\\text{lower}}\text{RAL} = \frac{2.36}{0.16} \times 0.0093 = \left.\begin{matrix}0.021948 \text{ cm}\\0.001488 \text{ cm}\end{matrix}\right\}$$

SCHEME OF OPERATION

We can now summarize the steps to be followed in the construction of mean and range charts to be used for the purpose of obtaining a continual check on the stability of a process.

1. Determine the mean value $\bar{\bar{x}}$ of the quantity that we are measuring; this is the mean value which the process could be expected to yield if it were functioning perfectly. The symbol $\bar{\bar{x}}$ indicates the mean of means of a large number of random samples during a stable period.

2. Determine the standard deviation σ of the measured quantity during this stable period. If this is not convenient, find the sample ranges during the same period, calculate the mean range $\bar{R}$, and estimate the standard deviation, using coefficient d of Table A-1.

3. Find the values of A_w and A_A from Table A-12 for the sample sizes used, and hence establish the warning and action limits $\bar{\bar{x}} \pm A_w \bar{R}$ and $\bar{\bar{x}} \pm A_A \bar{R}$, respectively. Plot these limits on a chart that also shows $\bar{\bar{x}}$. If any future observation falls outside the warning limits, further samples should be carefully watched, and if the trend persists, an investigation of the process should follow. An observation outside the action limits indicates that the process has gone out of control and remedial action is immediately necessary.

4. Find the upper and lower warning and action limits for range, using the D coefficients of Table A-13. Plot the mean range and the control limits. If any future sample range is found to be outside the action limits, the process is likely to have moved away from a stable position and steps must be taken to find the cause and remedy the situation.

5. If in establishing the control limits, we find an occasional value outside the action limits, the value is disregarded and a new mean and new control limits are calculated. There is little theoretical justification for this except that if the process is truly in control, a point outside the limits can be considered as not belonging to the population of items in control. Such a point is then considered to be an outlier and is rejected (cf. Chapter 11).

In general terms, if a process that was in control becomes unstable, we blame this on "assignable causes of variation," which we seek to remove. Such a cause may be identified as a specific part of the production process, but the possibility of poor sampling should not be ignored. Finally, it may be that the production process has changed (e.g., due to wear of some parts) and new control charts have to be set up.

Example. A manufacturer of a certain type of resistors decided to set up control charts for his product. Twenty-five samples of size $n = 4$ were taken. The means and range of each sample are shown in Table 20-1. Compute the limits for the mean and range charts for the process and check whether the process is in control. If so, estimate the standard deviation for the process. If the specification limits are 120 ± 8 ohms, is the process able to meet the specification?

TABLE 20-1

Sample	*Resistance in ohms*		*Sample*	*Resistance in ohms*	
	Mean $\bar{x}$	*Range* R		*Mean* $\bar{x}$	*Range* R
1	122	10	15	120	13
2	126	6	16	125	12
3	118	5	17	118	5
4	121	8	18	119	8
5	126	11	19	124	4
6	125	5	20	121	3
7	119	4	21	125	7
8	121	3	22	121	5
9	124	11	23	120	4
10	125	14	24	121	4
11	123	10	25	119	6
12	124	9			
13	122	7		$\sum \bar{x} = 3050$	$\sum R = 185$
14	121	11			

Hence

$$\bar{\bar{x}} = \frac{\sum \bar{x}}{n} = \frac{3050}{25} = 122 \text{ ohms}$$

$$\bar{R} = \frac{\sum R}{n} = \frac{185}{25} = 7.4 \text{ ohms}$$

The trial control limits are computed with the aid of Tables A-12 and A-13 as follows:

For mean chart:

$$\begin{aligned}{}^{\text{upper}}_{\text{lower}}\text{warning limits} &= \bar{\bar{x}} \pm A_w\bar{R} \\ &= 122.00 \pm 0.476 \times 7.4 \\ &= \left.\begin{matrix}125.52\\118.48\end{matrix}\right\}\text{ohms}\end{aligned}$$

$$\begin{aligned}{}^{\text{upper}}_{\text{lower}}\text{action limits} &= \bar{\bar{x}} \pm A_A\bar{R} \\ &= 122.00 \pm 0.750 \times 7.4 \\ &= \left.\begin{matrix}127.55\\116.45\end{matrix}\right\}\text{ohms}\end{aligned}$$

For range chart:

$${}^{\text{upper}}_{\text{lower}}\text{warning limits} = \left.\begin{matrix}D_{WU}\bar{R} = 1.93 \times 7.4 = 14.28\\ D_{WL}\bar{R} = 0.29 \times 7.4 = 2.15\end{matrix}\right\}\text{ohms}$$

$${}^{\text{upper}}_{\text{lower}}\text{action limits} = \left.\begin{matrix}D_{AU}\bar{R} = 2.58 \times 7.4 = 19.09\\ D_{AL}\bar{R} = 0.10 \times 7.4 = 0.74\end{matrix}\right\}\text{ohms}$$

These limits, as well as the appropriate mean values, are shown in Figs. 20-2 and 20-3. The plots of the mean and the range for each sample show that all points fall inside the action limits. Four of the sample means fall outside the warning limits, but the corresponding ranges are inside the warning limits for the range; this indicates that the situation is not serious but has to be watched. Therefore, we can assume for the time being that the process can be brought into control at this level.

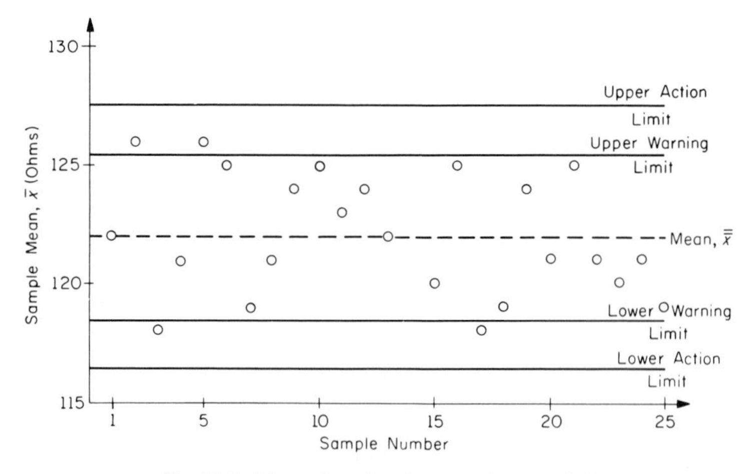

Fig. 20-2. Mean chart for the example on p. 373.

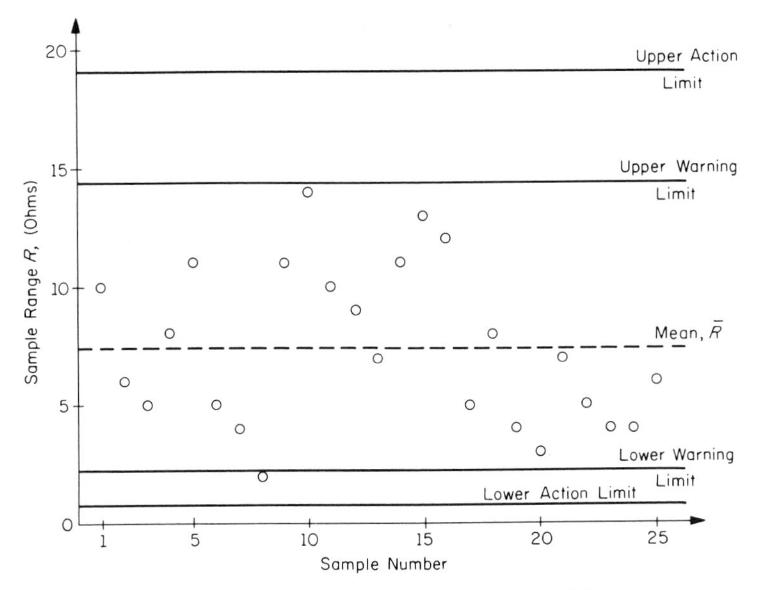

Fig. 20-3. Range chart for the example on p.373.

To obtain an estimate of the population standard deviation σ, we use Table A-1 for $n = 4$, $d = 0.4857$. Thus

$$s = \bar{R}d = 7.4 \times 0.4857 = 3.594 \text{ ohms}$$

The natural tolerance limits are

$$\bar{\bar{x}} \pm 3\sigma = 122 \pm 3 \times 3.594 = \left.\begin{matrix}111.22\\132.78\end{matrix}\right\}\text{ohms}$$

The specification limits are 120 ± 8 ohms, i.e., 112 to 128 ohms. The process does not, therefore, meet the specification.

CUMULATIVE SUM CHART

If we are interested only in changes that are large and, therefore, easily detectable on a control chart, the use of these simple-to-plot charts is adequate. However, if we want to detect at an early stage small changes that may be obscured by residual variation, there is another type of chart which is more efficient and still not laborious to use. This is the cumulative sum chart, often called the *cusum chart*, in which a cumulative sum of the differences between the process variable and the previously accepted mean value is plotted after each observation.

We can call this mean value the target T. Then, at any instant t, the cumulative sum of deviations about T is

$$S_t = \sum_{i=1}^{t} (x_i - T)$$

This value is plotted against time t, as shown in Fig. 20-4 for the data of Fig. 20-2, where, instead of time, successive samples are used. We can see that the slope of the cusum chart indicates the deviation of the process mean from the target, and such a change is visually easier to detect than the variation in the control chart.

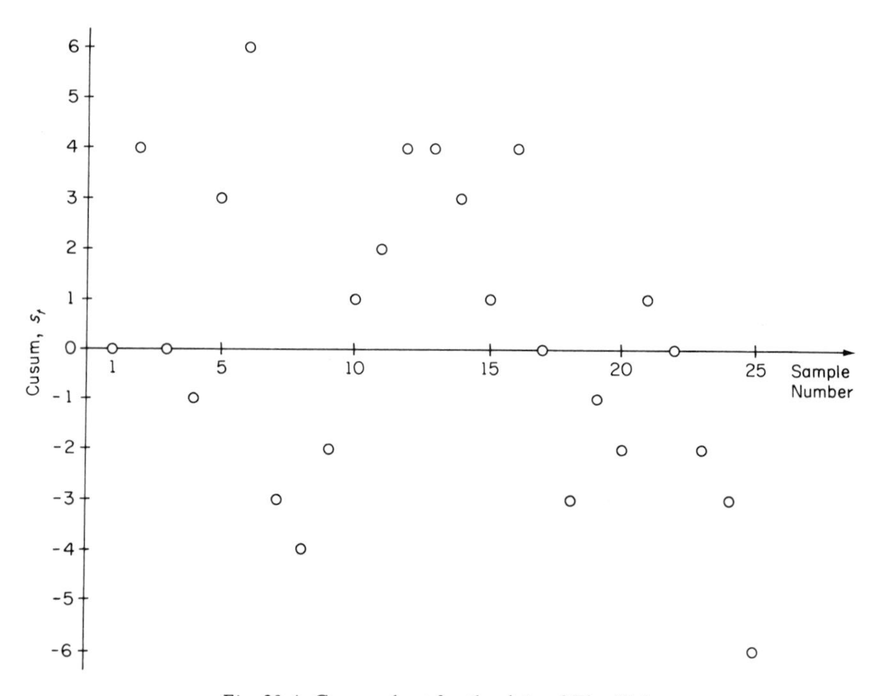

Fig. 20-4. Cusum chart for the data of Fig. 20-2.

Each successive value of S_t is obtained simply by adding the value of $(x - T)$ for the last observation. Now, if we are interested in the change in the mean between the kth and nth observations, we can write the mean value between $(k + 1)$ and n as

$$\bar{x} = \sum_{i=k+1}^{n} \frac{x_i}{n-k}$$

or

$$\bar{x} = T + \frac{\sum_{i=k+1}^{n} (x_i - T)}{n - k}$$

Thus

$$\bar{x} = T + \frac{S_n - S_k}{n - k}$$

i.e.,

$$\bar{x} = T + \frac{\text{change in cusum}}{\text{number of observations made}}$$

The fraction clearly represents the slope of the cusum, and it is important, therefore, to choose a convenient scale for the desired variation to be readily noticeable.

CHARTS USING ATTRIBUTES

In dealing with attributes, we are concerned with the binomial and Poisson distributions. The former is applicable when we are dealing with *defective* items, the latter when we count the number of *defects* per item.

If $\bar{p}$ is the mean proportion of defective items, then the mean number of defectives in a sample of size n is $n\bar{p}$. We may recall from Chapter 7 that the standard deviation of the number of defectives is $\sqrt{n\bar{p}q} = \sqrt{n\bar{p}(1 - \bar{p})}$. Hence we can write the action limits for the mean *number* of defectives as

$$\text{MAL} = n\bar{p} \pm 3.09\sqrt{n\bar{p}(1 - \bar{p})} \tag{20-12}$$

Instead of operating in terms of the number of defectives, we can express the control limits in terms of the *mean proportion* of defectives $\bar{p}$. The standard deviation of the mean proportion of defectives is

$$\frac{\sqrt{n\bar{p}q}}{n} = \sqrt{\frac{\bar{p}(1 - \bar{p})}{n}}$$

The action limits for the mean proportion of defectives are then

$$\text{proportion AL} = \bar{p} \pm 3.09\sqrt{\frac{\bar{p}(1 - \bar{p})}{n}} \tag{20-13}$$

If a negative limit is obtained in Eq. (20-12) and (20-13), it is replaced by zero, as a negative proportion or number of defectives is not possible.

When dealing with an expected number of *defects* in a material np, the Poisson distribution is applicable. From Eq. (8-6) the standard deviation is $\sqrt{np}$. Thus, in this case, the action limits for the number of defects are

$$\text{number AL} = np \pm 3.09\sqrt{np} \tag{20-14}$$

Example. Thirty successive samples, each consisting of 50 machine bolts, were checked by a "go–no-go" gage. The results are shown in Table 20-2.

Find the mean proportion of defectives and calculate the upper and lower action limits for such a mean proportion. Are any of the plotted points outside these limits? If so, adjust the mean proportion of defectives and the corresponding limits for future production.

TABLE 20-2

Sample	*Number defective*	*Proportion defective p*	*Sample*	*Number defective*	*Proportion defective p*
1	2	0.04	17	0	0.00
2	1	0.02	18	0	0.00
3	5	0.10	19	1	0.02
4	1	0.02	20	3	0.06
5	2	0.04	21	2	0.04
6	1	0.02	22	1	0.02
7	3	0.06	23	1	0.02
8	0	0.00	24	2	0.04
9	1	0.02	25	0	0.00
10	2	0.04	26	0	0.00
11	1	0.02	27	1	0.02
12	0	0.00	28	1	0.02
13	0	0.00	29	2	0.04
14	3	0.06	30	2	0.04
15	2	0.04			
16	2	0.04			$\sum p = 0.84$

The mean proportion of defectives is

$$\bar{p} = \frac{0.84}{30} = 0.028$$

The action limits are given by Eq. (20-13):

$$\text{proportion AL} = \bar{p} \pm 3.09\sqrt{\frac{\bar{p}(1-\bar{p})}{n}}$$

$$\text{proportion AL} = 0.028 \pm 3.09\sqrt{\frac{0.028 \times 0.972}{50}}$$

$$= (-0.044, 0.1001)$$

We substitute zero for the lower limit and write the limits as (0, 0.1001).

Figure 20-5 shows the control limits and a plot of the proportion defective in each sample. It can be seen that all the points fall inside the limits; therefore, we can use these limits as control limits for future production. Had

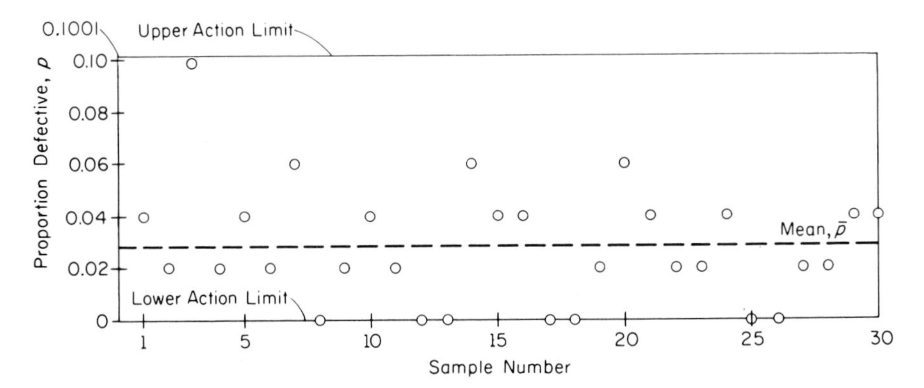

Fig. 20-5. Control chart for the proportion defective in the example on p. 378.

some points fallen outside, they could have been eliminated, and a new $\bar{p}$ and new limits should have been calculated from the remaining samples.

Example. In producing aluminum sheets for a new type of aircraft, it was required to set up control charts for the number of defects per sheet of a certain size. Twenty-five sample sheets were drawn and the number of defects per sheet was recorded. The results are shown in Table 20-3. On the basis of these results set up control limits for the number of defects for future production.

TABLE 20-3

Sample	*Number of defects per sheet*	*Sample*	*Number of defects per sheet*
1	1	14	1
2	0	15	2
3	2	16	0
4	1	17	0
5	0	18	8
6	0	19	2
7	2	20	0
8	2	21	0
9	0	22	1
10	1	23	0
11	1	24	1
12	2	25	1
13	0		
			$\sum = 28$

The expected number of defects per sheet is

$$np = \frac{28}{25} = 1.12$$

From Eq. (20-14) the action limits for the number of defects are

$$np \pm 3.09\sqrt{np} = 1.12 \pm 3.09 \times 1.059$$
$$= (-2.15, 4.39)$$

Substituting zero for the negative value, we write the limits as (0, 4.39). A plot of the control limits and of the number of defects in each sample

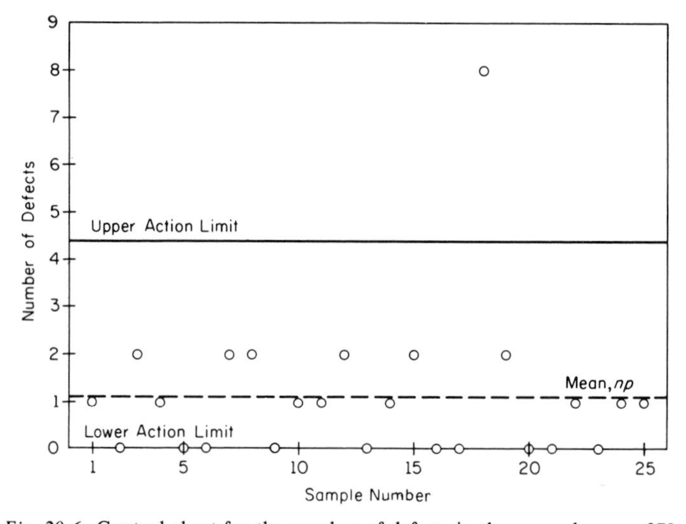

Fig. 20-6. Control chart for the number of defects in the example on p. 379.

(Fig. 20-6) shows that the 18th sheet falls outside the control limits. In this case it is best to base future work on a mean of defects np that does not include this particular sheet. We shall now recalculate np and the control limits, ignoring the 18th sheet.

$$\sum \text{number of defects} = 20$$

$$np = \frac{20}{24} = 0.833$$

Control limits are

$$0.833 \pm 3.09\sqrt{0.833}$$
$$= (-1.98, 3.65)$$

Thus the revised control limits are (0, 3.65).

It can be seen by inspection that none of the samples has a number of defects outside these new limits. Thus the new np and control limits can be used in future control work on the number of defects in the production of the aluminum sheets. We may add that these limits do not guarantee that a fixed percentage of the population will lie between them, as the probability of the number of defects falling between these limits depends on the mean expected number of defects np. For large n, np is approximately normally distributed, and hence we can assume that "almost" all the values will fall between the control limits.

ADVANTAGES OF QUALITY CONTROL

The notes on control charts given here are limited to basic principles, and a detailed study of the topic must be sought in one of the numerous books on quality control.

Quality control is used in all manufacturing processes, from pharmaceuticals to concrete products. Its special feature is that it replaces a negative form of inspection—i.e., discarding of defective items—by a more positive form. The process is watched continually and systematically so that a departure from a stable condition is diagnosed early; appropriate action can then be taken and the fault can be rectified before an excessive number of defectives has been manufactured. We must remember that quality control accepts the fact that some fraction of the items produced will be defective but makes it possible to ensure that the specified fraction is not exceeded.

Solved Problems

20-1. The mean range of shear strength of a certain kind of spot weld was determined from many tests on samples of 6 welds to be $\bar{R} = 35$ kN/m^2. Set up control limits for successive samples of size $n = 6$.

SOLUTION

From Table A-13, the factors for warning limits are $D_{WU} = 1.72$, $D_{WL} = 0.42$. For action limits the factors are $D_{AU} = 2.22$ and $D_{AL} = 0.21$. Hence

$$\begin{matrix}\text{upper}\\\text{lower}\end{matrix}\text{warning limits} = \left.\begin{matrix} D_{WU}\bar{R} = 1.72 \times 35 = 60.2 \text{ kN/m}^2 \\ D_{WL}\bar{R} = 0.42 \times 35 = 14.7 \text{ kN/m}^2 \end{matrix}\right\}$$

$$\begin{matrix}\text{upper}\\\text{lower}\end{matrix}\text{action limits} = \left.\begin{matrix} D_{AU}\bar{R} = 2.22 \times 35 = 77.7 \text{ kN/m}^2 \\ D_{AL}\bar{R} = 0.21 \times 35 = \ \ 7.4 \text{ kN/m}^2 \end{matrix}\right\}$$

20-2. Samples of 6 rings are taken at regular intervals from an assembly line for engine pistons. The inside diameters of the rings are measured, and the sample mean

$\bar{x}$ and the sample range R are determined. For the first 30 samples, the sums of the means and of the ranges were $\sum \bar{x} = 135.81$ cm and $\sum R = 0.063$ cm.

a. What are the control action limits for the mean and range charts?
b. What are the 3σ natural tolerance limits for the ring diameter, assuming that $\bar{\bar{x}}$ and $\bar{R}$ can be used to estimate the mean and standard deviation of the population?
c. Is the process able to produce rings with inside diameters within the specification limits of 4.525 ± 0.005 cm?
d. What proportion of rings will fall outside the specification limits if the process is in control with the calculated $\bar{\bar{x}}$ and derived σ?

SOLUTION

a. Compute

$$\bar{\bar{x}} = \frac{\sum \bar{x}}{n} = \frac{135.81}{30} = 4.527 \text{ cm}$$

$$\bar{R} = \frac{\sum R}{n} = \frac{0.063}{30} = 0.0021 \text{ cm}$$

From Table A-1, for $n = 6$, $d = 0.3945$. Hence

$$s = 0.0021 \times 0.3945 = 0.0008293$$

For the mean chart:

$$\text{action limits} = \bar{\bar{x}} \pm A_A \bar{R}$$

From Table A-12, for $n = 6$, $A_A = 0.498$. Hence

$$\text{action limits} = 4.527 \pm 0.498 \times 0.0021$$
$$= 4.52805, 4.52595 \text{ cm}$$

For the range chart:

$${}^{\text{upper}}_{\text{lower}}\text{action limit} = \begin{matrix} D_{AU}\bar{R} \\ D_{AL}\bar{R} \end{matrix}$$

From Table A-13, for $n = 6$, $D_{AU} = 2.22$ and $D_{AL} = 0.21$. Hence

$$\text{upper action limit} = 2.22 \times 0.0021$$
$$= 0.00466 \text{ cm}$$

$$\text{lower action limit} = 0.21 \times 0.0021$$
$$= 0.00044 \text{ cm}$$

b. As instructed, we assume that $\sigma = s$. Hence

$$\text{natural tolerance limits} = \bar{\bar{x}} \pm 3\sigma$$
$$= 4.527 \pm 3 \times 0.0008293$$
$$= 4.524, 4.529 \text{ cm}$$

c. The specification limits are $4.525 \pm 0.005 = 4.520, 4.530$ cm. Since the natural tolerance limits fall within the specification, the process is able to meet the specification.

d. We have $\bar{\bar{x}} = 4.527$ cm and $\sigma = 0.0008293$ cm.

The specification limits are 4.520 and 4.530 cm. Compute

$$z_1 = \frac{4.530 - 4.527}{0.0008293} = 3.62$$

From Table A-4, the area under the normal probability curve for such a value of z is approximately 0.49984. Also

$$z_2 = \frac{4.527 - 4.520}{0.0008293} = 8.45$$

This deviation is so large that the corresponding area under the normal probability curve can be taken as 0.5000. The total area between the specification limits is

$$0.50000 + 0.49984 = 0.99984$$

Hence the percentage outside these limits is

$$1 - 0.99984 = 0.00016 \qquad \text{or } 0.016 \text{ percent}$$

a very small percentage indeed.

20-3. A control chart for a new kind of plastics is to be initiated. Twenty-five samples of 100-m plastic sheets from the assembly line were inspected for flaws during a period of time. The results are given in Table 20-4.

TABLE 20-4

Sheet number	*Number of flaws per sheet*	*Sheet number*	*Number of flaws per sheet*
1	2	14	3
2	0	15	2
3	7	16	0
4	8	17	4
5	9	18	5
6	10	19	1
7	10	20	0
8	13	21	2
9	14	22	3
10	16	23	2
11	10	24	5
12	7	25	4
13	4		

Set up the necessary control chart.

SOLUTION

Compute the mean number of flaws

$$np = \frac{\sum \text{flaws}}{25}$$

$$= \frac{141}{25} = 5.64 \text{ flaws per sheet}$$

The control limits are

$$np \pm 3.09\sqrt{np}$$

$$= 5.64 \pm 3.09\sqrt{5.64}$$

$$= -1.70, 12.98$$

Thus the control limits are (0, 12.98). These limits and the mean number of flaws per sheet, as well as the individual number of flaws per sheet, are shown in Fig. 20-7. It can be seen that the samples numbered 8, 9, and 10 fall outside the upper control limit. Removing the value for these samples and recalculating, we get

$$np = \frac{98}{22} = 4.45 \text{ flaws per sheet}$$

$$\text{new control limits} = 4.45 \pm 3.09\sqrt{4.45}$$

$$= -2.07, 10.97$$

i.e., the control limits are (0, 10.97). By inspection, it is seen that none of the remaining sheets has flaws outside these limits.

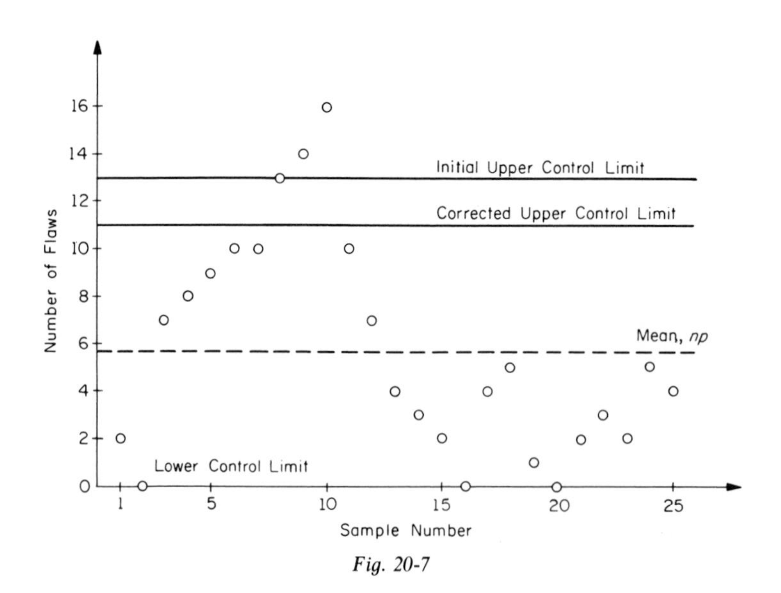

Fig. 20-7

A close look at Fig. 20-7 will show what appears to be an approximately cyclic pattern of variation in the number of flaws in successive sheets. In such cases it is advisable to check the production process before continuing production, since this trend is unlikely to be caused by chance alone.

Problems

20-1. From numerous tests it has been found that the mean strength and mean range of a certain timber are 3510 and 390 kN/m^2 respectively. Using these values, set up control limits for the mean chart when the sample size is 5.

20-2. Establish range and mean charts (containing warning and action limits) on the basis of the data for samples of size $n = 3$ in Table 20-5.

TABLE 20-5

Sample number	*Mean*	*Range*	*Sample number*	*Mean*	*Range*
1	11.97	0.5	13	17.87	8.7
2	14.87	6.3	14	14.97	0.1
3	15.35	7.5	15	14.60	9.8
4	15.72	6.6	16	14.12	7.7
5	11.12	4.9	17	13.21	7.5
6	11.06	6.7	18	12.86	1.2
7	11.97	9.9	19	12.38	9.8
8	12.27	6.9	20	16.99	6.5
9	12.85	0.1	21	18.35	7.6
10	13.23	8.3	22	18.52	4.0
11	16.12	5.1	23	14.13	8.4
12	16.61	3.2	24	14.60	9.8

20-3. Screws produced by an automatic machine are to be checked. A sample of 60 screws was checked by a go–no-go gage, and it was found that in 30 successive samples the following defective number of screws was obtained:

5, 4, 3, 1, 2, 0, 0, 0, 1, 2, 0, 1, 2, 1, 0, 1, 1,

0, 1, 0, 0, 1, 2, 0, 1, 0, 1, 0, 0, 1

a. Is this process in control?
b. Taking the process in its present condition, will it produce screws of which 5 percent or less are defective?

20-4. Thirty 20-m lengths of carpets were inspected, and the average number of defects per unit area was found to be 2.1. Compute the 2.5σ control limits. What are the probability limits corresponding to a probability of 0.3 percent of falling outside the upper control limits?

20-5. For the data on concrete strength given in Problem 4-7:

a. Estimate the standard deviation from the mean range.
b. Construct control charts for the mean and range, and plot the means and the ranges on the respective charts.
c. Are any of the plotted points outside the action limits? If so, adjust the mean of means and the mean range and the corresponding limits for future production.

20-6. The percentage content of compound A in a product was measured three times a day during 60 days. The 180 measurements, with a mean value of approximately 0.38, were tabulated as shown in Table 20-6 and then plotted in Figure 20-8.

TABLE 20-6

0.30	0.39	0.36	0.22	0.27	0.29	0.46	0.44	0.35
0.27	0.27	0.23	0.37	0.36	0.58	0.39	0.26	0.38
0.44	0.36	0.30	0.34	0.41	0.69	0.36	0.23	0.27
0.12	0.16	0.22	0.24	0.28	0.33	0.13	0.18	0.35
0.51	0.32	0.50	0.39	0.55	0.66	0.53	0.42	0.42
0.61	0.46	0.65	0.46	0.29	0.29	0.39	0.53	0.46
0.37	0.26	0.22	0.29	0.52	0.54	0.55	0.38	0.39
0.32	0.45	0.46	0.43	0.38	0.30	0.28	0.49	0.29
0.50	0.30	0.47	0.56	0.43	0.45	0.58	0.63	0.27
0.52	0.27	0.51	0.29	0.30	0.38	0.39	0.38	0.53
0.52	0.45	0.39	0.63	0.33	0.37	0.47	0.40	0.39
0.42	0.56	0.46	0.40	0.47	0.49	0.32	0.69	0.49
0.72	0.43	0.75	0.41	0.40	0.49	0.22	0.34	0.51
0.28	0.36	0.41	0.37	0.45	0.42	0.51	0.52	0.13
0.29	0.37	0.45	0.44	0.68	0.28	0.31	0.38	0.42
0.33	0.56	0.31	0.33	0.31	0.46	0.34	0.58	0.22
0.20	0.52	0.38	0.33	0.23	0.21	0.18	0.26	0.36
0.23	0.38	0.33	0.23	0.39	0.38	0.22	0.23	0.28
0.35	0.94	0.23	0.16	0.29	0.46	0.22	0.27	0.18
0.36	0.23	0.20	0.23	0.27	0.35	0.37	0.38	0.37

Since little can be gleaned from these data, it is proposed to plot a cusum chart using the mean as a reference.

Draw this cusum chart and comment on the pattern of fluctuation. Read the data horizontally.

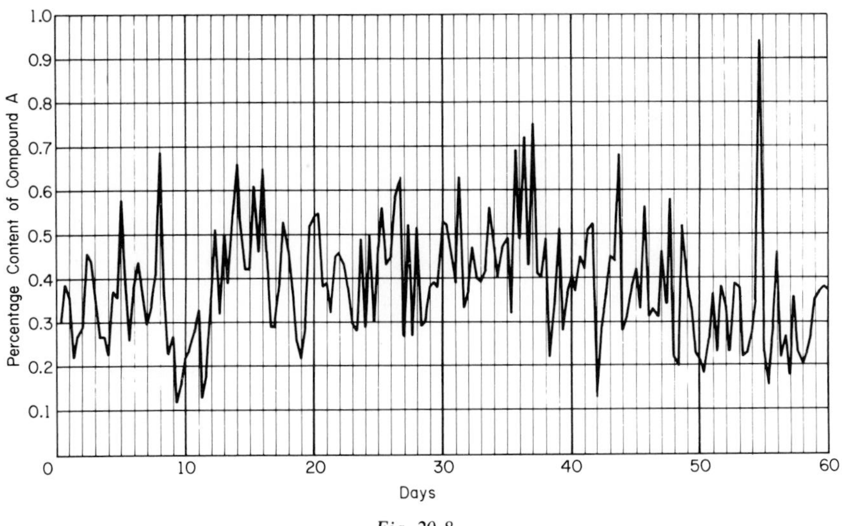

Fig. 20-8.

20-7. Data for the shear strength of spot welds has been maintained to construct mean and range control charts. From tests on 30 samples each of size $n = 4$, it is found that $\sum \bar{x} = 15{,}450$ N, and $\sum R = 1206$ N. Assuming the process is in control, determine:

a. the limits for the mean and range charts;
b. the standard deviation for the process.
c. If the minimum specifications for the spot weld is 400 N, what percentage of the welds does not meet the minimum specifications?

Chapter 21

Introduction to the Design of Experiments

The statistical methods discussed in the preceding chapters are generally used in the interpretation of experimental results. In some cases the planning of the experiment, in the broad sense of the word, may be outside our control, but in others we may be able to decide the procedure and details of the tests. This should be done in the light of the statistical analysis which will be subsequently applied; more information can then be extracted from the given experimental effort than when the statistical aspects of the program are considered only a posteriori.

The design of experiments is thus of considerable importance. In general, we require some advance knowledge of the variability of results. Such information may be available from previous work or alternatively should be obtained from pilot tests.

In the present chapter we shall consider only two topics: the choice of size of the sample and the design of experiments involving several variables. No more than an introduction will be given in either case.

CHOICE OF SAMPLE SIZE

As we have seen in the earlier chapters, our decisions about populations or batches (such as their acceptability under specification) are usually based on tests on samples. A problem that often arises is how many observations should be made so that the risk of making a wrong decision is acceptably small. Some risk is, of course, always present because of the random variation in results, but the risk becomes smaller as the number of observations increases. It is clear, however, that for reasons of economy of time and effort the testing should be kept to a minimum consistent with the maximum risk of a wrong decision, which we are prepared to accept. In practice, there is thus a certain optimum testing effort. In this book we cannot evaluate the cost of testing or the economic results of making a wrong decision; therefore, we shall consider our decision making on the basis of a specified risk of being wrong.[1]

The information required first of all is the distribution of the true mean value of the measured property of batches of the material.[2] Let us consider, for example, the strength of a plastic, and assume that in the past we received a sequence of batches of the plastic. From each batch k samples were taken and tested, and the mean of the k values was obtained. The means so obtained differ from one another for two reasons: They are estimates of different (batch) means and they are also affected by the testing error (arising from sampling and measuring).

If $\bar{x}_i$ is the mean strength obtained from tests on batch i, then

$$\bar{x}_i = \mu_i + \epsilon_i$$

where μ_i is the true mean strength of batch i, and ϵ_i is the testing error. Hence

$$\sigma_{\bar{x}}^2 = \sigma_\mu^2 + \frac{\sigma_T^2}{k}$$

where σ_μ^2 is the variance of μ, $\sigma_{\bar{x}}^2$ is the variance of the batch means available from the test results, and σ_T^2 is the variance of the testing error, assumed known.

In this manner σ_μ^2 can be readily estimated and the distribution of the underlying strength of the plastic can be determined. Assume that this is as shown in Fig. 21-1a. Suppose further that the specification calls for a minimum strength of 3000 kN/m^2. In practice, we obtain n test specimens of the

1. Details of the method described in the following pages and its further development are presented clearly in O. L. Davies, ed., *Statistical Methods in Research and Production* (Edinburgh, Scotland: Oliver & Boyd Ltd., 1958).

2. Since the tests refer usually to a batch of material, this term will be used in preference to a "finite population."

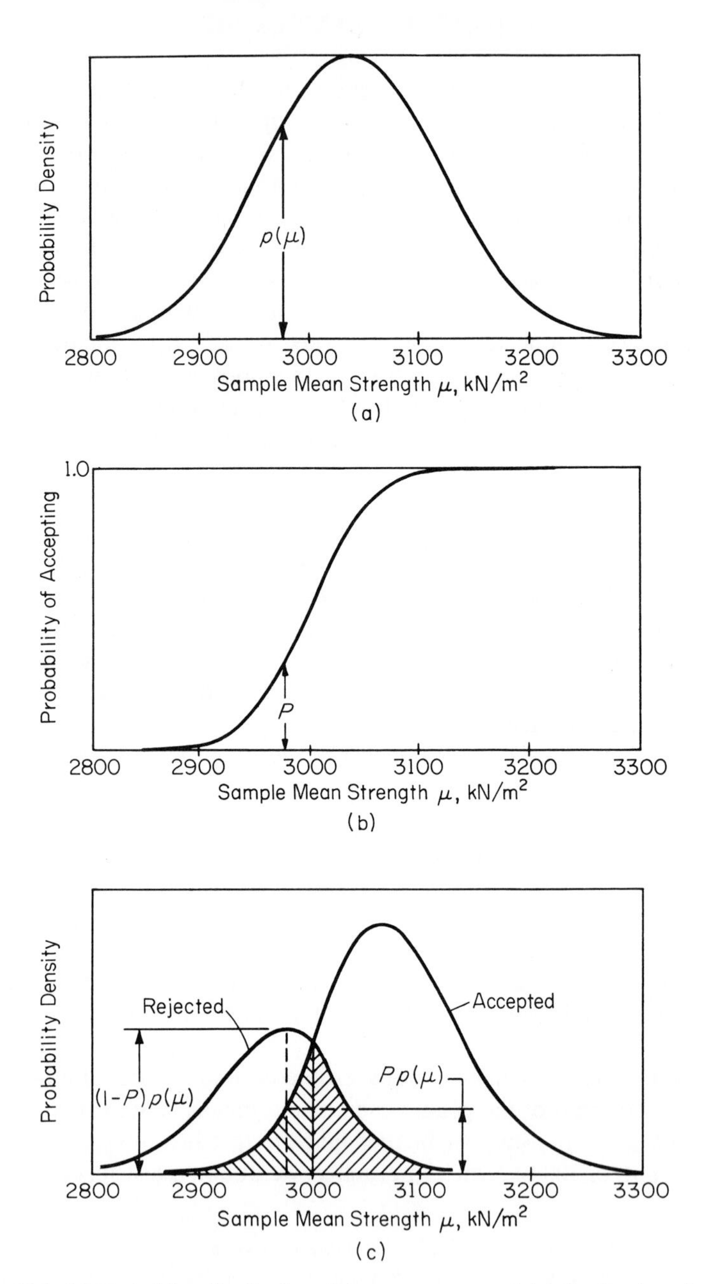

Fig. 21-1. (a) Underlying distribution; (b) power curve; and (c) distribution of accepte and rejected batches.

plastic and find their strengths. If the mean strength is greater than 3000 kN/m^2, we accept the batch of the plastic, and, if not, we reject it. The problem is: How large should n be?

To answer this, we have to consider the testing scheme, which is characterized by the *power curve*. This shows the probability of a plastic from a batch of any given strength being accepted as satisfactory; hence the power curve is related to the testing error. We shall recall from Chapter 13 that when we accept a value as being within a specified range, while in fact it is not, we are committing a Type II error. Figure 21-1b shows the shape of the power curve, also known as the *operating characteristic curve*.

The steepness of the power curve indicates the discriminating power of our testing scheme. The curve becomes steeper as the sample size increases and becomes vertical for $n = \infty$. In practice, we usually compromise between the advantage of increasing the discriminating power of our test and the increased cost of more extensive testing (which may be extremely high if the test is destructive).

The power curve can be used not only with a normal distribution, but also with a t distribution; G. P. Sillitto computed the number of observations needed in a t test in order to control the probabilities of Types I and II errors.[3]

CONSTRUCTION OF THE POWER CURVE

As we have said, the ordinate of the power curve represents the probability P that a batch of any strength μ will be accepted. In our testing scheme we perform n tests on samples from each batch and make our decision on the basis of whether the mean of our tests $\bar{x}$ is greater or smaller than 3000 kN/m^2. The results from the n tests will differ among themselves because of the testing error, which can be considered as the resultant of the sampling and measurement errors. Such testing error can be estimated from the differences among the n determinations.

The underlying distribution of the strength of the plastic is normal, and this in practice must be the case, since we are concerned with the distribution of means $\bar{x}$ (see the Central Limit Theorem, Chapter 10). Assuming that the distribution of testing errors is also normal, we can now obtain the probability P that for any given value of mean batch strength μ, the sample mean $\bar{x}$ will exceed 3000 and the batch will be accepted.

3. See W. Volk, *Applied Statistics for Engineers* (New York: McGraw-Hill Book Company, 1958).

For n observations in a sample, the standard deviation of the sample mean is $\sigma_T/\sqrt{n}$. Thus we calculate

$$z = \frac{3000 - \mu}{\sigma_T/\sqrt{n}} \tag{21-1}$$

and, using Table A-4, find the probability P of z having at most this value. [This is $0.5 + F(z)$.]

A plot of P against μ gives the power curve (Fig. 21-1b). This curve is descriptive of the testing scheme and depends only on σ_T^2 and n.

It is important to note that the power curve is determined by two points only; this will be illustrated in an example later in this chapter.

From Equation (21-1), we can see that there is a general relation between the absolute difference between the population mean and the sample mean, the standard deviation of the population, the sample size, and the probability of obtaining a difference not greater than the given one due to chance when in fact it is a "real" difference, i.e. of committing the Type II error. Thus

$$z = \frac{|\bar{x} - \mu|}{\sigma/\sqrt{n}}$$

It may be useful to consider this argument more fully. We recall that any testing scheme has an associated Type I error. The probability α of committing such an error is generally set at 5 percent or 1 percent, depending on the risk that we wish to take. Having specified the level of committing the Type I error for an acceptable testing scheme, the experimenter is faced with the problem of determining the sample size n for a desirable magnitude of the Type II error.

For a specific sample size, a plot of the probability of wrongly accepting a hypothesis (β) against the measured quantity $\bar{x}$ results in an operating characteristic curve (briefly called an OC curve). Such an OC curve passes through two points: $(\bar{x}_1, \beta)$ and $(\mu, 1 - \alpha)$. The former point arises from the fact that the testing scheme requires that we detect a level of the measured quantity $\bar{x}_1$ with an associated risk, β, of making the wrong decision. The latter point is found when we conclude that the measured quantity is μ when it really is $\bar{x}$, with the risk of a wrong decision in this case being $1 - \alpha$. The construction of the OC curve is illustrated in Fig. 21-2. This figure shows a typical OC curve for the test of the hypothesis that the mean of a normal distribution with a known standard deviation σ is μ against one-sided and two-sided alternatives. In a one-sided test, we hypothesize that $\bar{x}_1 \leq \mu$ against the alternative of $\bar{x}_1 > \mu$ (or a hypothesis of $\bar{x}_1 \geq \mu$ against the alternative of $\bar{x}_1 < \mu$); in a two-sided test, the hypothesis is $\bar{x}_1 = \mu$ against the alternative of $\bar{x}_1 \neq \mu$.

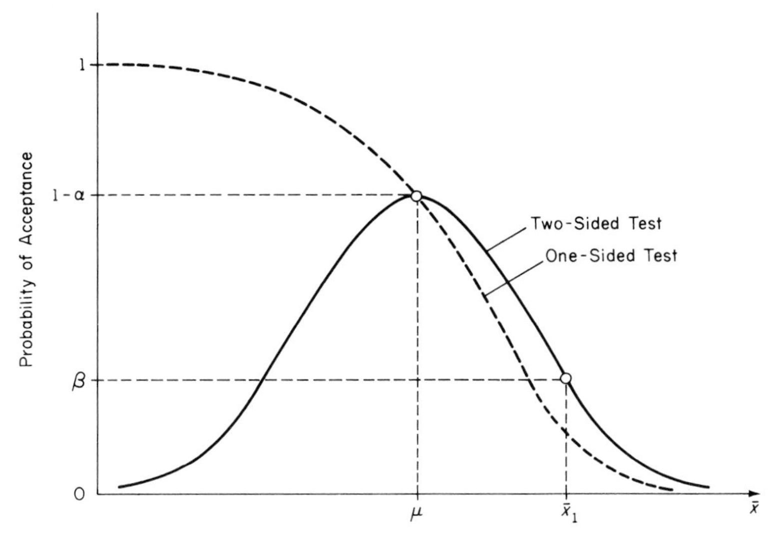

Fig. 21-2. The OC curves for one-sided and two-sided tests.

A more useful form of the OC curve is found when the probability of acceptance of the hypothesis is plotted against a dimensionless quantity λ, where $\lambda = |\bar{x} - \mu|/\sigma$, as shown in Fig. 21-3. A family of such curves for various values of sample size n is found in Figs. 21-4, 21-5, 21-6, and 21-7. The first two figures are for a two-sided normal test with a level of significance $\alpha = 0.05$ and $\alpha = 0.01$, respectively. Figures 21-6 and 21-7 are for a one-sided normal test with $\alpha = 0.05$ and $\alpha = 0.01$, respectively. For a given level of significance, the required sample size n is obtained by entering the appropriate figure with the point (λ, β) and reading out the value of n corresponding to the OC curve passing through this point. Other OC curves for the χ^2 test, F test, and t test have been derived by Ferris et al.[4]

Example. Past experience with a certain rope has shown its breaking strength to be 195 N with a standard deviation $\sigma = 10$ N. How many samples do we need in order to detect that the actual true mean strength of a consignment is 190 N? We are willing to take the risks: $\alpha = 0.05$ and $\beta = 0.1$.

4. C. D. Ferris, F. E. Grubbs, and C. L. Weaver, "Operating Characteristics for the Common Statistical Tests of Significance," *Annals of Mathematical Statistics*, vol. 17, (June 1946), p. 178.

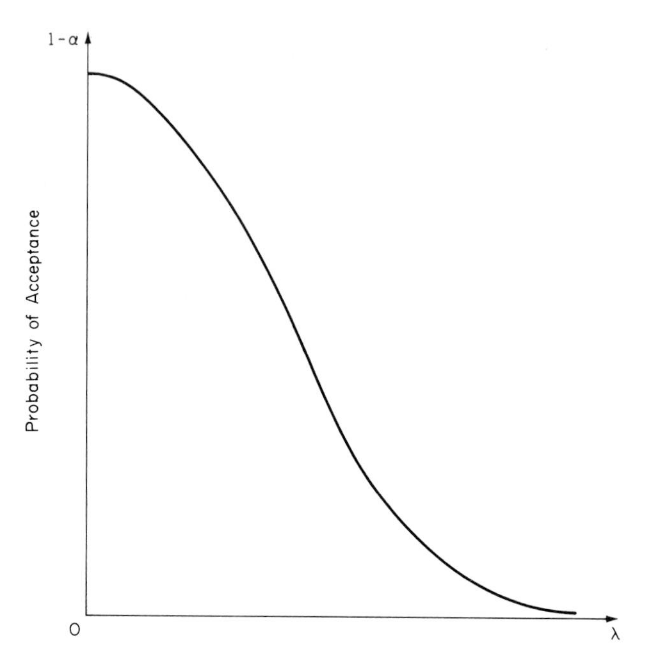

Fig. 21-3. Standardized OC curve for a fixed sample size n.

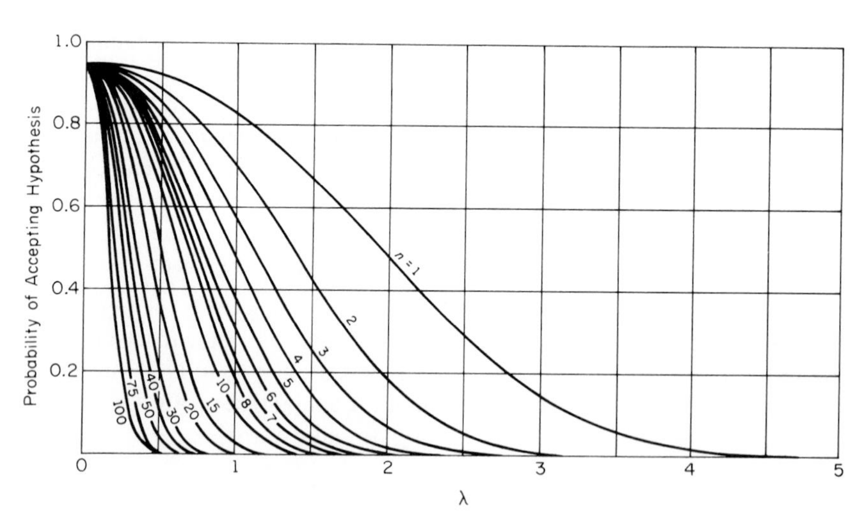

Fig. 21-4. The OC curves for different values of n for the two-sided normal test for a level of significance $\alpha = 0.05$. (From Charles L. Ferris, Frank E. Grubbs, and Chalmers L. Weaver, "Operating Characteristics for the Common Statistical Tests of Significance," *Annals of Mathematical Statistics*, vol. 17, June 1946.)

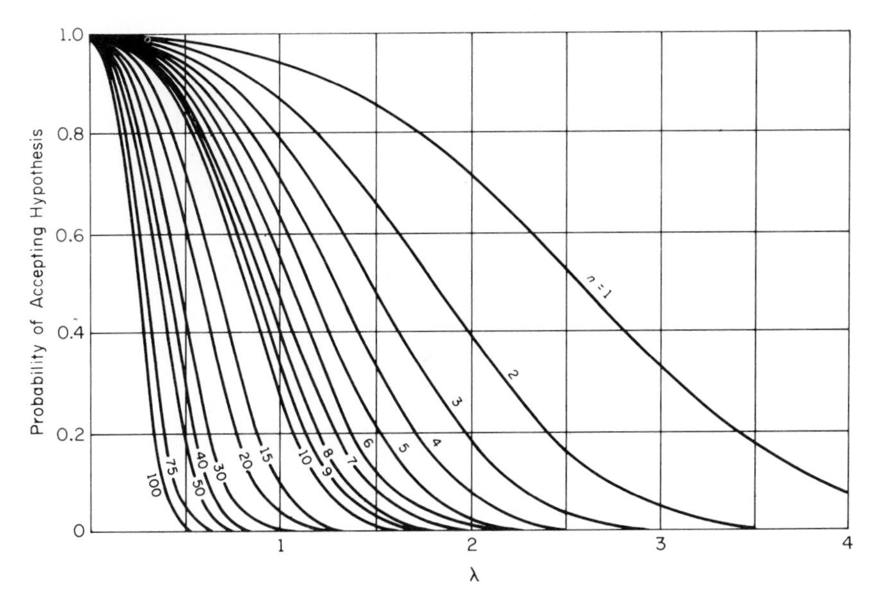

Fig. 21-5. The OC curves for different values of n for the two-sided normal test for a level of significance $\alpha = 0.01$. (From A. H. Bowker and G. J. Lieberman, *Engineering Statistics*, Prentice-Hall, Inc., Englewood Cliffs, N.J., 1959.)

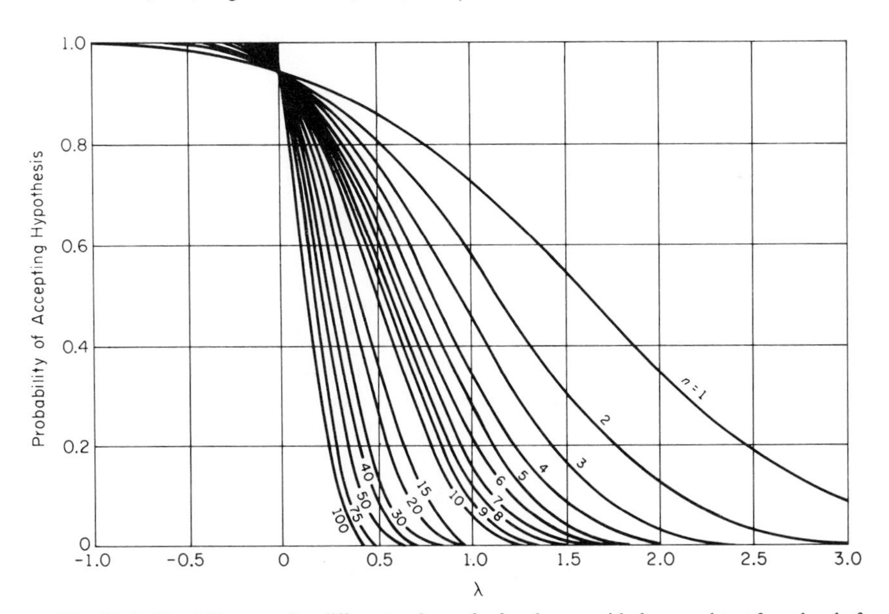

Fig. 21-6. The OC curves for different values of n for the one-sided normal test for a level of significance $\alpha = 0.05$. (From A. H. Bowker and G. J. Lieberman, *Engineering Statistics*, Prentice-Hall Inc., Englewood Cliffs, N.J., 1959.)

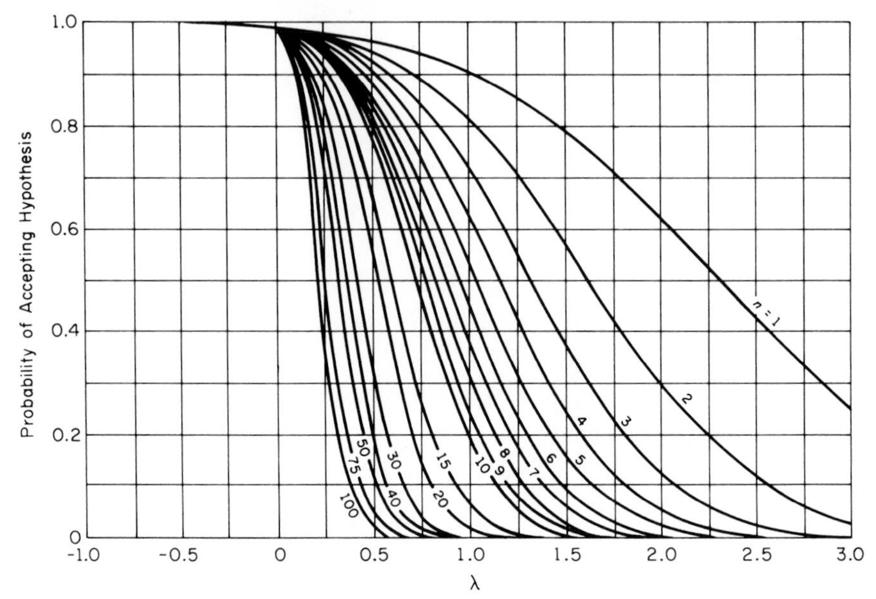

Fig. 21-7. The OC curves for different values of n for the one-sided normal test for a level of significance $\alpha = 0.01$. (From A. H. Bowker and G. J. Lieberman, *Engineering Statistics*, Prentice-Hall, Inc., Englewood Cliffs, N.J., 1959.)

It is clear that this is a one-sided test with the hypothesis that $\bar{x}_1 \leq 195$. In this case $\bar{x} = 190$ N. Therefore,

$$\lambda = \frac{|\bar{x} - \mu|}{\sigma}$$

$$= \frac{|190 - 195|}{10} = 0.5$$

Since the test is one-sided with $\alpha = 0.05$, we enter Fig. 21-6 with the point $(\lambda, \beta) = (0.5, 0.10)$ and find that $n = 35$. A value $\bar{x}_c$ between 190 N and 195 N becomes critical when we reject the hypothesis for observed values of $\bar{x}$ below it and accept the hypothesis for $\bar{x}$ values above it. This critical value is given by

$$\bar{x}_c = 195 - \frac{z\sigma}{\sqrt{n}}$$

Now, from Table A-4 for $\alpha = 0.05$ [i.e., $F(z) = 0.45$], $z = 1.645$. Hence

$$\bar{x}_c = 195 - 1.645 \times \frac{10}{\sqrt{35}} = 192.2 \text{ N}$$

The decision rule is then: Choose a random sample of 35 ropes and if the mean strength of these is less than 192.2 N, reject the hypothesis that $\mu = 195$ N; otherwise, accept the hypothesis.

DISTRIBUTION OF STRENGTH OF ACCEPTED AND REJECTED BATCHES

Referring to the previous example on the strength of plastic (p. 389), because the discrimination of the power curve is not perfect, it is to be expected that some of the accepted batches will actually have a strength below 3000 kN/m^2, and, conversely, some of the rejected batches will in fact have a strength above 3000 kN/m^2.

To determine the relative frequency with which batches of a strength μ are offered *and* accepted, we multiply the ordinate $p(\mu)$ of Fig. 21-1a by the ordinate P of Fig. 21-1b. The product $p(\mu)P$ then gives the distribution of strength of the accepted batches. This is plotted against μ in Fig. 21-1c.

Similarly, the distribution of strength of batches offered and rejected is given by the product $p(\mu)(1 - P)$. A plot of this is also given in Fig. 21-1c.

The area under the "accepted" curve, as a fraction of the sum of the areas under the two curves, represents the proportion of all the batches that are accepted. The remainder represents the proportion of rejected batches.

The proportion of cases in which we accept a batch of plastic, when in fact its strength is below 3000 kN/m^2, is represented by the area under the "accepted" curve to the left of the abscissa of 3000, as a fraction of the total area under both curves. Similarly, the proportion of cases in which we reject a batch, when in fact its strength exceeds 3000 kN/m^2, is represented by the area under the "rejected" curve to the right of the abscissa of 3000, as a fraction of the total area under both curves.

The acceptable level of the percentage of "wrong decisions" depends on the consequences. Obviously, the level would have to be considerably lower when we are dealing with human life than when we fail to discriminate, e.g., between the qualities of two products offered at the same price.

The method outlined here can also be used when full information about the underlying distribution is not available, but this is outside the scope of the present book.[5]

Example. We require limestone containing at least 92 percent calcium carbonate. From past experience the percentage of the carbonate in different

5. See O. L. Davies, ed., *Statistical Methods in Research and Production* (Edinburgh, Scotland: Oliver & Boyd Ltd., 1958).

batches is known to be normally distributed with a mean of 91 percent and a variance of 5 percent. The variance due to the testing error (i.e., sampling and measurement errors) obtained from a single test on each batch ($k = 1$) is known to be $\sigma_T^2 = 1$ (percent)^2. [NOTE: To avoid confusion, it is preferable not to work in percentages but in "units."]

Obtain power curves for samples (per batch of the rock) of size $n = 1, 4, 25$, and ∞. Then, for $n = 4$,

a. Obtain the distribution of the content of calcium carbonate in accepted and rejected batches.
b. Calculate the percentage of batches accepted.
c. Calculate the fraction of the accepted batches that, in fact, have a calcium carbonate content below 92 percent.

SOLUTION

We are given that the variance of the batch contents is 5 and that the variance due to testing is 1. Therefore, since k equals 1, the variance of the underlying distribution exclusive of testing error is $5 - 1 = 4$, or the standard deviation is 2.

Thus we can assume that the underlying distribution is normal with mean $= 91$ and $\sigma = 2$. This distribution is plotted in Fig. 21-8a. To construct the power curves for $n = 1, 4, 25$, and ∞, we need to calculate P for

$$z = \frac{92 - \mu}{\sigma_T/\sqrt{n}}$$

Thus

For $n = 1$: $z = (92 - \mu)$

For $n = 4$: $z = (92 - \mu) \times 2$

For $n = 25$: $z = (92 - \mu) \times 5$

For $n = \infty$: $z = \infty$ for $\mu > 92$ and $\mu < 92$

From Table A-4 we can find the probability P of z having at most the preceding values. [This is $0.5 - F(z)$ if $\mu < 92$, and it is $0.5 + F(z)$ for $\mu > 92$.] These values of P for various values of μ are shown in Tables 21-1 to 21-3 for the different values of n.

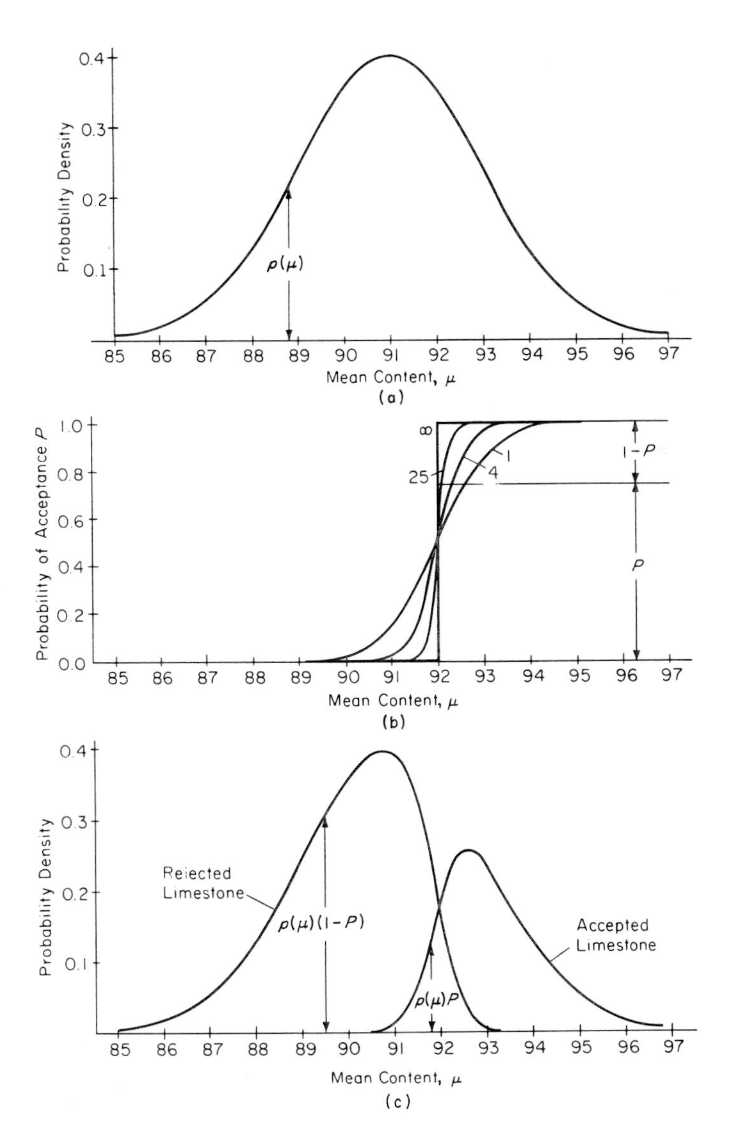

Fig. 21-8. (a) Underlying distribution; (b) power curves for $n = 1$, 4, 25, and ∞; (c) distribution of accepted and rejected batches of sample size $n = 4$.

For $n = 1$:

TABLE 21-1

μ	z	P
88.0	4.0	0.0000
88.4	3.6	0.0002
88.8	3.2	0.0007
89.2	2.8	0.0026
89.6	2.4	0.0082
90.0	2.0	0.0228
90.4	1.6	0.0548
90.8	1.2	0.1151
91.2	0.8	0.2119
91.6	0.4	0.3446
92.0	0.0	0.5000
92.4	−0.4	0.6554
92.8	−0.8	0.7881
93.2	−1.2	0.8849
93.6	−1.6	0.9452
94.0	−2.0	0.9772
94.4	−2.4	0.9918
94.8	−2.8	0.9974
95.2	−3.2	0.9993
95.6	−3.6	0.9998
96.0	−4.0	1.0000

For $n = 4$:

TABLE 21-2

μ	z	P
90.0	4.0	0.0000
90.4	3.2	0.0007
90.8	2.4	0.0082
91.2	1.6	0.0548
91.6	0.8	0.2119
92.0	0.0	0.5000
92.4	−0.8	0.7881
92.8	−1.6	0.9452
93.2	−2.4	0.9918
93.6	−3.2	0.9993
94.0	−4.0	1.0000

For $n = 25$:

TABLE 21-3

μ	z	P
91.0	5.0	0.0000
91.2	4.0	0.0000
91.4	3.0	0.0013
91.6	2.0	0.0228
91.8	1.0	0.1587
92.0	0.0	0.5000
92.2	−1.0	0.8413
92.4	−2.0	0.9772
92.6	−3.0	0.9987
92.8	−4.0	1.0000
93.0	−5.0	1.0000

For $n = \infty$:

$$z = \infty \text{ for any } \mu < 92 \qquad \text{whence } P = 0.0000$$

$$z = -\infty \text{ for any } \mu > 92 \quad \text{whence } P = 1.0000$$

The plot of P versus μ gives the power curve. These curves are shown in Fig. 21-8b for $n = 1$, 4, 25, and ∞. The power curve for $n = \infty$ is sometimes called the *ideal* power curve.

a. To obtain the distribution of the content of calcium carbonate in accepted and rejected batches, we have to find $p(\mu)P$ and $p(\mu)(1 - P)$ for various values of μ, corresponding to $n = 4$. These values are given in Table 21-4.

The plots of $p(\mu)P$ and $p(\mu)(1 - P)$ versus μ give the required distributions. These are shown in Fig. 21-8c.

b. The percentage of batches accepted is obtained from the ratio:

$$\frac{\sum p(\mu)P}{\sum p(\mu)} = \frac{1.5687}{4.9947} = 31.4 \text{ percent}$$

c. The fraction of the accepted batches that, in fact, have a calcium carbonate content below 92 percent is obtained by summing up the column $p(\mu)P$ from $\mu = 85.0$ to $\mu = 91.8$ (just below $\mu = 92$). This sum $= 0.1821$. Therefore, the required fraction $= 0.1821/1.5687 = 11.6$ percent. This is equivalent to 3.64 percent of all the accepted and rejected batches. From Fig. 21-8c it can be observed that the mean content of calcium carbonate in rejected batches is 90.8 percent and, in accepted batches, equals 92.6 percent.

TABLE 21-4

μ	$p(\mu)$	P	$p(\mu)P$	$p(\mu)(1-P)$
85.0	0.0044	0.0000	0.0000	0.0044
85.4	0.0079	0.0000	0.0000	0.0079
85.8	0.0136	0.0000	0.0000	0.0136
86.2	0.0224	0.0000	0.0000	0.0224
86.6	0.0355	0.0000	0.0000	0.0355
87.0	0.0540	0.0000	0.0000	0.0540
87.4	0.0790	0.0000	0.0000	0.0790
87.8	0.1109	0.0000	0.0000	0.1109
88.2	0.1497	0.0000	0.0000	0.1497
88.6	0.1942	0.0000	0.0000	0.1942
89.0	0.2420	0.0000	0.0000	0.2420
89.4	0.2897	0.0000	0.0000	0.2897
89.8	0.3332	0.0000	0.0000	0.3332
90.2	0.3683	0.0002	0.0001	0.3682
90.6	0.3910	0.0026	0.0010	0.3900
91.0	0.3989	0.0228	0.0091	0.3898
91.4	0.3910	0.1151	0.0450	0.3460
91.8	0.3683	0.3446	0.1269	0.2414
92.2	0.3332	0.6554	0.2184	0.1148
92.6	0.2897	0.8849	0.2564	0.0333
93.0	0.2420	0.9772	0.2365	0.0055
93.4	0.1942	0.9974	0.1937	0.0005
93.8	0.1497	0.9998	0.1497	0.0000
94.2	0.1109	1.0000	0.1109	0.0000
94.6	0.0790	1.0000	0.0790	0.0000
95.0	0.0540	1.0000	0.0540	0.0000
95.4	0.0355	1.0000	0.0355	0.0000
95.8	0.0224	1.0000	0.0224	0.0000
96.2	0.0136	1.0000	0.0136	0.0000
96.6	0.0079	1.0000	0.0079	0.0000
97.0	0.0044	1.0000	0.0044	0.0000
97.4	0.0024	1.0000	0.0024	0.0000
97.8	0.0012	1.0000	0.0012	0.0000
98.2	0.0006	1.0000	0.0006	0.0000
$\sum =$	4.9947		1.5687	3.4260

SIMPLIFIED CASE

In some cases a simpler approach to deciding on the size of the sample can be adopted. For example, this is the case in testing concrete, where we may be concerned with the maximum error of the mean strength determined by tests on a sample.

The t tests of Chapter 13 can be written as

$$t = \frac{|\mu - \bar{x}|}{s_d}$$

where μ is the population mean, $\bar{x}$ the sample mean, and s_d the standard deviation of the sample mean.

The value of t for the largest value of $|\mu - \bar{x}|$ corresponding to different levels of significance and numbers of degrees of freedom is given in Table A-8.

Now, $s_d = s/\sqrt{n}$, where s is the estimate of the population standard deviation from the sample and n the sample size.

In tests on the compressive strength of concrete, the coefficient of variation within a sample is under many circumstances $V = 5$ percent.[6] Hence

$$s = \frac{V}{100}\mu = 0.05\mu$$

Expressing $|\mu - \bar{x}|$ as a percentage of the population mean, we put

$$E = \frac{\mu - \bar{x}}{\mu} \times 100$$

Then,

$$t = \frac{\mu E \sqrt{n}}{V\mu}$$

whence

$$E = \frac{Vt}{\sqrt{n}} \tag{21-2}$$

Working at the 10 percent level of significance,[7] and using a sample size $n = 3$, we have, from Table A-8, $t = 2.920$ for $\nu = 3 - 1 = 2$. Then,

$$E = \frac{0.05 \times 2.920}{\sqrt{3}} \times 100 = 8.5 \text{ percent}$$

Thus, for a sample of three specimens, the error of the mean will exceed 8.5 percent in 10 percent of the cases.

6. W. A. Cordon, "Size and Number of Samples and Statistical Considerations in Sampling." ASTM, Special Technical Publication No. 169, 1955.
7. Ibid.

If V is based on a large number of tests, we reach in the limit t for an infinite number of degrees of freedom (cf. u of Chapter 13); at the 10 percent level of significance, we have $t = 1.645$. Then,

$$E = \frac{0.05 \times 1.645}{\sqrt{3}} \times 100 = 4.8 \text{ percent}$$

Thus the mean of three specimens will indicate the *average* strength of concrete with an "error" of 4.8 percent of the mean, exceeded in 10 percent of the cases.

We can now consider the problem of determining the size of the sample n when the "error" in the estimated strength of concrete is not to exceed 5 percent in 90 percent of our tests. The coefficient of variation is assumed to be 5 percent. From Eq. (21-2),

$$n = \left(\frac{Vt}{E}\right)^2 = \left(\frac{0.05 \times 1.645}{0.05}\right)^2 = 2.7 \tag{21-3}$$

The next higher integer is 3; i.e., we require a sample of three specimens. This is, indeed, the number commonly used.

SAMPLING OF ATTRIBUTES

A sampling plan similar to that outlined in the earlier part of this chapter can be used with attributes, such as defective items. In fact, our case of distribution of strength can be considered as that of the sampling of attributes, since each test can be classified as success (strength greater than 3000 kN/m^2) or failure (strength smaller than 3000 kN/m^2). In the general case, we expect a small number of failures, or defectives; therefore, we can use the Poisson distribution as a sufficiently good approximation. (This is justified when the proportion of defectives does not exceed 0.1.)

Suppose that to inspect a batch we draw a sample of 100 and accept the batch if the number of defectives does not exceed 2 but reject it if it does. Knowing the average number of defectives in the sample,[8] we can determine from the Poisson distribution (Fig. 8-1) the probability of our sample containing 0, 1, 2, 3 ... defectives. Conversely, from batches with different proportions of defectives (this being reflected in the average number of defectives in the samples), we can draw samples containing a specified number of defectives with a probability determined by the Poisson distribution. For example, if the batch being inspected contains 1.5 percent of defectives ($np = 1.5$), the probability of drawing a sample containing at least 3

8. This is directly related to the proportion of defectives in the material tested.

defectives is 0.20 (from Fig. 8-1). Thus one sample out of five will contain 3 or more defectives. We might use a sample with at least 3 defectives as a basis for rejection of the batch. We know, however, that the percentage of defectives is 1.5 percent so that we are wrong in our decision to reject in 20 percent of the cases.

Let us consider Fig. 8-1 further. We can observe that not more than 2 defectives in a sample would be obtained with a probability of 0.2 when the average number of defectives is as high as 4.3. We thus run a risk of 0.2 of accepting a batch containing 4.3 percent of defectives.

We can see that there is a definite probability of wrongly rejecting a batch and also of wrongly accepting a batch. These are, of course, the Type I and Type II errors, respectively. The former is simply the confidence level α, and the latter, β, determines the power curve associated with any sampling plan; see Chapter 13.

Instead of specifying the sample size and the acceptable number of defectives in it, we can define the curve by two points, (p_1, α) and (p_2, β), where α is the probability of rejecting a batch in which the proportion of defectives is p_1, β is the probability of accepting a batch in which the proportion of defectives is p_2, and $p_2 > p_1$.

We can plot the power curve as shown in Fig. 21-1b, from which we read off the probabilities of accepting batches of different qualities.

The risk of a Type I error can be made as small as we wish by increasing the acceptance interval, i.e., by increasing the acceptable number of defectives. We thus usually fix this risk, and to compare procedures we compare the Type II errors.

It is important to realize that the tacit assumption underlying the procedures outlined here is that a small number of defectives is admissible; otherwise, of course, a 100 percent inspection would be necessary, and there would be no question of sampling procedure.

We should further stress that the power curve alone provides no information about the quality of the material but only gives the probability of accepting a batch containing a specified proportion of defectives. To know the qualities of accepted and rejected material, we must know the underlying distribution of the material as manufactured (see Fig. 21-1a).

Finally, we should observe that information obtained from a sample of size n depends on the sample size only and is in no way related to the size of the batch from which the sample was drawn. Thus, provided that the batches are homogeneous, it is more economical to test large batches than small ones.

Example. (a) Construct the power curve for a sampling plan such that the risk of rejecting a batch containing not more than the specified proportion of defectives p is 0.1, when the sample size $n = 50$, and we accept samples when the number of defectives per sample is $r \leq 2$. (b) For approximately the same risk of wrong rejection, establish the decision rule when the sample size $n = 150$ and construct the power curve for this sampling plan.

SOLUTION

a. Figure 8-1 shows that for the probability of 0.1 of rejecting a batch with the average number of defectives per sample np, the number r of observed defectives which is *exceeded* is as follows:

r	0	1	2	3	4	5
np	0.1	0.53	1.10	1.54	2.4	3.1

Thus, if we reject samples with $r > 2$, $np = 1.10$, whence the proportion of defectives in the batch is 0.022. Since the probability of accepting a batch is then 0.9, we have a point on the power curve (0.022, 0.9).

To find a second point, consider, for example, $p = 0.08$. Then $np = 4$, and for our decisions rule $r \leq 2$, Fig. 8-1 gives the probability of accepting a batch containing this proportion of defectives as 0.24. Thus the second point on the power curve is (0.08, 0.24).

Some of the other points are (from Fig. 8-1):

np	6	5	3	2	1	0
Proportion of defectives in batch p	0.12	0.10	0.06	0.04	0.02	0.0
Probability of accepting the batch	0.06	0.12	0.41	0.67	0.92	1.0

b. When $n = 150$ and p is still 0.022, the average number of defects in a sample is $np = 150 \times 0.022 = 3.3$. However, bearing in mind that the actual number of defectives in a sample used as our decision rule must be an integer, we choose the nearest value of r, namely $r \leq 5$. For $np = 3.3$, this gives a probability of 0.12 of rejecting a batch containing not more than the specified proportion of defectives (i.e., a wrong rejection).

The power curve thus passes through the point (0.022, 0.88). From Fig. 8-1, the probability of accepting a batch containing a proportion of defectives $p = 0.06$ ($np = 9$) on the basis of our decision rule $r \leq 5$ is 0.12. Thus the second point on the power curve is (0.06, 0.12).

Some of the other points (from Fig. 8-1) are

np	10	7.5	6	4.5	2.4	1.5	0
Proportion of defectives in batch p	0.067	0.05	0.04	0.03	0.016	0.01	0.0
Probability of accepting the batch	0.070	0.25	0.44	0.70	0.962	0.995	1.0

The two power curves are shown in Fig. 21-9. We can see that although the risk of wrongly rejecting a batch is nearly the same with either sampling plan, the use of the larger sample ensures a better discrimination against acceptance of batches containing too great a proportion of defectives.

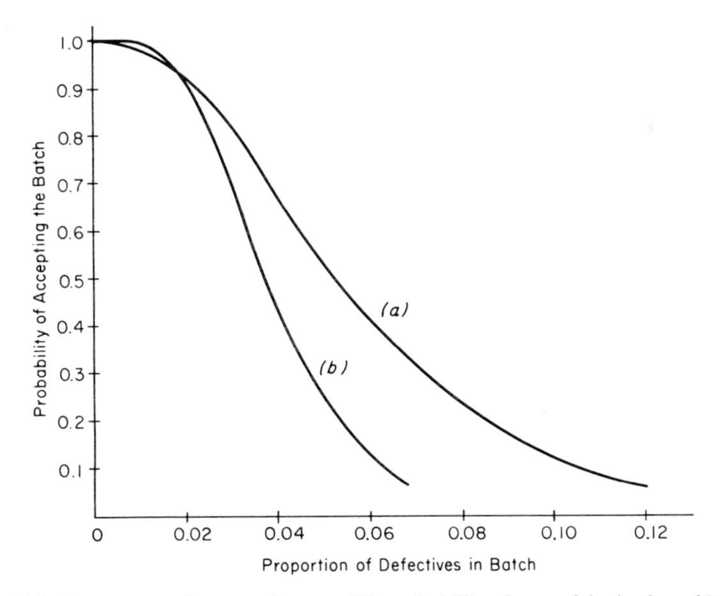

Fig. 21-9. Power curves for example on p. 405; probability of wrongful rejection of 0.10. (a) Decision rule: Accept batch when a sample of 50 contains not more than 2 defectives. (b) decision rule: Accept batch when a sample of 150 contains not more than 5 defectives.

RANDOMIZED BLOCKS

The first requirements of a properly designed experiment are that an unbiased estimate of error can be obtained and that the error be at a minimum. This can be achieved by the use of a randomized-block design.

Suppose that we wish to compare four methods of molding a plastic, and further that the material arrives in batches large enough only to permit four tests per batch. From our knowledge of engineering materials, we expect the batches to vary from one another. In statistical terminology of design of experiments, a unit such as our batch would be called a *block.*

The term has its origin in agricultural experiments, where it denoted a strip of land consisting of adjacent plots believed to be more similar to one another than plots chosen at random. This greater homogeneity of units within the block than of those chosen at random is the essential feature of a block.[9] However, the homogeneity must be related to the measurements of the dependent variable, and blocks should not be formed on the basis of irrelevant or unrelated variables. Thus blocks are normally formed on the basis of prior information about the units, but when dealing with random units, we can use blocks to control some sources of variation not associated with the units. For example, if we suspect a day-to-day variation in laboratory conditions, we may consider all the units tested the same day as a block. In the analysis of variance the day-to-day variation would be eliminated from the estimate of error.

Suppose further that we are going to use each method of molding, which we shall call *treatment*, four times, i.e., *replicate* four times. The question is: How should we arrange the tests?

It does not require any knowledge of statistics to realize that to assign each block to one treatment would be purposeless, for the observed differences might be those between blocks and not between treatments; we would have no information on the basis of which to decide which is the case. It is obvious, then, that the blocks have to be distributed between the treatments.

The distribution could be made at random, e.g., by assigning a letter *A*, *B*, *C*, or *D* to the four treatments, and writing *A* on four cards, *B* on another four cards, and so on, putting all 16 cards in a hat, and withdrawing four for each block. We would thus determine the treatment to which the four components of each block should be subjected. An actual drawing from a hat has produced the data in Table 21-5.

TABLE 21-5

Block	*Treatment*			
1	*D*	*C*	*A*	*A*
2	*B*	*A*	*C*	*C*
3	*D*	*D*	*B*	*D*
4	*B*	*B*	*A*	*C*

9. If all subjects are homogeneous, nothing is gained by the use of blocks.

This procedure would enable us to estimate the difference between the treatments, but the error of the mean value for each treatment would include the differences between blocks. Specifically, treatment D would be applied only to blocks 1 and 3, and no treatment would be applied to specimens from all four blocks. It is obvious, therefore, that we should apply each treatment to each block so that the mean value for each treatment is independent of the differences between the blocks. Our test program might thus be as shown in Table 21-6.

TABLE 21-6

Block		*Treatment*		
1	*A*	*B*	*C*	*D*
2	*A*	*B*	*C*	*D*
3	*A*	*B*	*C*	*D*
4	*A*	*B*	*C*	*D*

It is possible, however, that some uncontrolled variables are acting, and these could influence our results. For example, if only four treatments can be performed in a day, treatment A would be applied first, and, say, with a lower temperature of the machine, we might obtain a consistently "low" reading. Or, the last treatment might be done in a hurry, and the reading could be "high." To remove such a bias, we should randomize the treatments by determining the order for each block by drawing letters A, B, C, and D from a hat. As an example, the results in Table 21-7 have been obtained.

TABLE 21-7

Block		*Treatment*		
1	*D*	*C*	*A*	*B*
2	*A*	*D*	*B*	*C*
3	*A*	*B*	*C*	*D*
4	*C*	*B*	*A*	*D*

This, then, is the randomized block.

Since the units in each block are more homogeneous than units selected at random, the differences between blocks can be taken into account in the analysis of variance, with the result that the estimate of experimental error will be smaller than if randomized selection had been used.

The analysis of variance of the results of a randomized block experiment is best illustrated by means of an example. We may note that the number of blocks for any treatment must be the same so that all the observations in the testing program as designed are necessary. Should one be missing (through mishap, breakdown, etc.) it has to be replaced by an estimate. On the other hand, with a randomized-groups design, the analysis of variance can be applied for unequal numbers of experiments for each treatment.

Example. In order to establish the temperature of complete melting of cadmium and tin alloys with various percentages of cadmium, a large piece from each of five different alloys was obtained and was cut into four smaller pieces, each of which was tested for the complete melting state. Find the influence of the content of cadmium on the melting state of the alloys. What can be said about the homogeneity of each individual large piece of alloy? The data are given in Table 21-8. Use $\alpha = 1$ percent.

SOLUTION

a. Compute the row, column, and the grand totals as shown in Table 21-8.

TABLE 21-8

	Melting point, °C				
Block t (percent cadmium)	*Replications (pieces) r*				
	1	*2*	*3*	*4*	*Totals*
40	201	185	182	179	747
50	200	195	220	199	814
60	257	240	224	225	946
70	252	228	275	250	1005
80	280	275	277	260	1092
Totals	1190	1123	1178	1113	$G = 4604$

b. Obtain the correction factor:

$$C = \frac{G^2}{tr} = \frac{(4604)^2}{5 \times 4} = 1{,}059{,}840.80$$

c. Square every measurement and add:

$$(201)^2 + (185)^2 + \cdots + (277)^2 + (260)^2 = 1{,}082{,}234.00$$

d. Replications. Square the totals of each column, sum these squares, and divide by the number of measurements in each column:

$$\frac{(1190)^2 + (1123)^2 + (1178)^2 + (1113)^2}{5} = 1{,}060{,}736.40$$

e. Blocks. Square the totals of each row, sum these squares, and divide by the number of measurements in each row:

$$\frac{(747)^2 + (814)^2 + (946)^2 + (1005)^2 + (1092)^2}{4} = 1{,}079{,}502.50$$

The analysis of variance is shown in Table 21-9.

TABLE 21-9

Source of variation	*Sum of squares (s.s.)*	*Degrees of freedom* ν	*Mean square*
Replications	(d) − C = 895.60	$r - 1 = 3$	298.53
Blocks	(e) − C = 19,661.70	$t - 1 = 4$	4,915.43
Error	1,835.90	$(r - 1)(t - 1) = 12$	152.99
Totals	(c) − C = 22,393.20	$rt - 1 = 19$	

The number of degrees of freedom associated with replications is $(r - 1)$, since this is the number of classifications that can arbitrarily be assigned; similarly, there are $(t - 1)$ degrees of freedom associated with the blocks. The number of degrees of freedom associated with the residual error is obtained from the assumption that blocks and replications are independent. In other words, the variations present in the data are attributable to chance and unaccountable factors. Therefore, as explained in Chapter 12, the number of degrees of freedom in this case will be

$$rt - 1 - (r - 1) - (t - 1) = rt - t - r + 1 = (r - 1)(t - 1)$$

$$\begin{aligned} \text{error sum of squares} &= \text{total (s.s.)} - \text{replication (s.s.)} - \text{block (s.s.)} \\ &= 22{,}393.20 - 895.60 - 19{,}661.70 \\ &= 1835.90 \end{aligned}$$

$$\text{value of } F \text{ for blocks} = \frac{4915.43}{152.99} = 32.13$$

Referring to Table A-10, for $\nu_1 = 4$ and $\nu_2 = 12$, $F = 5.41$ at the 1 percent level of significance. We conclude, therefore, that the block effect is significant at that level, i.e., the cadmium content affects the melting state of the alloy. The value of F for replications $= 298.53/152.99 = 1.95$, which is not significant at the 5 percent level ($v_1 = 3$ and $v_2 = 12$). Therefore, we can state that there is no evidence of each of the large pieces of alloy being nonhomogeneous.

LATIN SQUARES

Let us now consider the situation in which we have, say, five treatments to be compared. We have 25 specimens but only 5 can be tested in any one day. We have no prior information about the specimen that would enable us to arrange them in blocks. We might suspect, however, that there may be some variation between observations made on different days, and to eliminate this effect, we would consider a group of 5 specimens tested on the same day as a block. We would, therefore, assign the specimens at random to each day, and of the day's block subject one to each treatment, A, B, C, D, and E. The procedure would be, for example, to number cards corresponding to the specimens as 1 to 25 and to make cards with letters A to E. We could then draw one number card and one letter card at a time, and this would determine which specimens are to be tested on the first day, and in what order. The letter card would then be replaced, and a further draw would determine the tests for the second day, etc.

The analysis of variance of these results would be the same as for the randomized blocks, and the day-to-day variation would be removed from the estimate of the experimental error.

It is possible, however, that the time of the day affects the results. To remove this error, we would arrange our testing program so that each treatment occurs not only once every day but also once at a particular time of the day. Denoting the five parts of a working day by numbers 1, 2, 3, 4, and 5, we would arrange the test program so that each treatment occurs once and only once in each row and each column, as shown in Table 21-10.

TABLE 21-10

	Part of the day				
Day	*1*	*2*	*3*	*4*	*5*
1	*A*	*B*	*C*	*D*	*E*
2	*E*	*A*	*B*	*C*	*D*
3	*D*	*E*	*A*	*B*	*C*
4	*C*	*D*	*E*	*A*	*B*
5	*B*	*C*	*D*	*E*	*A*

Such an arrangement is known as a Latin square. Since it is a square, the number of observations (specimens) must be equal to the square of the number of treatments. Thus, with a large number of treatments, the total testing effort is high, but a high reduction in errors is achieved as every row and every column is a complete replication. The experiment should be designed so that the differences among rows and columns represent major sources of variation.

Example. It is desired to know whether the rate of flow of fuels through different types of nozzles is affected by temperature. An experiment was carried out by six operators, A, B, ..., F, chosen at random, at six different temperatures on six different types of nozzles. The coded results are shown in Table 21-11. Use $\alpha = 1$ percent for temperature effects and $\alpha = 5$ percent for both nozzle and operator effects.

SOLUTION

First, the totals for each row, column, operator, and the grand total G are computed as shown in Table 21-11.

Second, the correction factor C is calculated:

$$C = \frac{G^2}{r^2} = \frac{(832)^2}{(6)^2} = 19{,}228.44$$

TABLE 21-11

	Volume of fuel through nozzle						
	Nozzle type						
Temperature, °C	*1*	*2*	*3*	*4*	*5*	*6*	*Totals*
0	*A* 24	*D* 20	*E* 22	*C* 17	*F* 12	*B* 18	113
5	*E* 20	*A* 15	*C* 18	*F* 11	*B* 19	*D* 10	93
10	*D* 16	*C* 22	*B* 24	*E* 18	*A* 13	*F* 15	108
20	*C* 24	*E* 32	*F* 27	*B* 22	*D* 30	*A* 24	159
40	*B* 26	*F* 29	*D* 28	*A* 32	*E* 30	*C* 27	172
80	*F* 33	*B* 34	*A* 30	*D* 28	*C* 29	*E* 33	187
Totals	143	152	149	128	133	127	$G = 832$
Operator	*A*	*B*	*C*	*D*	*E*	*F*	Total
Totals for operators	138	143	137	132	155	127	832

a. The sum of the squares of all measurements is then obtained:

$$(24)^2 + (20)^2 + (22)^2 + \cdots + (29)^2 + (33)^2 = 20{,}884$$

b. The sum of squares of row totals is obtained and then divided by the number of measurements in each row, i.e., 6:

$$\frac{(113)^2 + (93)^2 + (108)^2 + (159)^2 + (172)^2 + (187)^2}{6} = 20{,}486.00$$

c. Similarly, for the columns:

$$\frac{(143)^2 + (152)^2 + (149)^2 + (128)^2 + (133)^2 + (127)^2}{6} = 19{,}326.00$$

d. Similarly, for the operators:

$$\frac{(138)^2 + (143)^2 + (137)^2 + (132)^2 + (155)^2 + (127)^2}{6} = 19{,}306.67$$

We can now form the table of the analysis of variance (Table 21-12).

TABLE 21-12

Source of variation	*Sums of squares (s.s.)*	*Degrees of freedom* v	*Mean square*
Temperature (rows)	(b) $- C =$ 1257.56	$r - 1 = 5$	251.51
Nozzle type (columns)	(c) $- C =$ 97.56	$r - 1 = 5$	19.51
Operators (treatments)	(d) $- C =$ 78.23	$r - 1 = 5$	15.65
Residual error	222.21	$(r - 1)(r - 2) = 20$	11.11
Totals	(a) $- C =$ 1655.56	$r^2 - 1 = 35$	

The number of degrees of freedom v associated with temperature, nozzle type, and operators is obviously $r - 1$, all three variables having the same number of classifications, namely, 6. The number of degrees of freedom for the sum of squares of the residual error is, in this case, $(r)(r) - 1 - (r - 1) - (r - 1) - (r - 1) = r^2 - 3r + 2 = (r - 1)(r - 2)$.

The residual error of the sum of squares

$$= \text{total (s.s.)} - \text{rows (s.s.)} - \text{columns (s.s.)} - \text{treatments (s.s.)}$$

$$= 1655.56 - 1257.56 - 97.56 - 78.23$$

$$= 222.21$$

Each mean square is obtained by dividing the sums of squares by the corresponding degrees of freedom ν. The variance ratio F is then computed for the different effects:

Temperature:

$$F = \frac{\text{mean square of temperature variation}}{\text{mean square of residual error}} = \frac{251.51}{11.11} = 22.64$$

Nozzle type:

$$F = \frac{19.51}{11.11} = 1.76$$

Operator:

$$F = \frac{15.65}{11.11} = 1.41$$

From Table A-10, for $\nu_1 = 5$ and $\nu_2 = 20$, the F values are 4.10 at the 1 percent level of significance, 2.71 at the 5 percent level of significance, and 2.16 at the 10 percent level of significance.

We conclude that the temperature effect is significant at the 1 percent level of significance (in fact, it passes the 0.1 percent level).[10] However, both the nozzle type and the operator's effects do not reach the 10 percent level of significance. (They fall between the 10 percent and the 20 percent level.)[11] Thus we can state that no definite effect on the volume of fuel because of different nozzles or operators is established.

Let us pursue further the effect of temperature on the volume of fuel. The mean volumes for the various temperatures are

0°C	5°C	10°C	20°C	40°C	80°C
18.83	15.50	18.00	26.50	28.67	31.17

The estimated standard error of each of such means is $s_{\bar{x}} = \sqrt{s^2/n}$, where s^2 = residual error mean square = 11.11, and $n = r = 6$. Thus the estimated error of each mean, $s_{\bar{x}} = \sqrt{11.11/6} = 1.361$. For testing the difference between a pair of means, the standard error is

$$\sqrt{2}\, s_{\bar{x}} = \sqrt{2} \times 1.361 = 1.924$$

10. See R. A. Fisher and F. Yates, *Statistical Tables for Biological and Medical Research* (Edinburgh, Scotland: Oliver & Boyd Ltd., 1963).

11. Ibid.

From Table A-8, the value of t at the 5 percent level of significance, for $\nu = 20$, is $t = 2.086$. Thus we have for the 95 percent confidence limits: $\pm(2.086)(1.924) = \pm 4.01$; i.e., the difference between two means must be at least ± 4.01 in order to reach significance at this level.

$$\text{mean volume} = \frac{832}{6 \times 6} = 23.11$$

and the 95 percent confidence interval $= 23.11 \pm 4.01 = (27.12, 19.10)$.

By comparing the mean volumes for the various temperatures, we may be led to suspect that the temperature effects fall into 3 sets: (0°C, 5°C, 10°C), (20°C, 40°C), and (80°C), but further study of the problem is outside the scope of this book.

BALANCED INCOMPLETE BLOCKS

The above procedure is simple but can be applied only if each block is large enough to include all the treatments. Suppose, however, that each batch is large enough only to make four specimens, but we want to compare five treatments.

The most efficient procedure is to use five batches of four and to replicate each of the five treatments four times, the arrangement being that of a balanced incomplete block. (See Table 21-13.)

TABLE 21-13

	Batch number				
	1	*2*	*3*	*4*	*5*
Treatment	*A*	*A*	*A*	*A*	*B*
	B	*B*	*B*	*C*	*C*
	C	*C*	*D*	*D*	*D*
	D	*E*	*E*	*E*	*E*

We can note that:

1. Each treatment occurs once and only once in four of the batches.
2. Any specified pair of treatments occurs in three of the batches (e.g., A and B can be compared in batches 1, 2, and 3).
3. The two batches in which a direct comparison of treatments (e.g., A and B in batches 4 and 5) is not possible can provide a means of comparison for they each contain the remaining three treatments, C, D, and E. The average of these three can be considered as a "standard" against which the difference between A and B can be assessed.

Thus, although the blocks are incomplete, for none of them contains the full number of treatments, they are balanced because each treatment occurs to the same extent. The disadvantage of this method is that the number of replications necessary may be large, generally t treatments requiring $(\sqrt{t} + 1)$ replications.

MULTIPLE FACTOR EXPERIMENTS

The various blocks considered up to now are applicable only to experiments containing one independent variable, which we termed *treatment.* In many practical cases, we may have several independent variables. This problem is often approached by keeping all independent variables but one constant in each series of tests. Although this may appear to be a reasonable approach, it is not always the best one, for the variables (referred to as factors) must necessarily be kept constant at an arbitrary value. The influence on some property of a variation in the factor A when the factor B is kept constant at value B_1 may be different from the influence when $B = B_2$. To detect this type of behavior, we may use the *factorial design* of experiments,[12] which has the additional advantage of giving the greatest amount of information about the given system of variables for the given amount of work. Specifically, information can be obtained about the *interaction* of the variables, i.e., the dependence of the effect of one factor on the value of another. This may be of considerable importance in many engineering applications. As a simple example, let us consider tests on the durability of concrete, in which we may vary the water/cement ratio, the type of aggregate, and the type of admixture. Suppose that the tests shown in Table 21-14 have been made.

Test 1 may tell us which admixture is best from the standpoint of durability, but this finding may not be true when we use lightweight instead of normal aggregate or a water/cement ratio of 0.7 instead of 0.4. The same

TABLE 21-14

Test number	*Water/cement ratio*	*Type of aggregate*	*Type of admixture*
1	0.4	Normal weight	Variable
2	variable	Lightweight	*A*
3	0.7	Variable	*B*

12. An excellent treatment of this topic is given by K. A. Brownlee, *Industrial Experimentation* (London: H.M.S.O., 1960).

may apply to the influence of aggregate when different admixtures or water/cement ratios are used. Of course, none of these variables may interact, admixture A being better than B whatever the other factors, and if we *know* this to be the case we need not worry about interaction. If, however, the assumption is unwarranted, ignoring the interaction may lead to completely erroneous conclusions.

In some cases we may choose deliberately to sacrifice the precision of the estimates of interactions, especially those of higher order, and reduce the size of the block, with an increase in the precision with which the average effects of the factors are estimated. This is the principle of *confounding*.

It is important to note that factorial experiments can be applied only when the dependent variable is a function of the sums of the functions of the independent variables, namely,

$$S = F_1(x) + F_2(y) + F_3(z) + \cdots \tag{21-4}$$

This limitation is not as restrictive as might appear at first, since transformation can often reduce other forms to that of Eq. 21-4. For example,

$$S = x^b (\sin\, y) e^{fz}$$

can be transformed to

$$\log S = b \log x + \log \sin y + fz$$

CLASSICAL THREE-FACTOR EXPERIMENT

Let us consider the problem of the durability of concrete and denote a measure of this dependent variable by x, and the three independent variables by P, Q, and R. We assume that all the variables can be expressed quantitatively, e.g., the type of aggregate can be measured by its density, etc.

Suppose that we have in hand data on the durability of concrete when P, Q, and R have values P_1, Q_1, and R_1, respectively. We now want to find the effect of changing the values to P_2, Q_2, and R_2; we are thus concerned with only two levels of each variable.

Let us denote the effect on x of changing P from P_1 to P_2 by $(P_1 - P_2)_x$. To determine this, we observe, say, the value of x for P_1, Q_1, R_1 [denoted by $(P_1 Q_1 R_1)_x$] and the value of x for P_2, Q_1, R_1, i.e., $(P_2 Q_1 R_1)_x$. We have thus

$$(P_1 - P_2)_x = (P_1 Q_1 R_1)_x - (P_2 Q_1 R_1)_x$$

This tells us nothing about $(P_1 - P_2)_x$ when $R = R_2$, and to determine this we would have to find

$$(P_1 - P_2)_x = (P_1 Q_1 R_2)_x - (P_2 Q_1 R_2)_x$$

The procedure would be similar for $Q = Q_2$, and the appropriate combinations of the different values of Q and R.

The second important observation concerns the experimental error. If we make only one observation at P_1, Q_1, R_1 and one at P_2, Q_1, R_1, we cannot tell whether the difference between the observed values of x is real or is caused by the errors of sampling, testing, etc. To make an estimate of the experimental error, at least two replications of each experiment are necessary. Thus, in order to determine $(P_1 - P_2)_x$, $(Q_1 - Q_2)_x$, and $(R_1 - R_2)_x$, each at *one* level of the other two factors, we have to make, for example, each of the following experiments twice: P_1, Q_1, R_1; P_2, Q_1, R_1; P_1, Q_2, R_1; and P_1, Q_1, R_2. This is a total of eight observations.

FACTORIAL THREE-FACTOR EXPERIMENT

Let us now consider the factorial design of the experiment. Here we need to perform experiments for all the possible combinations of the three factors, namely, P_1, Q_1, R_1; P_1, Q_2, R_1; P_1, Q_1, R_2; P_2, Q_1, R_1; P_2, Q_2, R_1; P_2, Q_1, R_2; P_2, Q_2, R_2; and P_1, Q_2, R_2. The number required is thus eight, which is the same as in the classical case.

Brownlee[13] represents the above combinations as coordinates of points on a system of axes P, Q, and R, with P_1, Q_1, R_1 as origin (Fig. 21-10). Then

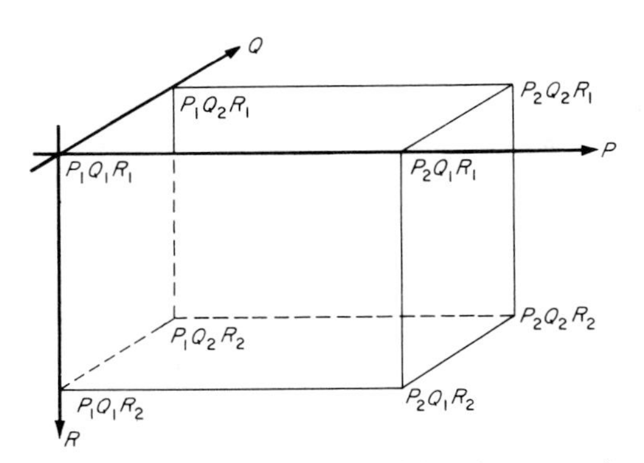

Fig. 21-10. Graphical representation of three-factor experiment.

$(P_1 - P_2)$, $(Q_1 - Q_2)$, and $(R_1 - R_2)$ are the edges of a rectangular parallelepiped. This enables us to visualize how to obtain an estimate of $(P_1 - P_2)_x$ for the various levels of the other two factors. Consider the plane containing the four points whose P coordinate is P_1. The average value of the dependent variable for these points characterizes P_1. Likewise, the aver-

13. Ibid.

age value for the plane containing the points for which $P = P_2$ characterizes P_2. The difference of the two averages is $(P_1 - P_2)_x$, i.e., the effect on the dependent variable of changing the value of P from P_1 to P_2.

Although the effects of the other two factors cancel out only approximately, the accuracy of our determination is twice as high as in the classical design, since each value is a mean of four observations instead of the two in the classical case (for the same total number of experiments). The full number of observations enters every comparison, even though each treatment contains only a limited number of observations.

But the main advantage of factorial design lies in the fact that it enables us to estimate the interactions between the factors. For example, in order to estimate the interaction between Q and R, we average the values of the dependent variable for the pairs of points differing in P only. These are

$$(Q_1 R_1)_x^P \qquad (Q_2 R_1)_x^P$$
$$(Q_1 R_2)_x^P \qquad (Q_2 R_2)_x^P$$

The top line gives $(Q_1 - Q_2)_x$ at $R = R_1$, and the bottom line gives $(Q_1 - Q_2)_x$ at $R = R_2$. Since each value is an average of two results, we can estimate the experimental error and hence determine whether $(Q_1 - Q_2)_x$ at $R = R_1$ is significantly different from $(Q_1 - Q_2)_x$ at $R = R_2$. If the difference is significant, there is interaction between Q and R. The same procedure can be used to determine the other interactions, and this gives us additional information on how each factor operates.

If the interactions are not significant, our results provide a better basis for generalized statements about the effects of each factor than if each factor had been tested with the others held at constant and arbitrary levels.

Example. An experiment was conducted to determine the pull-off force in newtons on glued parts of a piece of furniture. The completely randomized experiment was carried out by two operators at two temperatures and two humidities. There were $n = 10$ observations for each treatment. The data are shown in Table 21-15. Use $\alpha = 1$ percent.

SOLUTION

Total up the columns as shown. Then

a. Calculate the correction factor

$$C = \frac{G^2}{8n} = \frac{(2359)^2}{80} = 69{,}561.01$$

b. Sum the squares of all observations:

$$(9)^2 + (16)^2 + \cdots + (45)^2 + (51)^2 = 80{,}151.00$$

TABLE 21-15

Pull-off force, N (coded)								
	Temperature, cold (A_1)				*Temperature, hot* (A_2)			
	Humidity, 50 percent (B_1)		*Humidity, 90 percent* (B_2)		*Humidity, 50 percent* (B_1)		*Humidity, 90 percent* (B_2)	
	Operator C_1	*Operator* C_2	*Operator* C_1	*Operator* C_2	*Operator* C_1	*Operator* C_2	*Operator* C_1	*Operator* C_2
	9	20	27	31	35	41	42	50
	16	22	28	30	39	40	40	43
	20	16	19	26	32	35	35	47
	18	21	15	20	28	39	30	52
	24	15	17	27	25	29	21	30
	12	14	14	29	40	34	36	39
	15	22	27	17	33	28	47	54
	16	21	29	18	32	41	49	29
	14	19	25	27	28	45	50	45
	13	17	23	28	29	42	53	51
$\sum =$	157	187	224	253	321	374	403	440 $G = 2359$

c. Sum the squares of the totals of each column, and divide this sum by the number of individual observations, i.e., $n = 10$:

$$\frac{(157)^2 + (187)^2 + \cdots + (440)^2}{10} = 77{,}272.90$$

To test whether the treatment means differ significantly, we proceed as follows:

$$\text{sum of squares for total} = (\text{b}) - C = 80{,}151.00 - 69{,}561.01$$
$$= 10{,}589.99$$

$$\text{sum of squares for treatments} = (\text{c}) - C = 77{,}272.90 - 69{,}561.01$$
$$= 7711.89$$

(See Table 21-16).

The number of degrees of freedom associated with treatments is clearly $(t - 1)$. To obtain the number for within-treatment variation, we first reckon the total number of observations: tn. Now, for each treatment (column) the sum of the deviations about the column mean must be zero, which introduces a constraint. There are t columns and hence t constraints. Thus, by

TABLE 21-16

Source of variation	*Sum of squares (s.s.)*	v	*Mean square*
Treatments	7,711.89	$t - 1 = 7$	1,101.7
Within treatments	2,878.10	$t(n - 1) = 72$	39.97
Total	10,589.99	79	

definition of degrees of freedom, $tn - t$ or $t(n - 1)$ is the number of degrees of freedom associated with within-treatment variation.

$$\text{within-treatments sum of squares} = \text{total (s.s.)} - \text{treatments (s.s.)}$$
$$= 2878.10$$

Testing the treatments mean square for significance, we have

$$F = \frac{1101.7}{39.97} = 27.6$$

which is much greater than the value of F for $v_1 = 7$ and $v_2 = 72$, from Table A-10 at the 1 percent level of significance. We, therefore, conclude that the treatment means differ very significantly.

Partitioning the Treatments Sum of Squares

The treatments sum of squares, having 7 degrees of freedom, can be divided into 7 components, each with 1 degree of freedom, i.e.,

1. Sum of squares for A, the temperature effect.
2. Sum of squares for B, the humidity effect.
3. Sum of squares for C, the operator effect.
4. Interaction between the temperature effect and humidity effect, $A \times B$.
5. Interaction between the temperature effect and operator effect, $A \times C$.
6. Interaction between the humidity effect and operator effect, $B \times C$.
7. Interaction among the temperature, humidity, and operator effects, $A \times B \times C$.

The Main Effects *A*, *B*, and *C*

To compare A_1 with A_2, we have the sum of pull-off forces:

$$\text{for } A_1 = 157 + 187 + 224 + 253 = 821 \text{ N}$$

$$\text{for } A_2 = 321 + 374 + 403 + 440 = 1538 \text{ N}$$

For each of these the sum is based upon $4 \times 10 = 40$ measurements. Thus

$$\text{sum of squares for } A = \frac{(821)^2 + (1538)^2}{40} - C$$

$$= 75{,}987.13 - 69{,}561.01 = 6426.12$$

Similarly, to compare B_1 with B_2, we have the sum of pull-off forces:

$$\text{for } B_1 = 157 + 187 + 321 + 374 = 1039 \text{ N}$$

$$\text{for } B_2 = 224 + 253 + 403 + 440 = 1320 \text{ N}$$

$$\text{sum of squares for } B = \frac{(1039)^2 + (1320)^2}{40} - C$$

$$= 70{,}548.03 - 69{,}561.01 = 987.02$$

To compare C_1 with C_2, we take the sum of pull-off forces:

$$\text{for } C_1 = 157 + 224 + 321 + 403 = 1105 \text{ N}$$

$$\text{for } C_2 = 187 + 253 + 374 + 440 = 1254 \text{ N}$$

$$\text{sum of squares for } C = \frac{(1105)^2 + (1254)^2}{40} - C$$

$$= 69{,}838.53 - 69{,}561.01 = 277.52$$

The Interactions $A \times B$, $B \times C$, $A \times C$, $A \times B \times C$

The interactions sum of squares may be obtained from the formula

$$A \times B \text{ interaction sum of squares} = \frac{[(a + d) - (b + c)]^2}{(4)(n_1)}$$

where the factors a, b, c, and d are sums corresponding to the arrangement shown in the following table:

	B_1	B_2
A_1	a	b
A_2	c	d

and n_1 is the number of observations contributing to each of the above sums; i.e., $2 \times 10 = 20$ observations in our case. Similarly, the tables for interactions $A \times C$ and $B \times C$ are as follows:

	C_1	C_2
A_1	a	b
A_2	c	d

	C_1	C_2
B_1	a	b
B_2	c	d

Thus the sum for

$$A_1B_1 = a = 157 + 187 = 344$$

$$A_1B_2 = b = 224 + 253 = 477$$

$$A_2B_1 = c = 321 + 374 = 695$$

$$A_2B_2 = d = 403 + 440 = 843$$

Similarly, the sum for

$$A_1C_1 = a = 157 + 224 = 381$$

$$A_1C_2 = b = 187 + 253 = 440$$

$$A_2C_1 = c = 321 + 403 = 724$$

$$A_2C_2 = d = 374 + 440 = 814$$

Finally, the sum for

$$B_1C_1 = a = 157 + 321 = 478$$

$$B_1C_2 = b = 187 + 374 = 561$$

$$B_2C_1 = c = 224 + 403 = 627$$

$$B_2C_2 = d = 253 + 440 = 693$$

(See Table 21-17.)

TABLE 21-17

	B_1	B_2	$\sum =$
A_1	344	477	821
A_2	695	843	1538
$\sum =$	1039	1320	2359

	C_1	C_2	$\sum =$
A_1	381	440	821
A_2	724	814	1538
$\sum =$	1105	1254	2359

	C_1	C_2	$\sum =$
B_1	478	561	1039
B_2	627	693	1320
$\sum =$	1105	1254	2359

Substituting the numerical values of a, b, c, and d in the formula, we obtain for:

$$A \times B \text{ interaction sum of squares} = \frac{[(344 + 843) - (477 + 695)]^2}{(4)(20)}$$
$$= 2.81$$

$$A \times C \text{ interaction sum of squares} = \frac{[(381 + 814) - (440 + 724)]^2}{(4)(20)}$$
$$= 12.01$$

$$B \times C \text{ interaction sum of squares} = \frac{[(478 + 693) - (561 + 627)]^2}{(4)(20)}$$
$$= 3.61$$

The number of degrees of freedom associated with any interaction sum of squares equals the product of the degrees of freedom associated with the factors for which the interaction is being calculated; i.e., the number of degrees of freedom for $A \times B$ interaction sum of squares is the number of degrees of freedom for factor A times the number of degrees of freedom for factor B, namely, $1 \times 1 = 1$. Thus the number of degrees of freedom for each of $A \times B$, $A \times C$, $B \times C$, $A \times B \times C$ interaction sums of squares is 1.

Since the treatments sum of squares is known, then

$$\begin{aligned} A \times B \times C \text{ interaction sum of squares} &= \text{treatment sum of squares} \\ &\quad - (\text{sum of the sums of squares for } \\ &\quad A, B, C, A \times B, A \times C, \text{ and } B \times C) \\ &= 7711.89 - (6426.12 + 987.02 \\ &\quad + 277.52 + 2.81 + 12.01 + 3.61) \\ &= 2.80 \end{aligned}$$

The summary of the complete analysis of variance is given in Table 21-18.

The values of F were obtained by dividing each of the mean squares by the error mean square, 39.97. Since the four interaction mean squares were less than the error mean square, the F values for these were not calculated, and we can say that the four interactions are not significant.

From Table A-10, we find that for $v_1 = 1$ and $v_2 = 72$, $F = 7.00$ approximately, at the 1 percent level of significance. Thus the main effects, A and B, are highly significant. The effect of C is almost significant at the 1 percent level. It is definitely significant at the 5 percent level, where $F = 3.98$, approximately.

TABLE 21-18

	Source of variation	*Sum of squares*	ν	*Mean square*	*F*
A	Temperature	6,426.12	1	6,426.12	160.77
B	Humidity	987.02	1	987.02	24.69
C	Operator	277.52	1	277.52	6.94
$A \times B$	Temperature × humidity	2.81	1	2.81	
$A \times C$	Temperature × operator	12.01	1	12.01	
$B \times C$	Humidity × operator	3.61	1	3.61	
$A \times B \times C$	Temperature × humidity × operator	2.80	1	2.80	
Error	Within treatments	2,878.10	72	39.97	
Totals		10,589.99	79		

Meaning of the Main and Interaction Effects

The main effect of A represents a comparison between the means of force for cold temperature A_1 and for the hot temperature A_2, averaged over the two levels of B and the two levels of C. The mean for A_1 is obtained from the original data and is equal to $\frac{821}{40} = 20.525$ N. The mean for $A_2 = \frac{1538}{40} = 38.45$ N. Since the A mean square is significant, we can conclude that these two means differ significantly; i.e., glued parts of a furniture piece become stronger with increase in temperature within the range of the experiment. Similarly, for the humidity effect B, the mean force for $B_1 = \frac{1039}{40} = 25.975$ N, and the mean force for $B_2 = \frac{1320}{40} = 33.00$ N. Since the analysis of variance showed the mean square for B to be significant, we can state that the means of forces for B_1 and B_2 differ significantly; i.e., the glued parts of the furniture become stronger with increase in humidity, in the range of the experiment. The same procedure can be followed in comparing the means of force, $\frac{1105}{40}$ and $\frac{1254}{40}$, for the two operators C_1 and C_2, respectively. The difference in means is significant at the 5 percent level but not quite significant at the 1 percent level.

The meaning of the interaction effects being nonsignificant is that the difference between the means due to a main effect for one level is not significantly different from the difference between the means due to the same effect at a higher level. For example, the statement that $A \times C$ interaction mean square is not significant is interpreted to say that the difference between means of force due to A_1 and A_2 for the first level of C is not significantly different from the differences between the means of force due to A_1 and A_2 for the second level of C. With a nonsignificant $A \times C$ interaction, we can state that the A effect, the difference between the effect of A_1 and

A_2 on the force, is independent of C; i.e., we have approximately the same difference between the effect of A_1 and A_2 on the force regardless of the levels of C. Similar explanations can be made of the nonsignificance of $A \times B$, $B \times C$, and $A \times B \times C$.

Graphical Interpretation

We can also examine interaction effects, say $A \times B$, by taking factor B for the x-axis and plotting the means of force for each level of A; i.e., corresponding to B_1,

$$\text{mean for } A_1 = \tfrac{344}{20} = 17.2 \text{ N}$$

$$\text{mean for } A_2 = \tfrac{695}{20} = 34.75 \text{ N}$$

and corresponding to B_2:

$$\text{mean for } A_1 = \tfrac{477}{20} = 23.85 \text{ N}$$

$$\text{mean for } A_2 = \tfrac{843}{20} = 42.15 \text{ N}$$

These values are shown plotted in Fig. 21-11a. If the lines for A_1 and A_2 were exactly parallel, the $A \times B$ interaction would be zero. In fact, the lines are very nearly parallel (within the limits of random sampling), which shows

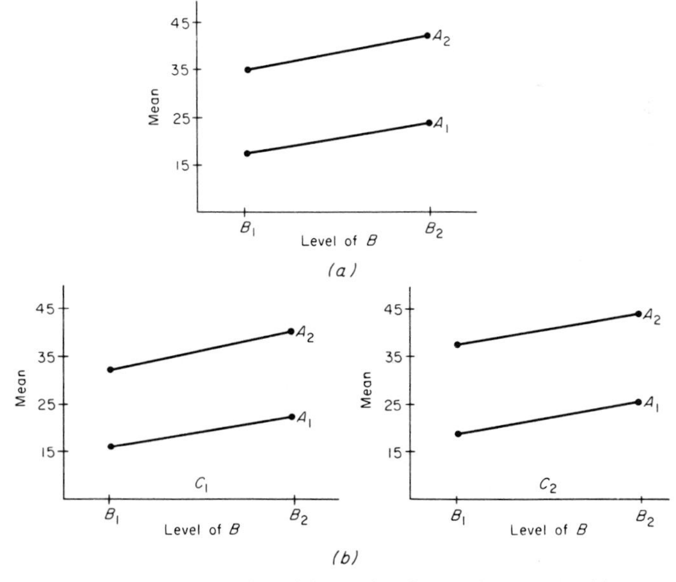

Fig. 21-11. Graphical representation of the results of example on p. 420. (a) Means for levels of A at each level of B; (b) means for levels of A at each level of B for C_1 and C_2, respectively.

that the $A \times B$ interaction is not significant. Similar graphical representation of the $A \times C$ and $B \times C$ interactions can be given.

For the nature of the $A \times B \times C$ interaction, we consider the $A \times B$ interaction separately for each level of C, as shown in Table 21-19. This table shows the means of force due to A and B for each level of C, as obtained from the original data.

TABLE 21-19

	C_1		C_2	
	B_1	B_2	B_1	B_2
A_1	15.7	22.4	18.7	25.3
A_2	32.1	40.3	37.4	44.0

The graphs for A_1 and A_2 versus B for levels C_1 and C_2, respectively, are shown in Fig. 21-11b. We notice that the forms of these graphs are similar, which confirms our finding of the nonsignificance of the $A \times B \times C$ interaction mean square. In other words, this nonsignificance means that the $A \times C$ interactions for the separate levels of B are of the same form, the $A \times B$ interactions for the separate levels of C are of the same form, and the $B \times C$ interactions for the separate levels of A are of the same form.

RANDOM NUMBERS

On several occasions we have achieved randomization by drawing numbered cards "from a hat." With a large number of items involved, this becomes tedious, and it may be more convenient to use a table of random numbers, such as Table A-14.

Numerous methods of using the table exist. For example, in order to arrange the numbers 1 to 13 in a random order, we can select any row, column, or diagonal in the table, and record the numbers 1 to 13 as they occur.

It may be quicker to read all the numbers in order and to divide each by 13, recording the remainder. Any remainder that has already been obtained is rejected. Since the remainders will be between 0 and 12 inclusive, we consider 0 as 13. We should note that since the highest possible two-digit multiple of 13 is 91, the remainders 1, 2, ..., 8 have a higher chance of occurring than others. To remove this bias, we ignore the numbers 92, 93, ..., 99.

As an alternative, we can use the divisor 20 instead of 13, rejecting any number that gives 0, 14, 15, 16, 17, 18, 19. Thus, using the fifth column of Table A-14, we record: 7, ~~14~~, 6, 9, ~~6~~, ~~9~~, 2, ~~9~~, ~~15~~, ~~7~~, ~~6~~, ~~19~~, ~~2~~, ~~0~~, ~~6~~, ~~15~~, 4, ~~17~~, ~~14~~, 10, 1, ~~6~~, 11, ~~19~~, ~~4~~, ~~16~~, 3, 8, ~~0~~, 5, ~~19~~, 12, 13; i.e., 7, 6, 9, 2, 4, 10, 1, 11, 3, 8, 5, 12, 13.

Solved Problem

21-1. The breaking strength of a fabric has been determined from numerous tests to be 250 N, with a standard deviation due to testing of 18 N. Recently, a new manufacturing process X was introduced, which seems to increase the strength of the fabric.

a. Find the criterion for rejecting the old manufacturing process A at a 1 percent level of significance when 36 specimens of the fabric are tested.
b. Using the criterion found in (a), and assuming that the standard deviation remains as a result of testing at 18 N, find the probability of accepting process A when process X has, in fact, improved on the mean strength to the value of 265 N.
c. Construct a power curve for the test of the hypothesis and interpret the graph.

SOLUTION

a. From Table A-4 for a one-tailed test, at a 1 percent level of significance (see Fig. 21-12a), the value of z corresponding to $F(z) = 0.49$ is $z = 2.33$. But,

$$z = \frac{\bar{x} - 250}{\sigma_T/\sqrt{n}} = \frac{\bar{x} - 250}{18/\sqrt{36}} = \frac{\bar{x} - 250}{3}$$
$$= 2.33$$

Hence

$$\bar{x} = 250 + 3 \times 2.33$$
$$= 257 \text{ N}$$

Therefore, the criterion is: Reject the hypothesis that process X is the same as process A if the mean breaking strength of 36 fabric specimens > 257 N. Otherwise, accept the hypothesis.

b. Figure 21-12b shows the two normal distributions corresponding to means of 250 and 265 N. The probability of accepting process A when the new mean breaking strength is actually 265 N (Type II error) is represented by the region marked β. To calculate the probability of committing a Type II error, we proceed as follows: Deviation of 265 from 257 in terms of $\sigma_{\bar{x}}$ is

$$\frac{265 - 257}{\sigma_T/\sqrt{n}} = \frac{8}{18/6} = 2.66$$

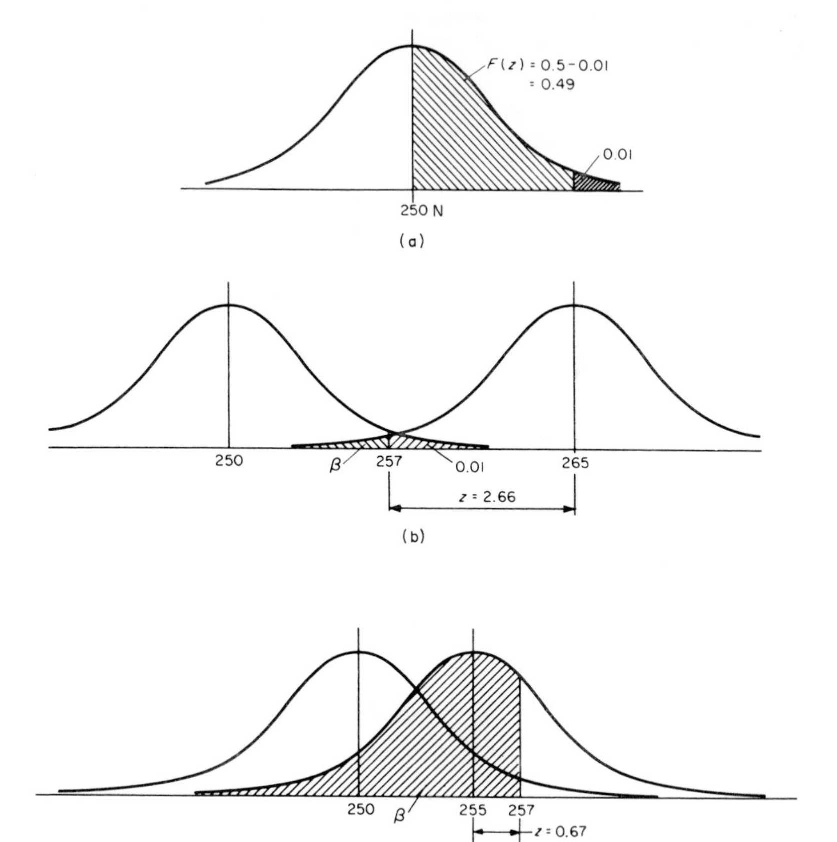

Fig. 21-12

Thus, using Table A-4, β = area under the right-hand normal curve to the left of $z = 2.66$, i.e., $0.5 - 0.4961 = 0.0039$ (small indeed).

c. In order to draw the power curve, we have to find β for various breaking strengths of the fabric manufactured by the new process X. Thus, if the breaking strength by the new process X is, say, 255 N, then its deviation from 257 N in terms of $\sigma_{\bar{x}}$ is

$$\frac{257 - 255}{18/\sqrt{36}} = 0.67$$

Using Table A-4, β = shaded area in Fig. 21-12c, that is, 0.5 + the area under the right-hand normal curve between $z = 0$ and $z = 0.67$. Thus

$$\beta = 0.5 + 0.2486 = 0.7486$$

Following the above procedure, we can compile the following table for β corresponding to the mean breaking strength μ.

μ	240	245	250	255	260	265	270
β	1.0000	1.0000	0.9901	0.7486	0.1587	0.0039	0.0000

Plotting μ versus $(1 - \beta)$ yields the power curve shown in Fig. 21-13. This curve

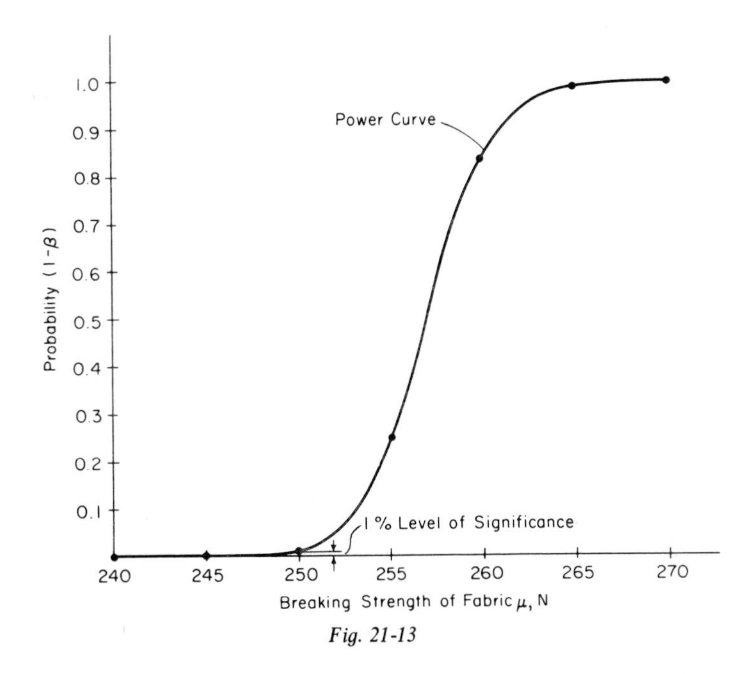

Fig. 21-13

indicates the power of the test to reject a false hypothesis. We notice from this curve that the probability of rejecting process A if the new breaking strength is less than 250 N is practically nil. On the other hand, we see that the curve rises sharply so that it is with almost a certainty that we reject the hypothesis of maintaining process A when the mean breaking strength is greater than 265 N. Note the point of inflection on the power curve at $\mu = 257$ N and $(1 - \beta) = 0.5$.

Problems

21-1. From previous data it was found that a certain chemical process produced a mean yield of 10 units with a standard deviation $\sigma_T = 0.3$. It is believed that by a newly developed process the mean yield can be increased.

a. If it is agreed to run 36 tests, design a decision rule for rejecting the old process at the 1 percent level of significance.
b. Following this decision rule, what is the probability of accepting the old process, when in fact the new process has increased the yield to 10.8? Assume the same standard deviation as before.
c. Construct the power curve for the process with $\sigma_T = 0.3$.
d. What sample size is necessary if the power of the test is to be 98 percent?

21-2. A project requires the use of components whose density has a coded value of at least 100. It is known from past experience that the standard deviation due to testing is 11. It is decided that the probability of accepting components with a density of 94 is to be 0.1, and the probability of rejecting those with a density of 112 is to be 0.1. Find a sample size n in order to avoid having either risk exceed the desired value. Hence draw a power curve for these decision rules.

21-3. The average production figures for four shifts measured during four years are as shown in Table 21-20. Establish the production trends within the working day and over the years. Use $\alpha = 1$ percent.

TABLE 21-20

	Shift			
Year	*12 p.m.–6 a.m.*	*6 a.m.–12 noon*	*12 noon–6 p.m.*	*6 p.m.–12 p.m.*
1	9.8	12.1	15.8	11.7
2	9.5	11.8	15.5	11.7
3	11.8	15.5	17.2	13.6
4	13.9	15.4	17.2	14.6

21-4. In order to decide on the choice of a clay for the manufacture of artificial aggregate, samples were obtained from three sources and were treated by two processes. The coded results of the performance of the clay are shown in Table 21-21. Test whether the performance depends on the source of clay or the process by which it was treated. The 5 percent level of significance is suggested as a basis for making decisions.

TABLE 21-21

	Source		
Process	*I*	*II*	*III*
	9	11	16
	12	13	15
A	13	13	17
	16	15	16
	15	16	16

	Source		
Process	*I*	*II*	*III*
	13	11	16
	15	13	17
B	17	14	15
	12	13	18
	15	16	18

21-5. To check the variation of gage blocks and of the sensitive devices measuring them, the manufacturer chose at random five gage blocks and five micrometers and asked five quality control engineers, *A*, *B*, *C*, *D*, and *E* to carry out the experiment. The 1-cm gage blocks were used with the coded results given in Table 21-22.

TABLE 21-22

	Micrometer				
Gage blocks	*1*	*2*	*3*	*4*	*5*
1	*B* 95	*E* 101	*D* 109	*C* 99	*A* 95
2	*C* 97	*A* 104	*B* 104	*E* 98	*D* 105
3	*D* 98	*B* 99	*C* 105	*A* 95	*E* 99
4	*E* 101	*C* 98	*A* 99	*D* 102	*B* 97
5	*A* 100	*D* 102	*E* 97	*B* 103	*C* 101

Assume there is no interaction present.

a. Estimate the component of the variance due to micrometers.
b. Estimate the component of the variance due to gage blocks.
c. Estimate the component of the variance due to engineers.
d. Test for micrometer and gage block effects at the 1 percent level of significance.
e. Test for engineer effects at the 5 percent level of significance.

21-6. A steel company manufactures a particular steel with an average ultimate strength of 356 MN/m^2 and a standard deviation of 4.8 MN/m^2. The manufacturer wishes to investigate whether a change in the composition of the steel alloy increases the strength. The company will take a 10 percent risk of not detecting an increase in the strength by as much as 3.4 MN/m^2; furthermore, it wants to reach the conclusion of no change in the strength with 0.99 probability ($\alpha = 0.01$). Note that the particular change in the steel alloy is assumed not to influence the standard deviation, nor to cause a decrease in the strength.

a. Find the required sample size n from the appropriate OC curve.
b. If the mean strength from the sample in (a) is found to be 361 MN/m^2, test the hypothesis that the strength is unaffected by the change in the steel alloy composition.

Appendix A

Bessel's Correction

Consider a sample of size n drawn from a population with a mean μ and standard deviation σ.

Let x_i be an observation in the sample. Then

$$\begin{aligned} x_i - \mu &= (x_i - \bar{x}) + (\bar{x} - \mu) \\ &= (x_i - \bar{x}) - \epsilon \end{aligned}$$

where $\epsilon = \mu - \bar{x}$ is the "error" or deviation of the sample mean, $\bar{x}$. Squaring, we obtain

$$(x_i - \mu)^2 = (x_i - \bar{x})^2 + \epsilon^2 - 2\epsilon(x_i - \bar{x})$$

For all the observations in the sample, we sum for i from 1 to n and obtain

$$\sum (x_i - \mu)^2 = \sum (x_i - \bar{x})^2 + n\epsilon^2 - 2\epsilon \sum (x_i - \bar{x})$$

But

$$\sum (x_i - \bar{x}) = 0 \quad \text{by definition of } \bar{x}$$

Therefore,

$$\sum (x_i - \mu)^2 = \sum (x_i - \bar{x})^2 + n\epsilon^2$$

If we repeat this calculation for a large number of samples, the mean value of the left-hand side of the above equation will (by definition of σ^2) tend to $n\sigma^2$. Similarly, the mean value of $n\epsilon^2 = n(\mu - \bar{x})^2$ will tend to n times the variance of $\bar{x}$, since ϵ represents the deviation of the sample mean from the population mean. Thus

$$n\epsilon^2 \rightarrow n\left(\frac{\sigma^2}{n}\right)$$

whence

$$n\sigma^2 \rightarrow \sum (x_i - \bar{x})^2 + \sigma^2$$

or

$$\sum (x_i - \bar{x})^2 \rightarrow (n - 1)\sigma^2$$

Thus

$$\frac{\sum (x_i - \bar{x})^2}{n - 1} \rightarrow \sigma^2$$

In other words, for a large number of random samples, the mean value of $[\sum (x_i - \bar{x})^2]/(n - 1)$ tends to σ^2, i.e., it is an unbiased estimate of the variance of the population. The estimate is denoted by s^2. Thus

$$s^2 = \frac{\sum (x_i - \bar{x})^2}{n - 1} \qquad \text{(4-3)}$$

Since the variance of the sample (taken as a finite population with a mean $\bar{x}$) σ^2 is given by

$$\sigma^2 = \frac{\sum (x_i - \bar{x})^2}{n}$$

Bessel's correction is

$$\frac{s^2}{\sigma^2} = \frac{n}{n - 1}$$

Appendix B

Derivation of $P_r = e^{-\mu}\mu^r/r!$ from the Poisson Process

Figure B-1 shows the random occurrence of events marked by x on a time scale. Let there be λ events on average in a unit time interval; then there will be on average λt events in a time interval of length t.

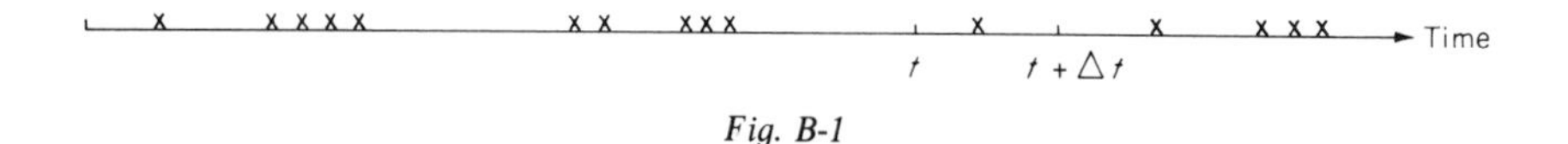

Fig. B-1

It is assumed that (a) the probability of observing one event in a very small interval of time Δt is $\lambda \, \Delta t$; (b) the probability of more than one event occurring in Δt is comparatively very small and, therefore, can be neglected; and (c) events in nonoverlapping time intervals are independent. Let

$P_r(t)$ = probability of observing exactly r events in a time interval from zero to time t, denoted $(0, t)$

Consider the time interval $(0, t + \Delta t)$ as split into two adjacent and nonoverlapping intervals $(0, t)$ and $(t, t + \Delta t)$. Now, r events can occur within the interval $(0, t + \Delta t)$ in any of the following ways:

r events occur in $(0, t)$ and 0 events occur in $(t, t + \Delta t)$

$r - 1$ events occur in $(0, t)$ and 1 event occurs in $(t, t + \Delta t)$

$r - 2$ events occur in $(0, t)$ and 2 events occur in $(t, t + \Delta t)$,

and so on. It follows, then, that for $r \geq 1$

$$\begin{aligned} P_r(t + \Delta t) = P_r(t) &\times \text{probability of no event in } (t, t + \Delta t) \\ + P_{r-1}(t) &\times \text{probability of 1 event in } (t, t + \Delta t) \\ + P_{r-2}(t) &\times \text{probability of 2 events in } (t, t + \Delta t) \\ \vdots \qquad & \qquad \vdots \qquad\qquad \vdots \qquad\qquad \vdots \\ + P_0(t) &\times \text{probability of } r \text{ events in } (t, t + \Delta t) \end{aligned}$$

From the assumptions made earlier, we then have

$$P_r(t + \Delta t) = P_r(t) \times (1 - \lambda\, \Delta t) + P_{r-1}(t) \times \lambda\, \Delta t + 0 \cdots$$

Rearranging yields

$$\frac{P_r(t + \Delta t) - P_r(t)}{\Delta t} = -\lambda[P_r(t) - P_{r-1}(t)]$$

In the limit, as $\Delta t \to 0$, we obtain

$$\frac{d}{dt}[P_r(t)] = -\lambda[P_r(t) - P_{r-1}(t)] \tag{a}$$

It can rigorously be shown[1] that the solution of the above differential equation, for $r = 1, 2, \ldots,$ is

$$P_r(t) = \frac{e^{-\lambda t}(\lambda t)^r}{r!} \tag{b}$$

using the initial condition that $P_r(0) = 0$. It can easily be verified by substitution that Eq. (b) is the solution of Eq. (a).

For $r = 0$ we can show by similar reasoning that

$$P_0(t + \Delta t) = P_0(t) \times (1 - \lambda\, \Delta t) + 0$$

1. T. C. Fry, Probability and Its Engineering Uses, (New York: Van Nostrand Reinhold Company, 1965) p. 217.

from which

$$\frac{dP_0(t)}{P_0(t)} = -\lambda\,dt$$

Integrating and using the initial condition that $P_0(0) = 1$,

$$P_0(t) = e^{-\lambda t} \tag{c}$$

Using the definition that $0! = 1$, we can combine Eqs. (b) and (c) to write

$$P_r(t) = \frac{e^{-\lambda t}(\lambda t)^r}{r!}$$

for $r = 0, 1, 2, \ldots$. Putting the average $\lambda t = \mu$, we find that

$$P_r = \frac{e^{-\mu}\mu^r}{r!} \tag{8-1}$$

Appendix C

Mean and Standard Deviation of Terms of Expansion of $\left(\frac{1}{2} + \frac{1}{2}\right)^{2n}$

Let the abscissae for which the expansion gives the ordinates vary in steps of Δx. We can then tabulate the values as shown in Table C-1. Hence

$$\text{mean} = \mu = \frac{\sum f_i x_i}{\sum f_i} = 2np\ \Delta x$$

Since $p = \frac{1}{2}$,

$$\mu = n\ \Delta x \tag{9-6}$$

Now

$$\sum [f_i x_i^2 - f_i x_i(\Delta x)] = \sum f_i x_i^2 - (\Delta x) \sum f_i x_i$$

Hence

$$\begin{aligned}\sum f_i x_i^2 &= (\Delta x) \sum f_i x_i + (2n - 1)(2n)(\Delta x)^2 p^2 (q + p)^{2n-2} \\ &= 2np(\Delta x)^2 + (2n - 1)(2n)(\Delta x)^2 p^2\end{aligned}$$

TABLE C-1

x_i	*Probability* f_i	$f_i x_i$	$f_i x_i^2$	$f_i x_i^2 - f_i x_i(\Delta x)$
0	q^{2n}	0	0	0
Δx	$2nq^{2n-1}p$	$2n\ \Delta x q^{2n-1}p$	$2n(\Delta x)^2 q^{2n-1}p$	0
$2\ \Delta x$	$\frac{2n(2n-1)}{2!} q^{2n-2}p^2$	$2n\ \Delta x(2n-1)q^{2n-2}p^2$	$4n(\Delta x)^2(2n-1)q^{2n-2}p^2$	$2n(\Delta x)^2(2n-1)q^{2n-2}p^2$
$\vdots$	$\vdots$	$\vdots$	$\vdots$	$\vdots$
$(2n-1)\ \Delta x$	$2nqp^{2n-1}$	$(2n-1)\ \Delta x(2n)qp^{2n-1}$	$(2n-1)^2(\Delta x)^2(2n)qp^{2n-1}$	$(2n-2)(2n)(\Delta x)^2(2n-1)qp^{2n-1}$
$2n\ \Delta x$	p^{2n}	$2n\ \Delta x p^{2n}$	$4n^2(\Delta x)^2 p^{2n}$	$2n(\Delta x)^2(2n-1)p^{2n}$
	$\sum f_i = (q+p)^{2n} = 1$	$\sum f_i x_i = 2np\ \Delta x(q+p)^{2n-1} = 2np\ \Delta x$		$\sum [f_i x_i^2 - f_i x_i(\Delta x)] = (2n-1)(2n)(\Delta x)^2 p^2 (q+p)^{2n-2}$

Since $p = q = \frac{1}{2}$,

$$\sum f_i x_i^2 = n(\Delta x)^2 - \frac{n}{2}(\Delta x)^2 + n^2(\Delta x)^2$$

$$= \mu^2 + \frac{n}{2}(\Delta x)^2$$

From Eq. (4-4),

$$\sigma^2 = \frac{\sum f_i x_i^2}{\sum f_i} - \mu^2$$

$$= \mu^2 + \frac{n}{2}(\Delta x)^2 - \mu^2$$

or

$$\sigma^2 = \frac{n}{2}(\Delta x)^2 \tag{9-7}$$

Appendix D

Evaluation of $\int_{-\infty}^{+\infty} e^{-(X^2/2\sigma^2)}\, dX$

To evaluate the integral

$$I = \int_{-\infty}^{+\infty} e^{-(X^2/2\sigma^2)}\, dX$$

let

$$I_1 = \int_{-a}^{+a} e^{-(X^2/2\sigma^2)}\, dX$$

Using an arbitrary variable y, we can also write:

$$I_1 = \int_{-a}^{+a} e^{-(y^2/2\sigma^2)}\, dy$$

Hence

$$I_1^2 = \int_{-a}^{+a} \int_{-a}^{+a} e^{-(X^2+y^2)/2\sigma^2}\, dX\, dy$$

As a physical interpretation of this integral, we can imagine that it represents the volume under a surface of height $e^{-(X^2+y^2)/2\sigma^2}$ on a square base of side $2a$ (Fig. D-1).

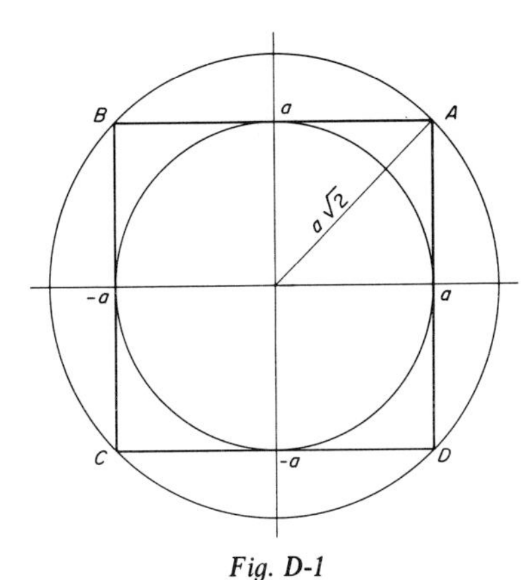

Fig. D-1

Let the square be $ABCD$ as shown in the figure. If we inscribe and circumscribe the square $ABCD$ by two circles, then the integral I_1^2 will be intermediate in value between the integrals I_3 and I_4, corresponding, respectively, to the volumes with the two circles as a base. Using polar coordinates, we find that

$$I_3 = \int_0^{+a} \int_0^{2\pi} e^{-r^2/2\sigma^2} r \, d\theta \, dr$$

and

$$I_4 = \int_0^{a\surd 2} \int_0^{2\pi} e^{-r^2/2\sigma^2} r \, d\theta \, dr$$

Hence

$$\int_0^{2\pi} d\theta \int_0^{a} re^{-r^2/2\sigma^2} \, dr < I_1^2 < \int_0^{2\pi} d\theta \int_0^{a\surd 2} re^{-r^2/2\sigma^2} \, dr$$

However, as $a \to \infty$, the integrals on the right-hand side and the left-hand side converge to

$$2\pi \int_0^{\infty} \sigma^2 \frac{r}{\sigma^2} e^{-r^2/2\sigma^2} \, dr = 2\pi\sigma^2[-e^{-r^2/2\sigma^2}]_0^{\infty}$$

$$= 2\pi\sigma^2$$

Hence

$$I^2 = \lim_{a \to \infty} I_1^2 = 2\pi\sigma^2$$

or

$$\int_{-\infty}^{+\infty} e^{-X^2/2\sigma^2}\, dX = \sigma\sqrt{2\pi} \qquad (9\text{-}12)$$

Appendix E

Relation Between t and r

Equation (15-32) gives the test for the significance of slope as

$$t = \frac{b}{s_b}$$

The correlation coefficient is given by Eq. (16-8) as

$$r = \frac{\sum XY}{\sqrt{\sum X^2 \sum Y^2}}$$

Now, from Eq. (15-18),

$$b = \frac{\sum XY}{\sum X^2}$$

and from Eq. (15-24),

$$s_b^2 = \frac{s_{y|x}^2}{\sum X^2}$$

Also, from Eq. (15-22),

$$s_{y|x}^2 = \frac{\sum \epsilon_i^2}{n-2}$$

$$= \frac{\sum (Y - \hat{Y})^2}{n-2}$$

Since

$$\hat{Y} = bX$$

we can expand:

$$\sum (Y - \hat{Y})^2 = \sum Y^2 - 2b \sum XY + b^2 \sum X^2$$

$$= \left(\sum Y^2\right)\left[1 + \frac{b(b \sum X^2 - 2 \sum XY)}{\sum Y^2}\right]$$

$$= \left(\sum Y^2\right)\left[1 + \frac{(\sum XY)(\sum XY - 2 \sum XY)}{\sum X^2 \sum Y^2}\right]$$

$$= \left(\sum Y^2\right)\left[1 - \frac{(\sum XY)^2}{\sum X^2 \sum Y^2}\right]$$

Hence

$$s_b^2 = \frac{\sum Y^2}{(n-2) \sum X^2}\left[1 - \frac{(\sum XY)^2}{\sum X^2 \sum Y^2}\right]$$

and

$$t = \frac{\sum XY(n-2)^{1/2}(\sum X^2)^{1/2}}{(\sum X^2)\left\{\sum Y^2\left[1 - \frac{(\sum XY)^2}{\sum X^2 \sum Y^2}\right]\right\}^{1/2}}$$

$$= \frac{r\sqrt{n-2}}{\sqrt{1-r^2}}$$

Appendix F

Proof that the Regression Plane Contains the Centroidal Point

Let $(\bar{y}, \bar{x}_1, \bar{x}_2)$ be the centroid of all observations. Dividing the first of Eqs. (17-5) by n, we obtain

$$b_0 + b_1 \frac{\sum x_1}{n} + b_2 \frac{\sum x_2}{n} = \frac{\sum y}{n}$$

Hence the definition of centroid

$$b_0 + b_1 \bar{x}_1 + b_2 \bar{x}_2 = \bar{y}$$

which proves that $(\bar{y}, \bar{x}_1, \bar{x}_2)$ lies on the plane.

Tables

TABLE A-1

Range Coefficient d

Number of observations n	*Coefficient* d	*Number of observations* n	*Coefficient* d
2	0.8862	14	0.2935
3	0.5908	15	0.2880
4	0.4857	16	0.2831
5	0.4299	17	0.2787
6	0.3945	18	0.2747
7	0.3698	19	0.2711
8	0.3512	20	0.2677
9	0.3367	24	0.2567
10	0.3249	50	0.2223
11	0.3152	100	0.1994
12	0.3069	1000	0.1543
13	0.2998		

This table gives values of d in equation $s = \bar{R}d$ for the estimate of the standard deviation from mean range $\bar{R}$. This method of estimating s is valid only when the underlying variate is approximately normally distributed.

This table is reproduced by permission of the author and the publishers from a paper "On the Extreme Individuals and the Range of Samples Taken from a Normal Population," by L. H. C. Tippett, *Biometrika*, vol. 17, 1925, pp. 364–387 and from Table 22 of *Biometrika Tables for Statisticians*, vol. 1 (London: Cambridge University Press, 1954).

TABLE A-2

Values of $e^{-\mu}$

μ	*0.00*	*0.01*	*0.02*	*0.03*	*0.04*	*0.05*	*0.06*	*0.07*	*0.08*	*0.09*
0.0	1.000	0.990	0.980	0.970	0.961	0.951	0.942	0.932	0.923	0.914

μ	*0.0*	*0.1*	*0.2*	*0.3*	*0.4*	*0.5*	*0.6*	*0.7*	*0.8*	*0.9*
0.0	1.000	0.905	0.819	0.741	0.670	0.607	0.549	0.497	0.449	0.407
1.0	0.368	0.333	0.301	0.273	0.247	0.223	0.202	0.183	0.165	0.150
2.0	0.135	0.122	0.111	0.100	0.091	0.082	0.074	0.067	0.061	0.055
3.0	0.050	0.045	0.041	0.037	0.033	0.030	0.027	0.025	0.022	0.020
4.0	0.018	0.017	0.015	0.014	0.012	0.011	0.010	0.0091	0.0082	0.0074
5.0	0.0067									

Intermediate values of $e^{-\mu}$ can be obtained by making use of the values in the upper part of the table. For example, the value of $e^{-1.11} = e^{-1.1} \times e^{-0.01} = 0.333 \times 0.990 = 0.330$.

TABLE A-3

Ordinates of the Normal Curve

$$f(z) = \frac{1}{\sqrt{2\pi}} e^{-z^2/2}$$

z	*.00*	*.01*	*.02*	*.03*	*.04*	*.05*	*.06*	*.07*	*.08*	*.09*
.0	.3989	.3989	.3989	.3988	.3986	.3984	.3982	.3980	.3977	.3973
.1	.3970	.3965	.3961	.3956	.3951	.3945	.3939	.3932	.3925	.3918
.2	.3910	.3902	.3894	.3885	.3876	.3867	.3857	.3847	.3836	.3825
.3	.3814	.3802	.3790	.3778	.3765	.3752	.3739	.3725	.3712	.3697
.4	.3683	.3668	.3653	.3637	.3621	.3605	.3589	.3572	.3555	.3538
.5	.3521	.3503	.3485	.3467	.3448	.3429	.3410	.3391	.3372	.3352
.6	.3332	.3312	.3292	.3271	.3251	.3230	.3209	.3187	.3166	.3144
.7	.3123	.3101	.3079	.3056	.3034	.3011	.2989	.2966	.2943	.2920
.8	.2897	.2874	.2850	.2827	.2803	.2780	.2756	.2732	.2709	.2685
.9	.2661	.2637	.2613	.2589	.2565	.2541	.2516	.2492	.2468	.2444
1.0	.2420	.2396	.2371	.2347	.2323	.2299	.2275	.2251	.2227	.2203
1.1	.2179	.2155	.2131	.2107	.2083	.2059	.2036	.2012	.1989	.1965
1.2	.1942	.1919	.1895	.1872	.1849	.1826	.1804	.1781	.1758	.1736
1.3	.1714	.1691	.1669	.1647	.1626	.1604	.1582	.1561	.1539	.1518
1.4	.1497	.1476	.1456	.1435	.1415	.1394	.1374	.1354	.1334	.1315
1.5	.1295	.1276	.1257	.1238	.1219	.1200	.1182	.1163	.1145	.1127
1.6	.1109	.1092	.1074	.1057	.1040	.1023	.1006	.0989	.0973	.0957
1.7	.0940	.0925	.0909	.0893	.0878	.0863	.0848	.0833	.0818	.0804
1.8	.0790	.0775	.0761	.0748	.0734	.0721	.0707	.0694	.0681	.0669
1.9	.0656	.0644	.0632	.0620	.0608	.0596	.0584	.0573	.0562	.0551
2.0	.0540	.0529	.0519	.0508	.0498	.0488	.0478	.0468	.0459	.0449
2.1	.0440	.0431	.0422	.0413	.0404	.0396	.0387	.0379	.0371	.0363
2.2	.0355	.0347	.0339	.0332	.0325	.0317	.0310	.0303	.0297	.0290
2.3	.0283	.0277	.0270	.0264	.0258	.0252	.0246	.0241	.0235	.0229
2.4	.0224	.0219	.0213	.0208	.0203	.0198	.0194	.0189	.0184	.0180
2.5	.0175	.0171	.0167	.0163	.0158	.0154	.0151	.0147	.0143	.0139
2.6	.0136	.0132	.0129	.0126	.0122	.0119	.0116	.0113	.0110	.0107
2.7	.0104	.0101	.0099	.0096	.0093	.0091	.0088	.0086	.0084	.0081
2.8	.0079	.0077	.0075	.0073	.0071	.0069	.0067	.0065	.0063	.0061
2.9	.0060	.0058	.0056	.0055	.0053	.0051	.0050	.0048	.0047	.0046
3.0	.0044	.0043	.0042	.0041	.0039	.0038	.0037	.0036	.0035	.0034
3.1	.0033	.0032	.0031	.0030	.0029	.0028	.0027	.0026	.0025	.0025
3.2	.0024	.0023	.0022	.0022	.0021	.0020	.0020	.0019	.0018	.0018
3.3	.0017	.0017	.0016	.0016	.0015	.0015	.0014	.0014	.0013	.0013
3.4	.0012	.0012	.0012	.0011	.0011	.0010	.0010	.0010	.0009	.0009
3.5	.0009									
3.6	.0006									
3.7	.0004									
3.8	.0003									
3.9	.0002									

The above ordinates give the probability density for $z = (x - \mu)/\sigma$ deviations from the mean μ (i.e., $z = 0$). To fit a normal frequency curve to observed data consisting of n observations, multiply the ordinate from the table for any value of z by n/σ. To fit a normal probability curve, multiply the ordinate by $1/\sigma$.

The values for z up to 3.0 are taken from Table II of Fisher and Yates: *Statistical Tables for Biological, Agricultural, and Medical Research*, published by Oliver & Boyd Ltd., Edinburgh, Scotland, by permission of the authors and publishers. The values for the range $z = 3.1$ to $z = 3.9$ are reproduced by permission of the author and publishers from Table A-2 of *Methods of Statistical Analysis* by Cyril H. Goulden, 2nd ed. (New York: John Wiley & Sons, Inc., 1960).

TABLE A-4

Areas under the Normal Curve

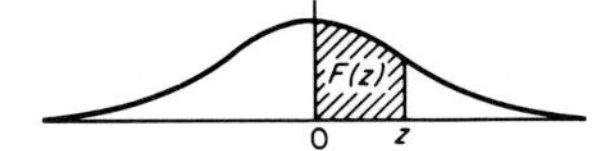

$$F(z) = \int_0^z \frac{1}{\sqrt{2\pi}} e^{-z^2/2}\, dz$$

z	.00	.01	.02	.03	.04	.05	.06	.07	.08	.09
.0	.0000	.0040	.0080	.0120	.0159	.0199	.0239	.0279	.0319	.0359
.1	.0398	.0438	.0478	.0517	.0557	.0596	.0636	.0675	.0714	.0753
.2	.0793	.0832	.0871	.0910	.0948	.0987	.1026	.1064	.1103	.1141
.3	.1179	.1217	.1255	.1293	.1331	.1368	.1406	.1443	.1480	.1517
.4	.1554	.1591	.1628	.1664	.1700	.1736	.1772	.1808	.1844	.1879
.5	.1915	.1950	.1985	.2019	.2054	.2088	.2123	.2157	.2190	.2224
.6	.2257	.2291	.2324	.2357	.2389	.2422	.2454	.2486	.2518	.2549
.7	.2580	.2611	.2642	.2673	.2704	.2734	.2764	.2794	.2823	.2852
.8	.2881	.2910	.2939	.2967	.2995	.3023	.3051	.3078	.3106	.3133
.9	.3159	.3186	.3212	.3238	.3264	.3289	.3315	.3340	.3365	.3389
1.0	.3413	.3438	.3461	.3485	.3508	.3531	.3554	.3577	.3599	.3621
1.1	.3643	.3665	.3686	.3708	.3729	.3749	.3770	.3790	.3810	.3830
1.2	.3849	.3869	.3888	.3907	.3925	.3944	.3962	.3980	.3997	.4015
1.3	.4032	.4049	.4066	.4082	.4099	.4115	.4131	.4147	.4162	.4177
1.4	.4192	.4207	.4222	.4236	.4251	.4265	.4279	.4292	.4306	.4319
1.5	.4332	.4345	.4357	.4370	.4382	.4394	.4406	.4418	.4430	.4441
1.6	.4452	.4463	.4474	.4485	.4495	.4505	.4515	.4525	.4535	.4545
1.7	.4554	.4564	.4573	.4582	.4591	.4599	.4608	.4616	.4625	.4633
1.8	.4641	.4649	.4656	.4664	.4671	.4678	.4686	.4693	.4699	.4706
1.9	.4713	.4719	.4726	.4732	.4738	.4744	.4750	.4756	.4762	.4767
2.0	.4772	.4778	.4783	.4788	.4793	.4798	.4803	.4808	.4812	.4817
2.1	.4821	.4826	.4830	.4834	.4838	.4842	.4846	.4850	.4854	.4857
2.2	.4861	.4865	.4868	.4871	.4875	.4878	.4881	.4884	.4887	.4890
2.3	.4893	.4896	.4898	.4901	.4904	.4906	.4909	.4911	.4913	.4916
2.4	.4918	.4920	.4922	.4925	.4927	.4929	.4931	.4932	.4934	.4936
2.5	.4938	.4940	.4941	.4943	.4945	.4946	.4948	.4949	.4951	.4952
2.6	.4953	.4955	.4956	.4957	.4959	.4960	.4961	.4962	.4963	.4964
2.7	.4965	.4966	.4967	.4968	.4969	.4970	.4971	.4972	.4973	.4974
2.8	.4974	.4975	.4976	.4977	.4977	.4978	.4979	.4980	.4980	.4981
2.9	.4981	.4982	.4983	.4983	.4984	.4984	.4985	.4985	.4986	.4986
3.0	.4987	.4987	.4987	.4988	.4988	.4989	.4989	.4989	.4990	.4990
3.1	.4990	.4991	.4991	.4991	.4992	.4992	.4992	.4992	.4993	.4993
3.2	.4993	.4993	.4994	.4994	.4994	.4994	.4994	.4995	.4995	.4995
3.3	.4995	.4995	.4996	.4996	.4996	.4996	.4996	.4996	.4996	.4997
3.4	.4997	.4997	.4997	.4997	.4997	.4997	.4997	.4997	.4998	.4998
⋮	⋮									
4.0	.499968									
5.0	.4999997									

This table gives the probability of a random value of a normal variate falling *in* the range $z = 0$ to $z = z$ (in the *shaded area in the figure*). The probability of the same variate having a deviation greater than z is given by $0.5 -$ probability from the table for the given z. The table refers to a single tail of the normal distribution; therefore, the probability of a normal variate falling in the range $\pm z = 2 \times$ probability from the table for the given z. The probability of a variate falling outside the range $\pm z$ is $1 - 2 \times$ probability from the table for given z.

The values in this table were obtained by permission of authors and publishers from C. E. Weatherburn, *Mathematical Statistics* (London: Cambridge University Press, 1957) (for $z = 0$ to $z = 3.1$); C. H. Richardson, *An Introduction to Statistical Analysis* (New York: Harcourt Brace & Jovanovich, Inc., 1944) (for $z = 3.2$ to $z = 3.4$); A. H. Bowker and G. J. Lieberman, *Engineering Statistics* (Englewood Cliffs, N. J.: Prentice-Hall, Inc., 1959)(for $z = 4.0$ and 5.0).

TABLE A-5

Values of Extreme Deviate $\frac{|x_m - \bar{x}|}{s_e}$ Not Rejected as an Outlier

ν \ n	3	4	5	6	7	8	9
	5 percent level						
10	2.02	2.29	2.49	2.63	2.75	2.85	2.93
11	1.99	2.26	2.44	2.58	2.70	2.79	2.87
12	1.97	2.22	2.40	2.54	2.65	2.75	2.83
13	1.95	2.20	2.38	2.51	2.62	2.71	2.79
14	1.93	2.18	2.35	2.48	2.59	2.68	2.76
15	1.92	2.16	2.33	2.46	2.56	2.65	2.73
16	1.90	2.14	2.31	2.44	2.54	2.63	2.70
17	1.89	2.13	2.30	2.42	2.52	2.61	2.68
18	1.88	2.12	2.28	2.41	2.51	2.59	2.66
19	1.87	2.11	2.27	2.39	2.49	2.58	2.65
20	1.87	2.10	2.26	2.38	2.48	2.56	2.63
24	1.84	2.07	2.23	2.35	2.44	2.52	2.59
30	1.82	2.04	2.20	2.31	2.40	2.48	2.55
40	1.80	2.02	2.17	2.28	2.37	2.44	2.51
60	1.78	1.99	2.14	2.25	2.33	2.41	2.47
120	1.76	1.97	2.11	2.21	2.30	2.37	2.43
∞	1.74	1.94	2.08	2.18	2.27	2.33	2.39
	1 percent level						
10	2.76	3.05	3.25	3.39	3.50	3.59	3.67
11	2.71	3.00	3.19	3.33	3.44	3.53	3.61
12	2.67	2.95	3.14	3.28	3.39	3.48	3.55
13	2.63	2.91	3.10	3.24	3.34	3.43	3.51
14	2.60	2.87	3.06	3.20	3.30	3.39	3.47
15	2.57	2.84	3.02	3.16	3.27	3.35	3.43
16	2.55	2.81	3.00	3.13	3.24	3.32	3.39
17	2.52	2.79	2.97	3.10	3.21	3.29	3.36
18	2.50	2.77	2.95	3.08	3.18	3.27	3.34
19	2.49	2.75	2.92	3.06	3.16	3.24	3.31
20	2.47	2.73	2.91	3.04	3.14	3.22	3.29
24	2.43	2.68	2.85	2.97	3.07	3.15	3.22
30	2.38	2.62	2.79	2.91	3.01	3.08	3.15
40	2.34	2.57	2.73	2.85	2.94	3.02	3.08
60	2.30	2.52	2.68	2.79	2.88	2.95	3.01
120	2.25	2.48	2.62	2.73	2.82	2.89	2.95
∞	2.22	2.43	2.57	2.68	2.76	2.83	2.88

(continued)

TABLE A-5—*Continued*

ν \ n	1 percent level						
	3	4	5	6	7	8	9
10	3.54	3.84	4.04	4.17	4.28	4.35	4.40
11	3.49	3.80	3.99	4.12	4.23	4.30	4.36
12	3.45	3.75	3.94	4.07	4.19	4.26	4.31
13	3.41	3.71	3.90	4.03	4.14	4.22	4.28
14	3.38	3.67	3.86	4.00	4.10	4.18	4.24
15	3.35	3.64	3.83	3.96	4.06	4.15	4.21
16	3.32	3.61	3.80	3.93	4.03	4.12	4.18
17	3.29	3.58	3.77	3.90	4.00	4.09	4.15
18	3.27	3.55	3.74	3.88	3.98	4.06	4.12
19	3.25	3.53	3.72	3.85	3.95	4.03	4.10
20	3.23	3.51	3.70	3.83	3.93	4.01	4.08
24	3.16	3.44	3.62	3.75	3.85	3.93	4.00
30	3.08	3.36	3.53	3.66	3.76	3.84	3.90
40	3.01	3.27	3.44	3.57	3.66	3.74	3.81
60	2.93	3.19	3.35	3.47	3.56	3.64	3.70
120	2.85	3.10	3.26	3.37	3.46	3.53	3.59
∞	2.78	3.01	3.17	3.28	3.36	3.43	3.48

An outlier is rejected if its value exceeds the tabulated value at the requisite level of significance.

x_m = greatest or smallest value that can be expected in a sample of size n at the given level of significance.

s_e = estimate of standard deviation from a sample with ν degrees of freedom.

This table is adapted by permission of the author and publishers from a paper "Tables of Percentage Points of the 'Studentized' Extreme Deviate from the Sample Mean" by K. R. Nair, *Biometrika*, vol. 39, 1952, pp. 189–191.

A slight revision of the above values has been suggested by H. A. David in a paper "Revised Upper Percentage Points of the Extreme Studentized Deviate from the Sample Mean," *Biometrika*, vol. 43, 1956, pp. 449–451.

TABLE A-6

Chauvenet's Criterion for Rejection of Outliers

Sample size n	$\frac{\lvert x_m - \bar{x} \rvert}{s}$	*Sample size* n	$\frac{\lvert x_m - \bar{x} \rvert}{s}$
2	1.15	20	2.24
3	1.38	25	2.33
4	1.53	30	2.39
5	1.64	40	2.50
6	1.73	50	2.58
7	1.80	60	2.64
8	1.86	80	2.74
9	1.91	100	2.81
10	1.96	150	2.94
11	2.00	200	3.02
12	2.04	300	3.14
13	2.07	400	3.23
14	2.10	500	3.29
16	2.15	600	3.34
18	2.20	1000	3.48

An outlier may be rejected if the actual value of $\lvert x_m - \bar{x} \rvert / s$ exceeds the tabulated value.

x_m = value of outlier, $\bar{x}$ = sample mean, s^2 = estimate of variance from sample, and n = sample size.

TABLE A-7
Distribution of χ^2

Degrees of freedom ν	Probability of a deviation greater than χ^2											
	0.99	*0.95*	*0.90*	*0.80*	*0.70*	*0.50*	*0.30*	*0.20*	*0.10*	*0.05*	*0.01*	*0.001*
1	.000157	.00393	.0158	.0642	.148	.455	1.074	1.642	2.706	3.841	6.635	10.827
2	.0201	.103	.211	.446	.713	1.386	2.408	3.219	4.605	5.991	9.210	13.815
3	.115	.352	.584	1.005	1.424	2.366	3.665	4.642	6.251	7.815	11.345	16.268
4	.297	.711	1.064	1.649	2.195	3.357	4.878	5.989	7.779	9.488	13.277	18.465
5	.554	1.145	1.610	2.343	3.000	4.351	6.064	7.289	9.236	11.070	15.086	20.517
6	.872	1.635	2.204	3.070	3.828	5.348	7.231	8.558	10.645	12.592	16.812	22.457
7	1.239	2.167	2.833	3.822	4.671	6.346	8.383	9.803	12.017	14.067	18.475	24.322
8	1.646	2.733	3.490	4.594	5.527	7.344	9.524	11.030	13.362	15.507	20.090	26.125
9	2.088	3.325	4.168	5.380	6.393	8.343	10.656	12.242	14.684	16.919	21.666	27.877
10	2.558	3.940	4.865	6.179	7.267	9.342	11.781	13.442	15.987	18.307	23.209	29.588
11	3.053	4.575	5.578	6.989	8.148	10.341	12.899	14.631	17.275	19.675	24.725	31.264
12	3.571	5.226	6.304	7.807	9.034	11.340	14.011	15.812	18.549	21.026	26.217	32.909
13	4.107	5.892	7.042	8.634	9.926	12.340	15.119	16.985	19.812	22.362	27.688	34.528
14	4.660	6.571	7.790	9.467	10.821	13.339	16.222	18.151	21.064	23.685	29.141	36.123
15	5.229	7.261	8.547	10.307	11.721	14.339	17.322	19.311	22.307	24.996	30.578	37.697
16	5.812	7.962	9.312	11.152	12.624	15.338	18.418	20.465	23.542	26.296	32.000	39.252

17	6.408	8.672	10.085	12.002	13.531	16.338	19.511	21.615	24.769	27.587	33.409	40.790
18	7.015	9.390	10.865	12.857	14.440	17.338	20.601	22.760	25.989	28.869	34.805	42.312
19	7.633	10.117	11.651	13.716	15.352	18.338	21.689	23.900	27.204	30.144	36.191	43.820
20	8.260	10.851	12.443	14.578	16.266	19.377	22.775	25.038	28.412	31.410	37.566	45.315
21	8.897	11.501	13.240	15.445	17.182	20.377	23.858	26.171	29.615	32.671	38.932	46.797
22	9.542	12.338	14.041	16.314	18.101	21.337	24.939	27.301	30.813	33.924	40.289	48.268
23	10.196	13.091	14.848	17.187	19.021	22.337	26.018	28.429	32.007	35.172	41.638	49.728
24	10.856	13.848	15.659	18.062	19.943	23.337	27.096	29.553	33.196	36.415	42.980	51.179
25	11.524	14.611	16.473	18.940	20.867	24.337	28.172	30.675	34.382	37.652	44.314	52.620
26	12.198	15.379	17.292	19.820	21.792	25.336	29.246	31.795	35.563	38.885	45.642	54.052
27	12.879	16.151	18.114	20.703	22.719	26.336	30.319	32.912	36.741	40.113	46.963	55.476
28	13.565	16.928	18.939	21.588	23.647	27.336	31.391	34.027	37.916	41.337	48.278	56.893
29	14.256	17.708	19.768	22.475	24.577	28.336	32.461	35.139	39.087	42.557	49.588	58.302
30	14.953	18.493	20.599	23.364	25.508	29.336	33.530	36.250	40.256	43.773	50.892	59.703

This table gives the probability α of a variate falling in the shaded area of the figure, i.e., outside the range 0 to χ^2 for a given number of degrees of freedom ν. For larger values of ν, the expression $\sqrt{2\chi^2} - \sqrt{2\nu - 1}$ may be used as a normal deviate with unit standard deviation, remembering that the probability for χ^2 corresponds to that of a single tail of the normal curve. For example, let $\chi^2 = 147.92$, $\nu = 113$; then $z = \sqrt{295.84} - \sqrt{225} = 2.2$, corresponding to an area $= 0.4861$ from Table A-4. Therefore, the probability of a variate exceeding z is $0.5000 - 0.4861 = 0.0139$, which is highly significant.

This table is taken from Table IV of Fisher and Yates, *Statistical Tables for Biological, Agricultural, and Medical Research*, published by Oliver & Boyd, Ltd., Edinburgh, Scotland, by permission of the authors and publishers.

TABLE A-8

Distribution of t:

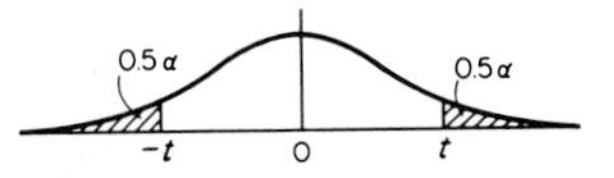

Degrees of freedom v	*Probability* α			
	0.10	*0.05*	*0.01*	*0.001*
1	6.314	12.706	63.657	636.619
2	2.920	4.303	9.925	31.598
3	2.353	3.182	5.841	12.941
4	2.132	2.776	4.604	8.610
5	2.015	2.571	4.032	6.859
6	1.943	2.447	3.707	5.959
7	1.895	2.365	3.499	5.405
8	1.860	2.306	3.355	5.041
9	1.833	2.262	3.250	4.781
10	1.812	2.228	3.169	4.587
11	1.796	2.201	3.106	4.437
12	1.782	2.179	3.055	4.318
13	1.771	2.160	3.012	4.221
14	1.761	2.145	2.977	4.140
15	1.753	2.131	2.947	4.073
16	1.746	2.120	2.921	4.015
17	1.740	2.110	2.898	3.965
18	1.734	2.101	2.878	3.922
19	1.729	2.093	2.861	3.883
20	1.725	2.086	2.845	3.850
21	1.721	2.080	2.831	3.819
22	1.717	2.074	2.819	3.792
23	1.714	2.069	2.807	3.767
24	1.711	2.064	2.797	3.745
25	1.708	2.060	2.787	3.725
26	1.706	2.056	2.779	3.707
27	1.703	2.052	2.771	3.690
28	1.701	2.048	2.763	3.674
29	1.699	2.045	2.756	3.659
30	1.697	2.042	2.750	3.646
40	1.684	2.021	2.704	3.551
60	1.671	2.000	2.660	3.460
120	1.658	1.980	2.617	3.373
∞	1.645	1.960	2.576	3.291

This table gives the values of t corresponding to various values of the probability α (level of significance) of a random variable falling inside the shaded areas in the figure, for a given number of degrees of freedom v available for the estimation of error. For a one-sided test the confidence limits are obtained for $\alpha/2$.

This table is taken from Table III of Fisher and Yates, *Statistical Tables for Biological, Agricultural, and Medical Research*, published by Oliver & Boyd Ltd., Edinburgh, Scotland, by permission of the authors and publishers.

TABLE A-9

Significance of a Difference Between Two Means with Different Variances

	ν_1	θ 0°	15°	30°	45°	60°	75°	90°
5 percent level	6	2.447	2.440	2.435	2.435	2.435	2.440	2.447
	8	2.447	2.430	2.398	2.364	2.331	2.310	2.306
$\nu_2 = 6$	12	2.447	2.423	2.367	2.301	2.239	2.193	2.179
	24	2.447	2.418	2.342	2.247	2.156	2.088	2.064
	∞	2.447	2.413	2.322	2.201	2.082	1.993	1.960
	6	2.306	2.310	2.331	2.364	2.398	2.430	2.447
	8	2.306	2.300	2.294	2.292	2.294	2.300	2.306
$\nu_2 = 8$	12	2.306	2.292	2.262	2.229	2.201	2.183	2.179
	24	2.306	2.286	2.236	2.175	2.118	2.077	2.064
	∞	2.306	2.281	2.215	2.128	2.044	1.982	1.960
	6	2.179	2.193	2.239	2.301	2.367	2.423	2.447
	8	2.179	2.183	2.201	2.229	2.262	2.292	2.306
$\nu_2 = 12$	12	2.179	2.175	2.169	2.167	2.169	2.175	2.179
	24	2.179	2.168	2..142	2.112	2.085	2.069	2.064
	∞	2.179	2.163	2.120	2.064	2.011	1.973	1.960
	6	2.064	2.088	2.156	2.247	2.342	2.418	2.447
	8	2.064	2.077	2.118	2.175	2.236	2.286	2.306
$\nu_2 = 24$	12	2.064	2.069	2.085	2.112	2.142	2.168	2.179
	24	2.064	2.062	2.058	2.056	2.058	2.062	2.064
	∞	2.064	2.056	2.035	2.009	1.983	1.966	1.960
	6	1.960	1.993	2.082	2.201	2.322	2.413	2.447
	8	1.960	1.982	2.044	2.128	2.215	2.281	2.306
$\nu_2 = \infty$	12	1.960	1.973	2.011	2.064	2.120	2.163	2.179
	24	1.960	1.966	1.983	2.009	2.035	2.056	2.064
	∞	1.960	1.960	1.960	1.960	1.960	1.960	1.960
1 percent level	6	3.707	3.654	3.557	3.514	3.557	3.654	3.707
	8	3.707	3.643	3.495	3.363	3.307	3.328	3.355
$\nu_2 = 6$	12	3.707	3.636	3.453	3.246	3.104	3.053	3.055
	24	3.707	3.631	3.424	3.158	2.938	2.822	2.797
	∞	3.707	3.626	3.402	3.093	2.804	2.627	2.576
	6	3.355	3.328	3.307	3.363	3.495	3.643	3.707
	8	3.355	3.316	3.239	3.206	3.239	3.316	3.355
$\nu_2 = 8$	12	3.355	3.307	3.192	3.083	3.032	3.039	3.055
	24	3.355	3.301	3.158	2.988	2.862	2.805	2.797
	∞	3.355	3.295	3.132	2.916	2.723	2.608	2.576
	6	3.055	3.053	3.104	3.246	3.453	3.636	3.707
	8	3.055	3.039	3.032	3.083	3.192	3.307	3.355
$\nu_2 = 12$	12	3.055	3.029	2.978	2.954	2.978	3.029	3.055
	24	3.055	3.020	2.938	2.853	2.803	2.793	2.797
	∞	3.055	3.014	2.909	2.775	2.661	2.595	2.576

(*continued*)

TABLE A-9—*Continued*

	v_1	θ 0°	15°	30°	45°	60°	75°	90°
	6	2.797	2.822	2.938	3.158	3.424	3.631	3.707
	8	2.797	2.805	2.862	2.988	3.158	3.301	3.355
$v_2 = 24$	12	2.797	2.793	2.803	2.853	2.938	3.020	3.055
	24	2.797	2.785	2.759	2.747	2.759	2.785	2.797
	∞	2.797	2.777	2.726	2.664	2.613	2.585	2.576
	6	2.576	2.627	2.804	3.093	3.402	3.626	3.707
	8	2.576	2.608	2.723	2.916	3.132	3.295	3.355
$v_2 = \infty$	12	2.576	2.595	2.661	2.775	2.909	3.014	3.055
	24	2.576	2.585	2.613	2.664	2.726	2.777	2.797
	∞	2.576	2.576	2.576	2.576	2.576	2.576	2.576

This table gives values of d for known values of v_1, v_2, and θ, where $\tan \theta = s_1/s_2$, and v_1 and v_2 are the corresponding numbers of degrees of freedom. If the difference of means exceeds $d\sqrt{s_1^2 + s_2^2}$, then it is significant at the specified level.

This table is taken from Table VI of Fisher and Yates, *Statistical Tables for Biological, Agricultural, and Medical Research*, published by Oliver & Boyd Ltd., Edinburgh, Scotland, by permission of the authors and publishers.

TABLE A-10

Distribution of Variance Ratio F (5 Percent Level of Significance)

ν_2 \ ν_1	1	2	3	4	5	6	7	8	9	10	12	15	20	24	30	40	60	120	∞
1	161.45	199.50	215.71	224.58	230.16	233.99	236.77	238.88	240.54	241.88	243.91	245.95	248.01	249.05	250.09	251.14	252.20	253.25	254.32
2	18.51	19.00	19.16	19.25	19.30	19.33	19.35	19.37	19.38	19.40	19.41	19.43	19.45	19.45	19.46	19.47	19.48	19.49	19.50
3	10.13	9.55	9.28	9.12	9.01	8.94	8.89	8.85	8.81	8.79	8.74	8.70	8.66	8.64	8.62	8.59	8.57	8.55	8.53
4	7.71	6.94	6.59	6.39	6.26	6.16	6.09	6.04	6.00	5.96	5.91	5.86	5.80	5.77	5.75	5.72	5.69	5.66	5.63
5	6.61	5.79	5.41	5.19	5.05	4.95	4.88	4.82	4.77	4.74	4.68	4.62	4.56	4.53	4.50	4.46	4.43	4.40	4.36
6	5.99	5.14	4.76	4.53	4.39	4.28	4.21	4.15	4.10	4.06	4.00	3.94	3.87	3.84	3.81	3.77	3.74	3.70	3.67
7	5.59	4.74	4.35	4.12	3.97	3.87	3.79	3.73	3.68	3.64	3.57	3.51	3.44	3.41	3.38	3.34	3.30	3.27	3.23
8	5.32	4.46	4.07	3.84	3.69	3.58	3.50	3.44	3.39	3.35	3.28	3.22	3.15	3.12	3.08	3.04	3.01	2.97	2.93
9	5.12	4.26	3.86	3.63	3.48	3.37	3.29	3.23	3.18	3.14	3.07	3.01	2.94	2.90	2.86	2.83	2.79	2.75	2.71
10	4.96	4.10	3.71	3.48	3.33	3.22	3.14	3.07	3.02	2.98	2.91	2.84	2.77	2.74	2.70	2.66	2.62	2.58	2.54
11	4.84	3.98	3.59	3.36	3.20	3.09	3.01	2.95	2.90	2.85	2.79	2.72	2.65	2.61	2.57	2.53	2.49	2.45	2.40
12	4.75	3.89	3.49	3.26	3.11	3.00	2.91	2.85	2.80	2.75	2.69	2.62	2.54	2.51	2.47	2.43	2.38	2.34	2.30
13	4.67	3.81	3.41	3.18	3.03	2.92	2.83	2.77	2.71	2.67	2.60	2.53	2.46	2.42	2.38	2.34	2.30	2.25	2.21
14	4.60	3.74	3.34	3.11	2.96	2.85	2.76	2.70	2.65	2.60	2.53	2.46	2.39	2.35	2.31	2.27	2.22	2.18	2.13
15	4.54	3.68	3.29	3.06	2.90	2.79	2.71	2.64	2.59	2.54	2.48	2.40	2.33	2.29	2.25	2.20	2.16	2.11	2.07
16	4.49	3.63	3.24	3.01	2.85	2.74	2.66	2.59	2.54	2.49	2.42	2.35	2.28	2.24	2.19	2.15	2.11	2.06	2.01
17	4.45	3.59	3.20	2.96	2.81	2.70	2.61	2.55	2.49	2.45	2.38	2.31	2.23	2.19	2.15	2.10	2.06	2.01	1.96
18	4.41	3.55	3.16	2.93	2.77	2.66	2.58	2.51	2.46	2.41	2.34	2.27	2.19	2.15	2.11	2.06	2.02	1.97	1.92
19	4.38	3.52	3.13	2.90	2.74	2.63	2.54	2.48	2.42	2.38	2.31	2.23	2.16	2.11	2.07	2.03	1.98	1.93	1.88
20	4.35	3.49	3.10	2.87	2.71	2.60	2.51	2.45	2.39	2.35	2.28	2.20	2.12	2.08	2.04	1.99	1.95	1.90	1.84
21	4.32	3.47	3.07	2.84	2.68	2.57	2.49	2.42	2.37	2.32	2.25	2.18	2.10	2.05	2.01	1.96	1.92	1.87	1.81
22	4.30	3.44	3.05	2.82	2.66	2.55	2.46	2.40	2.34	2.30	2.23	2.15	2.07	2.03	1.98	1.94	1.89	1.84	1.78
23	4.28	3.42	3.03	2.80	2.64	2.53	2.44	2.37	2.32	2.27	2.20	2.13	2.05	2.00	1.96	1.91	1.86	1.81	1.76
24	4.26	3.40	3.01	2.78	2.62	2.51	2.42	2.36	2.30	2.25	2.18	2.11	2.03	1.98	1.94	1.89	1.84	1.79	1.73
25	4.24	3.39	2.99	2.76	2.60	2.49	2.40	2.34	2.28	2.24	2.16	2.09	2.01	1.96	1.92	1.87	1.82	1.77	1.71
26	4.23	3.37	2.98	2.74	2.59	2.47	2.39	2.32	2.27	2.22	2.15	2.07	1.99	1.95	1.90	1.85	1.80	1.75	1.69
27	4.21	3.35	2.96	2.73	2.57	2.46	2.37	2.31	2.25	2.20	2.13	2.06	1.97	1.93	1.88	1.84	1.79	1.73	1.67
28	4.20	3.34	2.95	2.71	2.56	2.45	2.36	2.29	2.24	2.19	2.12	2.04	1.96	1.91	1.87	1.82	1.77	1.71	1.65
29	4.18	3.33	2.93	2.70	2.55	2.43	2.35	2.28	2.22	2.18	2.10	2.03	1.94	1.90	1.85	1.81	1.75	1.70	1.64
30	4.17	3.32	2.92	2.69	2.53	2.42	2.33	2.27	2.21	2.16	2.09	2.01	1.93	1.89	1.84	1.79	1.74	1.68	1.62
40	4.08	3.23	2.84	2.61	2.45	2.34	2.25	2.18	2.12	2.08	2.00	1.92	1.84	1.79	1.74	1.69	1.64	1.58	1.51
60	4.00	3.15	2.76	2.53	2.37	2.25	2.17	2.10	2.04	1.99	1.92	1.84	1.75	1.70	1.65	1.59	1.53	1.47	1.39
120	3.92	3.07	2.68	2.45	2.29	2.18	2.09	2.02	1.96	1.91	1.83	1.75	1.66	1.61	1.55	1.50	1.43	1.35	1.25
∞	3.84	3.00	2.60	2.37	2.21	2.10	2.01	1.94	1.88	1.83	1.75	1.67	1.57	1.52	1.46	1.39	1.32	1.22	1.00

Continued

TABLE A-10 (*Continued*) (1 Percent Level of Significance)

ν_2 \ ν_1	1	2	3	4	5	6	7	8	9	10	12	15	20	24	30	40	60	120	∞
1	4,052.4	4,999.5	5,403.3	5,624.6	5,763.7	5,859.0	5,928.3	5,981.6	6,022.5	6,055.8	6,106.3	6,157.3	6,208.7	6,234.6	6,260.7	6.286.8	6,313.0	6,339.4	6,366.0
2	98.50	99.00	99.17	99.25	99.30	99.33	99.36	99.37	99.39	99.40	99.42	99.43	99.45	99.46	99.47	99.47	99.48	99.49	99.50
3	34.12	30.82	29.46	28.71	28.24	27.91	27.67	27.49	27.34	27.23	27.05	26.87	26.69	26.60	26.50	26.41	26.32	26.22	26.12
4	21.20	18.00	16.69	15.98	15.52	15.21	14.98	14.80	14.66	14.55	14.37	14.20	14.02	13.93	13.84	13.74	13.65	13.56	13.46
5	16.26	13.27	12.06	11.39	10.97	10.67	10.46	10.29	10.16	10.05	9.89	9.72	9.55	9.47	9.38	9.29	9.20	9.11	9.02
6	13.74	10.92	9.78	9.15	8.75	8.47	8.26	8.10	7.98	7.87	7.72	7.56	7.40	7.31	7.23	7.14	7.06	6.97	6.88
7	12.25	9.55	8.45	7.85	7.46	7.19	6.99	6.84	6.72	6.62	6.47	6.31	6.16	6.07	5.99	5.91	5.82	5.74	5.65
8	11.26	8.65	7.59	7.01	6.63	6.37	6.18	6.03	5.91	5.81	5.67	5.52	5.36	5.28	5.20	5.12	5.03	4.95	4.86
9	10.56	8.02	6.99	6.42	6.06	5.80	5.61	5.47	5.35	5.26	5.11	4.96	4.81	4.73	4.65	4.57	4.48	4.40	4.31
10	10.04	7.56	6.55	5.99	5.64	5.39	5.20	5.06	4.94	4.85	4.71	4.56	4.41	4.33	4.25	4.17	4.08	4.00	3.91
11	9.65	7.21	6.22	5.67	5.32	5.07	4.89	4.74	4.63	4.54	4.40	4.25	4.10	4.02	3.94	3.86	3.78	3.69	3.60
12	9.33	6.93	5.95	5.41	5.06	4.82	4.64	4.50	4.39	4.30	4.16	4.01	3.86	3.78	3.70	3.62	3.54	3.45	3.36
13	9.07	6.70	5.74	5.21	4.86	4.62	4.44	4.30	4.19	4.10	3.96	3.82	3.66	3.59	3.51	3.43	3.34	3.25	3.17
14	8.86	6.51	5.56	5.04	4.70	4.46	4.28	4.14	4.03	3.94	3.80	3.66	3.51	3.43	3.35	3.27	3.18	3.09	3.00
15	8.68	6.36	5.42	4.89	4.56	4.32	4.14	4.00	3.89	3.80	3.67	3.52	3.37	3.29	3.21	3.13	3.05	2.96	2.87
16	8.53	6.23	5.29	4.77	4.44	4.20	4.03	3.89	3.78	3.69	3.55	3.41	3.26	3.18	3.10	3.02	2.93	2.84	2.75
17	8.40	6.11	5.18	4.67	4.34	4.10	3.93	3.79	3.68	3.59	3.46	3.31	3.16	3.08	3.00	2.92	2.83	2.75	2.65
18	8.29	6.01	5.09	4.58	4.25	4.01	3.84	3.71	3.60	3.51	3.37	3.23	3.08	3.00	2.92	2.84	2.75	2.66	2.57
19	8.18	5.93	5.01	4.50	4.17	3.94	3.77	3.63	3.52	3.43	3.30	3.15	3.00	2.92	2.84	2.76	2.67	2.58	2.49
20	8.10	5.85	4.94	4.43	4.10	3.87	3.70	3.56	3.46	3.37	3.23	3.09	2.94	2.86	2.78	2.69	2.61	2.52	2.42
21	8.02	5.78	4.87	4.37	4.04	3.81	3.64	3.51	3.40	3.31	3.17	3.03	2.88	2.80	2.72	2.64	2.55	2.46	2.36
22	7.95	5.72	4.82	4.31	3.99	3.76	3.59	3.45	3.35	3.26	3.12	2.98	2.83	2.75	2.67	2.58	2.50	2.40	2.31
23	7.88	5.66	4.76	4.26	3.94	3.71	3.54	3.41	3.30	3.21	3.07	2.93	2.78	2.70	2.62	2.54	2.45	2.35	2.26
24	7.82	5.61	4.72	4.22	3.90	3.67	3.50	3.36	3.26	3.17	3.03	2.89	2.74	2.66	2.58	2.49	2.40	2.31	2.21
25	7.77	5.57	4.68	4.18	3.86	3.63	3.46	3.32	3.22	3.13	2.99	2.85	2.70	2.62	2.54	2.45	2.36	2.27	2.17
26	7.72	5.53	4.64	4.14	3.82	3.59	3.42	3.29	3.18	3.09	2.96	2.82	2.66	2.58	2.50	2.42	2.33	2.23	2.13
27	7.68	5.49	4.60	4.11	3.78	3.56	3.39	3.26	3.15	3.06	2.93	2.78	2.63	2.55	2.47	2.38	2.29	2.20	2.10
28	7.64	5.45	4.57	4.07	3.75	3.53	3.36	3.23	3.12	3.03	2.90	2.75	2.60	2.52	2.44	2.35	2.26	2.17	2.06
29	7.60	5.42	4.54	4.04	3.73	3.50	3.33	3.20	3.09	3.00	2.87	2.73	2.57	2.49	2.41	2.33	2.23	2.14	2.03
30	7.56	5.39	4.51	4.02	3.70	3.47	3.30	3.17	3.07	2.98	2.84	2.70	2.55	2.47	2.39	2.30	2.21	2.11	2.01
40	7.31	5.18	4.31	3.83	3.51	3.29	3.12	2.99	2.89	2.80	2.66	2.52	2.37	2.29	2.20	2.11	2.02	1.92	1.80
60	7.08	4.98	4.13	3.65	3.34	3.12	2.95	2.82	2.72	2.63	2.50	2.35	2.20	2.12	2.03	1.94	1.84	1.73	1.60
120	6.85	4.79	3.95	3.48	3.17	2.96	2.79	2.66	2.56	2.47	2.34	2.19	2.03	1.95	1.86	1.76	1.66	1.53	1.38
∞	6.63	4.61	3.78	3.32	3.02	2.80	2.64	2.51	2.41	2.32	2.18	2.04	1.88	1.79	1.70	1.59	1.47	1.32	1.00

If the computed statistic $F = s_1^2/s_2^2$, with the larger s in the numerator, exceeds the tabulated value of F at the specified level with ν_1, ν_2 degrees of freedom, we reject the null hypothesis that $\sigma_1 = \sigma_2$ (see the figure).

Table adapted by permission of authors and publishers from E. L. Crow, F. A. Davis, and M. W. Maxfield, *Statistics Manual* (New York: Dover Publications, Inc., 1960).

TABLE A-11

Values of Correlation Coefficient r

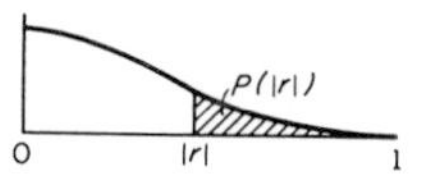

	5 Percent level of significance				1 percent level of significance				
	Total number of variables				Total number of variables				
ν	2	3	4	5	2	3	4	5	ν
1	.997	.999	.999	.999	1.000	1.000	1.000	1.000	1
2	.950	.975	.983	.987	.990	.995	.997	.998	2
3	.878	.930	.950	.961	.959	.976	.983	.987	3
4	.811	.881	.912	.930	.917	.949	.962	.970	4
5	.754	.836	.874	.898	.874	.917	.937	.949	5
6	.707	.795	.839	.867	.834	.886	.911	.927	6
7	.666	.758	.807	.838	.798	.855	.885	.904	7
8	.632	.726	.777	.811	.765	.827	.860	.882	8
9	.602	.697	.750	.786	.735	.800	.836	.861	9
10	.576	.671	.726	.763	.708	.776	.814	.840	10
11	.553	.648	.703	.741	.684	.753	.793	.821	11
12	.532	.627	.683	.722	.661	.732	.773	.802	12
13	.514	.608	.664	.703	.641	.712	.755	.785	13
14	.497	.590	.646	.686	.623	.694	.737	.768	14
15	.482	.574	.630	.670	.606	.677	.721	.752	15
16	.468	.559	.615	.655	.590	.662	.706	.738	16
17	.456	.545	.601	.641	.575	.647	.691	.724	17
18	.444	.532	.587	.628	.561	.633	.678	.710	18
19	.433	.520	.575	.615	.549	.620	.665	.698	19
20	.423	.509	.563	.604	.537	.608	.652	.685	20
21	.413	.498	.552	.592	.526	.596	.641	.674	21
22	.404	.488	.542	.582	.515	.585	.630	.663	22
23	.396	.479	.532	.572	.505	.574	.619	.652	23
24	.388	.470	.523	.562	.496	.565	.609	.642	24
25	.381	.462	.514	.553	.487	.555	.600	.633	25
26	.374	.454	.506	.545	.478	.546	.590	.624	26
27	.367	.446	.498	.536	.470	.538	.582	.615	27
28	.361	.439	.490	.529	.463	.530	.573	.606	28
29	.355	.432	.482	.521	.456	.522	.565	.598	29
30	.349	.426	.476	.514	.449	.514	.558	.591	30
35	.325	.397	.445	.482	.418	.481	.523	.556	35
40	.304	.373	.419	.455	.393	.454	.494	.526	40
45	.288	.353	.397	.432	.372	.430	.470	.501	45
50	.273	.336	.379	.412	.354	.410	.449	.479	50

(continued)

TABLE A-11—*Continued*

v	*5 Percent level of significance*				*1 percent level of significance*				v
	Total number of variables				*Total number of variables*				
	2	*3*	*4*	*5*	*2*	*3*	*4*	*5*	
60	.250	.308	.348	.380	.325	.377	.414	.442	60
70	.232	.286	.324	.354	.302	.351	.386	.413	70
80	.217	.269	.304	.332	.283	.330	.362	.389	80
90	.205	.254	.288	.315	.267	.312	.343	.368	90
100	.195	.241	.274	.300	.254	.297	.327	.351	100
125	.174	.216	.246	.269	.228	.266	.294	.316	125
150	.159	.198	.225	.247	.208	.244	.270	.290	150
200	.138	.172	.196	.215	.181	.212	.234	.253	200
300	.113	.141	.160	.176	.148	.174	.192	.208	300
400	.098	.122	.139	.153	.128	.151	.167	.180	400
500	.088	.109	.124	.137	.115	.135	.150	.162	500
1000	.062	.077	.088	.097	.081	.096	.106	.116	1000

The critical value of r at a given level of significance, total number of variables, and degrees of freedom v, is read from the table. If the computed $|r|$ exceeds the critical value, then the null hypothesis that there is no association between the variables is rejected at the given level. The test is an equal-tails test, since we are usually interested in either positive or negative correlation. The shaded portion of the figure is the stipulated probability as a level of significance.

Table reproduced with the permission of the authors and publisher from E. L. Crow, F. A. Davis, and M. W. Maxfield, *Statistical Manual* (New York: Dover Publications, Inc., 1960).

TABLE A-12

Control Chart Limits for Mean

[Factor A_w corresponds to 95 percent probability ($z = 1.96$);
factor A_A corresponds to 99.8 percent probability ($z = 3.09$)]

Sample size n	2	3	4	5	6	7	8	9	10	11	12
Warning factor A_w	1.229	0.668	0.476	0.377	0.316	0.274	0.244	0.220	0.202	0.186	0.174
Action factor A_A	1.937	1.054	0.750	0.594	0.498	0.432	0.384	0.347	0.317	0.294	0.274

To obtain limits for a given sample size *n*, multiply the mean range $\bar{R}$ by the appropriate value of A_w and A_A; then add to and subtract from mean $\bar{\bar{x}}$.

Table adapted from Table G of O. L. Davies, ed., *Statistical Methods in Research and Production* (Edinburgh, Scotland, Oliver & Boyd Ltd., 1958), by permission of the Imperial Chemical Industries, Ltd., and the publishers.

TABLE A-13

Control Chart Limits for Range

[Factors D_{WU} and D_{WL} correspond to 95 percent probability;
factors D_{AU} and D_{AL} correspond to 99.8 percent probability]

Sample size n	2	3	4	5	6	7	8	9	10	11	12
Upper warning factor D_{WU}	2.81	2.17	1.93	1.81	1.72	1.66	1.62	1.58	1.56	1.53	1.51
Lower warning factor D_{WL}	0.04	0.18	0.29	0.37	0.42	0.46	0.50	0.52	0.54	0.56	0.58
Upper action factor D_{AU}	4.12	2.99	2.58	2.36	2.22	2.12	2.04	1.99	1.94	1.90	1.87
Lower action factor D_{AL}	0.00	0.04	0.10	0.16	0.21	0.26	0.29	0.32	0.35	0.38	0.40

To obtain limits, multiply mean range $\bar{R}$ by the appropriate value of *D*.

Table adapted from Table G of O. L. Davies (ed.), *Statistical Methods in Research and Production* (Edinburgh, Scotland, Oliver & Boyd Ltd., 1958), by permission of the Imperial Chemical Industries, Ltd., and the publishers.

TABLE A-14

Random Numbers

53 74 23 99 67	61 32 28 69 84	94 62 67 86 24	98 33 41 19 95	47 53 53 38 09
63 38 06 86 54	99 00 65 26 94	02 82 90 23 07	79 62 67 80 60	75 91 12 81 19
35 30 58 21 46	06 72 17 10 94	25 21 31 75 96	49 28 24 00 49	55 65 79 78 07
63 43 36 82 69	65 51 18 37 88	61 38 44 12 45	32 92 85 88 65	54 34 81 85 35
98 25 37 55 26	01 91 82 81 46	74 71 12 94 97	24 02 71 37 07	03 92 18 66 75
02 63 21 17 69	71 50 80 89 56	38 15 70 11 48	43 40 45 86 98	00 83 26 91 03
64 55 22 21 82	48 22 28 06 00	61 54 13 43 91	82 78 12 23 29	06 66 24 12 27
85 07 26 13 89	01 10 07 82 04	59 63 69 36 03	69 11 15 83 80	13 29 54 19 28
58 54 16 24 15	51 54 44 82 00	62 61 65 04 69	38 18 65 18 97	85 72 13 49 21
34 85 27 84 87	61 48 64 56 26	90 18 48 13 26	37 70 15 42 57	65 65 80 39 07
03 92 18 27 46	57 99 16 96 56	30 33 72 85 22	84 64 38 56 98	99 01 30 98 64
62 95 30 27 59	37 75 41 66 48	86 97 80 61 45	23 53 04 01 63	45 76 08 64 27
08 45 93 15 22	60 21 75 46 91	98 77 27 85 42	28 88 61 08 84	69 62 03 42 73
07 08 55 18 40	45 44 75 13 90	24 94 96 61 02	57 55 66 83 15	73 42 37 11 61
01 85 89 95 66	51 10 19 34 88	15 84 97 19 75	12 76 39 43 78	64 63 91 08 25
72 84 71 14 35	19 11 58 49 26	50 11 17 17 76	86 31 57 20 18	95 60 78 46 75
88 78 28 16 84	13 52 53 94 53	75 45 69 30 96	73 89 65 70 31	99 17 43 48 76
45 17 75 65 57	28 40 19 72 12	25 12 74 75 67	60 40 60 81 19	24 62 01 61 16
96 76 28 12 54	22 01 11 94 25	71 96 16 16 88	68 64 36 74 45	19 59 50 88 92
43 31 67 72 30	24 02 94 08 63	38 32 36 66 02	69 36 38 25 39	48 03 45 15 22
50 44 66 44 21	66 06 58 05 62	68 15 54 35 02	42 35 48 96 32	14 52 41 52 48
22 66 22 15 86	26 63 75 41 99	58 42 36 72 24	58 37 52 18 51	03 37 18 39 11
96 24 40 14 51	23 22 30 88 57	95 67 47 29 83	94 69 40 06 07	18 16 36 78 86
31 73 91 61 19	60 20 72 93 48	98 57 07 23 69	65 95 39 69 58	56 80 30 19 44
78 60 73 99 84	43 89 94 36 45	56 69 47 07 41	90 22 91 07 12	78 35 34 08 72
84 37 90 61 56	70 10 23 98 05	85 11 34 76 60	76 48 45 34 60	01 64 18 39 96
36 67 10 08 23	98 93 35 08 86	99 29 76 29 81	33 34 91 58 93	63 14 52 32 52
07 28 59 07 48	89 64 58 89 75	83 85 62 27 89	30 14 78 56 27	86 63 59 80 02
10 15 83 87 60	79 24 31 66 56	21 48 24 06 93	91 98 94 05 49	01 47 59 38 00
55 19 68 97 65	03 73 52 16 56	00 53 55 90 27	33 42 29 38 87	22 13 88 83 34
53 81 29 13 39	35 01 20 71 34	62 33 74 82 14	53 73 19 09 03	56 54 29 56 93
51 86 32 68 92	33 98 74 66 99	40 14 71 94 58	45 94 19 38 81	14 44 99 81 07
35 91 70 29 13	80 03 54 07 27	96 94 78 32 66	50 95 52 74 33	13 80 55 62 54
37 71 67 95 13	20 02 44 95 94	64 85 04 05 72	01 32 90 76 14	53 89 74 60 41
93 66 13 83 27	92 79 64 64 72	28 54 96 53 84	48 14 52 98 94	56 07 93 89 30
02 96 08 45 65	13 05 00 41 84	93 07 54 72 59	21 45 57 09 77	19 48 56 27 44
49 83 43 48 35	82 88 33 69 96	72 36 04 19 76	47 45 15 18 60	82 11 08 95 97
84 60 71 62 46	40 80 81 30 37	34 39 23 05 38	25 15 35 71 30	88 12 57 21 77
18 17 30 88 71	44 91 14 88 47	89 23 30 63 15	56 34 20 47 89	99 82 93 24 98
79 69 10 61 78	71 32 76 95 62	87 00 22 58 40	92 54 01 75 25	43 11 71 99 31
75 93 36 57 83	56 20 14 82 11	74 21 97 90 65	96 42 68 63 86	74 54 13 26 94
38 30 92 29 03	06 28 81 39 38	62 25 06 84 63	61 29 08 93 67	04 32 92 08 09
51 28 50 10 34	31 57 75 95 80	51 97 02 74 77	76 15 48 49 44	18 55 63 77 09
21 31 38 86 24	37 79 81 53 74	73 24 16 10 33	52 83 90 94 76	70 47 14 54 36
29 01 23 87 88	58 02 39 37 67	42 10 14 20 92	16 55 23 42 45	54 96 09 11 06
95 33 95 22 00	18 74 72 00 18	38 79 58 69 32	81 76 80 26 92	82 80 84 25 39
90 84 60 79 80	24 36 59 87 38	82 07 53 89 35	96 35 23 79 18	05 98 90 07 35
46 40 62 98 82	54 97 20 56 95	15 74 80 08 32	16 46 70 50 80	67 72 16 42 79
20 31 89 03 43	38 46 82 68 72	32 14 82 99 70	80 60 47 18 97	63 49 30 21 30
71 59 73 05 50	08 22 23 71 77	91 01 93 20 49	82 96 59 26 94	66 39 67 98 60

This table is taken from Table 33 of Fisher and Yates: *Statistical Tables for Biological, Agricultural, and Medical Research*, published by Oliver & Boyd Ltd., Edinburgh, Scotland, by permission of the authors and publishers.

TABLE A-15

Cumulative Terms of Binomial Distribution

		p									
n	*r′*	.05	.10	.15	.20	.25	.30	.35	.40	.45	.50
2	1	.0975	.1900	.2775	.3600	.4375	.5100	.5775	.6400	.6975	.7500
	2	.0025	.0100	.0225	.0400	.0625	.0900	.1225	.1600	.2025	.2500
3	1	.1426	.2710	.3859	.4880	.5781	.6570	.7254	.7840	.8336	.8750
	2	.0072	.0280	.0608	.1040	.1562	.2160	.2818	.3520	.4252	.5000
	3	.0001	.0010	.0034	.0080	.0156	.0270	.0429	.0640	.0911	.1250
4	1	.1855	.3439	.4780	.5904	.6836	.7599	.8215	.8704	.9085	.9375
	2	.0140	.0523	.1095	.1808	.2617	.3483	.4370	.5248	.6090	.6875
	3	.0005	.0037	.0120	.0272	.0508	.0837	.1265	.1792	.2415	.3125
	4	.0000	.0001	.0005	.0016	.0039	.0081	.0150	.0256	.0410	.0625
5	1	.2262	.4095	.5563	.6723	.7627	.8319	.8840	.9222	.9497	.9688
	2	.0226	.0815	.1648	.2627	.3672	.4718	.5716	.6630	.7438	.8125
	3	.0012	.0086	.0266	.0579	.1035	.1631	.2352	.3174	.4069	.5000
	4	.0000	.0005	.0022	.0067	.0156	.0308	.0540	.0870	.1312	.1875
	5	.0000	.0000	.0001	.0003	.0010	.0024	.0053	.0102	.0185	.0312
6	1	.2649	.4686	.6229	.7379	.8220	.8824	.9246	.9533	.9723	.9844
	2	.0328	.1143	.2235	.3447	.4661	.5798	.6809	.7667	.8364	.8906
	3	.0022	.0158	.0473	.0989	.1694	.2557	.3529	.4557	.5585	.6562
	4	.0001	.0013	.0059	.0170	.0376	.0705	.1174	.1792	.2553	.3438
	5	.0000	.0001	.0004	.0016	.0046	.0109	.0223	.0410	.0692	.1094
	6	.0000	.0000	.0000	.0001	.0002	.0007	.0018	.0041	.0083	.0156
7	1	.3017	.5217	.6794	.7903	.8665	.9176	.9510	.9720	.9848	.9922
	2	.0444	.1497	.2834	.4233	.5551	.6706	.7662	.8414	.8976	.9375
	3	.0038	.0257	.0738	.1480	.2436	.3529	.4677	.5801	.6836	.7734
	4	.0002	.0027	.0121	.0333	.0706	.1260	.1998	.2898	.3917	.5000
	5	.0000	.0002	.0012	.0047	.0129	.0288	.0556	.0963	.1529	.2266
	6	.0000	.0000	.0001	.0004	.0013	.0038	.0090	.0188	.0357	.0625
	7	.0000	.0000	.0000	.0000	.0001	.0002	.0006	.0016	.0037	.0078
8	1	.3366	.5695	.7275	.8322	.8999	.9424	.9681	.9832	.9916	.9961
	2	.0572	.1869	.3428	.4967	.6329	.7447	.8309	.8936	.9368	.9648
	3	.0058	.0381	.1052	.2031	.3215	.4482	.5722	.6846	.7799	.8555
	4	.0004	.0050	.0214	.0563	.1138	.1941	.2936	.4059	.5230	.6367
	5	.0000	.0004	.0029	.0104	.0273	.0580	.1061	.1737	.2604	.3633
	6	.0000	.0000	.0002	.0012	.0042	.0113	.0253	.0498	.0885	.1445
	7	.0000	.0000	.0000	.0001	.0004	.0013	.0036	.0085	.0181	.0352
	8	.0000	.0000	.0000	.0000	.0000	.0001	.0002	.0007	.0017	.0039
9	1	.3698	.6126	.7684	.8658	.9249	.9596	.9793	.9899	.9954	.9980
	2	.0712	.2252	.4005	.5638	.6997	.8040	.8789	.9295	.9615	.9805
	3	.0084	.0530	.1409	.2618	.3993	.5372	.6627	.7682	.8505	.9102
	4	.0006	.0083	.0339	.0856	.1657	.2703	.3911	.5174	.6386	.7461
	5	.0000	.0009	.0056	.0196	.0489	.0988	.1717	.2666	.3786	.5000
	6	.0000	.0001	.0006	.0031	.0100	.0253	.0536	.0994	.1658	.2539
	7	.0000	.0000	.0000	.0003	.0013	.0043	.0112	.0250	.0498	.0898
	8	.0000	.0000	.0000	.0000	.0001	.0004	.0014	.0038	.0091	.0195
	9	.0000	.0000	.0000	.0000	.0000	.0000	.0001	.0003	.0008	.0020
10	1	.4013	.6513	.8031	.8926	.9437	.9718	.9865	.9940	.9975	.9990
	2	.0861	.2639	.4557	.6242	.7560	.8507	.9140	.9536	.9767	.9893
	3	.0115	.0702	.1798	.3222	.4744	.6172	.7384	.8327	.9004	.9453
	4	.0010	.0128	.0500	.1209	.2241	.3504	.4862	.6177	.7340	.8281
	5	.0001	.0016	.0099	.0328	.0781	.1503	.2485	.3669	.4956	.6230
	6	.0000	.0001	.0014	.0064	.0197	.0473	.0949	.1662	.2616	.3770
	7	.0000	.0000	.0001	.0009	.0035	.0106	.0260	.0548	.1020	.1719
	8	.0000	.0000	.0000	.0001	.0004	.0016	.0048	.0123	.0274	.0547
	9	.0000	.0000	.0000	.0000	.0000	.0001	.0005	.0017	.0045	.0107
	10	.0000	.0000	.0000	.0000	.0000	.0000	.0000	.0001	.0003	.0010

(*continued*)

TABLE A-15—*Continued*

		p									
n	*r'*	.05	.10	.15	.20	.25	.30	.35	.40	.45	.50
12	1	.4596	.7176	.8578	.9313	.9683	.9862	.9943	.9978	.9992	.9998
	2	.1184	.3410	.5565	.7251	.8416	.9150	.9576	.9804	.9917	.9968
	3	.0196	.1109	.2642	.4417	.6093	.7472	.8487	.9166	.9579	.9807
	4	.0022	.0256	.0922	.2054	.3512	.5075	.6533	.7747	.8655	.9270
	5	.0002	.0043	.0239	.0726	.1576	.2763	.4167	.5618	.6956	.8062
	6	.0000	.0005	.0046	.0194	.0544	.1178	.2127	.3348	.4731	.6128
	7	.0000	.0001	.0007	.0039	.0143	.0386	.0846	.1582	.2607	.3872
	8	.0000	.0000	.0001	.0006	.0028	.0095	.0255	.0573	.1117	.1938
	9	.0000	.0000	.0000	.0001	.0004	.0017	.0056	.0153	.0356	.0730
	10	.0000	.0000	.0000	.0000	.0000	.0002	.0008	.0028	.0079	.0193
	11	.0000	.0000	.0000	.0000	.0000	.0000	.0001	.0003	.0011	.0032
	12	.0000	.0000	.0000	.0000	.0000	.0000	.0000	.0000	.0001	.0002
15	1	.5367	.7941	.9126	.9648	.9866	.9953	.9984	.9995	.9999	.1.0000
	2	.1710	.4510	.6814	.8329	.9198	.9647	.9858	.9948	.9983	.9995
	3	.0362	.1841	.3958	.6020	.7639	.8732	.9383	.9729	.9893	.9963
	4	.0055	.0556	.1773	.3518	.5387	.7031	.8273	.9095	.9576	.9824
	5	.0006	.0127	.0617	.1642	.3135	.4845	.6481	.7827	.8796	.9408
	6	.0001	.0022	.0168	.0611	.1484	.2784	.4357	.5968	.7392	.8491
	7	.0000	.0003	.0036	.0181	.0566	.1311	.2452	.3902	.5478	.6964
	8	.0000	.0000	.0006	.0042	.0173	.0500	.1132	.2131	.3465	.5000
	9	.0000	.0000	.0001	.0008	.0042	.0152	.0422	.0950	.1818	.3036
	10	.0000	.0000	.0000	.0001	.0008	.0037	.0124	.0338	.0769	.1509
	11	.0000	.0000	.0000	.0000	.0001	.0007	.0028	.0093	.0255	.0592
	12	.0000	.0000	.0000	.0000	.0000	.0001	.0005	.0019	.0063	.0176
	13	.0000	.0000	.0000	.0000	.0000	.0000	.0001	.0003	.0011	.0037
	14	.0000	.0000	.0000	.0000	.0000	.0000	.0000	.0000	.0001	.0005
	15	.0000	.0000	.0000	.0000	.0000	.0000	.0000	.0000	.0000	.0000
20	1	.6415	.8784	.9612	.9885	.9968	.9992	.9998	1.0000	1.0000	1.0000
	2	.2642	.6083	.8244	.9308	.9757	.9924	.9979	.9995	.9999	1.0000
	3	.0755	.3231	.5951	.7939	.9087	.9645	.9879	.9964	.9991	.9998
	4	.0159	.1330	.3523	.5886	.7748	.8929	.9556	.9840	.9951	.9987
	5	.0026	.0432	.1702	.3704	.5852	.7625	.8818	.9490	.9811	.9941
	6	.0003	.0113	.0673	.1958	.3828	.5836	.7546	.8744	.9447	.9793
	7	.0000	.0024	.0219	.0867	.2142	.3920	.5834	.7500	.8701	.9423
	8	.0000	.0004	.0059	.0321	.1018	.2277	.3990	.5841	.7480	.8684
	9	.0000	.0001	.0013	.0100	.0409	.1133	.2376	.4044	.5857	.7483
	10	.0000	.0000	.0002	.0026	.0139	.0480	.1218	.2447	.4086	.5881
	11	.0000	.0000	.0000	.0006	.0039	.0171	.0532	.1275	.2493	.4119
	12	.0000	.0000	.0000	.0001	.0009	.0051	.0196	.0565	.1308	.2517
	13	.0000	.0000	.0000	.0000	.0002	.0013	.0060	.0210	.0580	.1316
	14	.0000	.0000	.0000	.0000	.0000	.0003	.0015	.0065	.0214	.0577
	15	.0000	.0000	.0000	.0000	.0000	.0000	.0003	.0016	.0064	.0207
	16	.0000	.0000	.0000	.0000	.0000	.0000	.0000	.0003	.0015	.0059
	17	.0000	.0000	.0000	.0000	.0000	.0000	.0000	.0000	.0003	.0013
	18	.0000	.0000	.0000	.0000	.0000	.0000	.0000	.0000	.0000	.0002
	19	.0000	.0000	.0000	.0000	.0000	.0000	.0000	.0000	.0000	.0000
	20	.0000	.0000	.0000	.0000	.0000	.0000	.0000	.0000	.0000	.0000

This table gives the probability of an event succeeding at least r' times out of n trials, when the probability of success in each trial is p. Linear interpolation will be accurate at most to two decimal places.

The table is reproduced in condensed form by permission of the publishers CRC Handbook of Tables for Applied Engineering Science, 2nd ed. (Cleveland: The Chemical Rubber Co., 1972).

TABLE A-16

Significance Points for the Absolute Value of the Smaller Sum of Signed Ranks (R) Obtained from Paired Observations

n	*5 Percent*	*2 Percent*	*1 Percent*
6	0	—	—
7	2	0	—
8	4	2	0
9	6	3	2
10	8	5	3
11	11	7	5
12	14	10	7
13	17	13	10
14	21	16	13
15	25	20	16
16	30	24	20
17	35	28	23
18	40	33	28
19	46	38	32
20	52	43	38
21	59	49	43
22	66	56	49
23	73	62	55
24	81	69	61
25	89	77	68

The values in this table were obtained by kind permission of the American Cyanimid Company.

TABLE A-17

5 percent Critical Points of Rank Sums (T)

n_2 \ n_1	2	3	4	5	6	7	8	9	10	11	12	13	14	15
4			10											
5		6	11	17										
6		7	12	18	26									
7		7	13	20	27	36								
8	3	8	14	21	29	38	49							
9	3	8	15	22	31	40	51	63						
10	3	9	15	23	32	42	53	65	78					
11	4	9	16	24	34	44	55	68	81	96				
12	4	10	17	26	35	46	58	71	85	99	115			
13	4	10	18	27	37	48	60	73	88	103	119	137		
14	4	11	19	28	38	50	63	76	91	106	123	141	160	
15	4	11	20	29	40	52	65	79	94	110	127	145	164	185
16	4	12	21	31	42	54	67	82	97	114	131	150	169	
17	5	12	21	32	43	56	70	84	100	117	135	154		
18	5	13	22	33	45	58	72	87	103	121	139			
19	5	13	23	34	46	60	74	90	107	124				
20	5	14	24	35	48	62	77	93	110					
21	6	14	25	37	50	64	79	95						
22	6	15	26	38	51	66	82							
23	6	15	27	39	53	68								
24	6	16	28	40	55									
25	6	16	28	42										
26	7	17	29											
27	7	17												
28	7													

The values in this table were obtained by permission from Colin White, "The Use of Ranks in a Test of Significance for Comparing Two Treatments," *Biometrics*, vol. 8, 1952, p. 37.

TABLE A-18

Factors for Converting to Imperial (British) Units

SI	*Imperial*	*SI*	*Imperial*
Length		*Force*	
1 millimeter (mm)	0.0394 inch (in)	1 newton (N)	0.225 pound force (lbf)
1 meter (m)	3.28 feet (ft)		
1 kilometer (km)	0.621 mile (mile)		
		Pressure, Stress	
Area		1 N/m^2 (= pascal, Pa)	0.0209 lbf/ft^2
1 square millimeter (mm^2)	1.55 × 10^{-3} square inch (in^2)	1 N/mm^2 (=1 MN/m^2)	145 lbf/in^2 (p.s.i.)
1 square meter (m^2)	10.76 square feet (ft^2)	1 N/mm^2	295 in. Hg
1 hectare (ha)	2.47 acres (acre)	1 millibar (mb)	0.0295 in. Hg
Volume		*Work, Energy*	
1 cubic millimeter (mm^3)	0.0610 × 10^{-3} cubic inches (in^3)	1 joule (J)	0.738 ft · lbf
		Moment, Torque	
1 cubic meter (m^3)	35.3 cubic feet (ft^3)	1 N · m	0.738 lbf · ft
1 liter (dm^3)	0.220 gallons (gal)		
	0.264 U.S. gallons (gal)	*Power*	
		1 watt (W) (=1 J/s)	1.34 × 10^{-3} horsepower (hp)
Mass		*Heat Quantity*	
1 gram (g)	0.0353 ounce (oz)	1 joule (J)	0.948 × 10^{-3} (Btu)
1 kilogram (kg)	2.20 pounds (lb)	1 watt (W)	3.41 Btu/h
1 ton (t) (1000 kg)	2.20 kips (kip)		
		Conductivity and U Value (Transmittance)	
Density			
1 kg/m^3	0.0624 lb/ft^3	1 W/m°C	6.93 Btu · in/ft^2 h°F
1 t/m^3	0.0361 lb/in^3	1 W/m^2 °C	0.176 Btu/ft^2 h°F
Flexural Stiffness		*Flow*	
1 N · mm^2	0.348 × 10^{-3} lbf · in^2	1 l/s	2.12 ft^3/min
Second Moment of Area		*Viscosity (kinematic)*	
1 mm^4	2.40 × 10^{-6} in^4	1 cm^2/s (=1 stoke, St)	1.076 × 10^{-3} ft^2/s
Velocity		*Concrete Mixes*	
1 m/s	3.28 ft/s	1 kg/m^3	1.69 lb/yd^3
1 km/h	0.911 ft/s		
	0.621 mile/h (m.p.h.)		

Answers to Problems

Chapter 2

2-1. (d) 21 years, 15 years.
2-2. (b) 69.6. (c) 42.7.
2-3. (b) 33.8 percent. (c) 71 percent.
2-5. (a) 0.75 percent (b) 0.885 (c) 0, 1, 2, ..., 12
2-6. (c) 46.7 percent for class width of 0.002 cm starting at 1.724 cm.

Chapter 3

3-1. 978, 1152; the set has no mode.
3-2. (c) 0.5354 cm.
3-3. (a) 13.75 cm. (b) 13.80 cm; error is $\simeq$0.34 percent.
3-4. 57.75, 54.6, 52.5.
3-5. 10.3 percent.
3-6. 10.56 percent, 10.5 percent, 10.6 percent, 10.66 percent.
3-7. (b) 231.1 h, 186.25 h, 50 h
3-8. 2.53, 1, 2

Chapter 4

4-1. 1403.2 N, 13.68 N, 42 N.
4-2. 133,884, 365.9, 37.4 percent. One cannot estimate the standard deviation from the range since it is not possible to calculate the mean range, $\bar{R}$, and the given values are far from being normally distributed.
4-3. 0.022 cm, (a) 0.00525 cm, $V = 0.982$ percent. (b) 0.00529 cm, $V = 0.989$ percent, s from mean range $= 0.00542$ cm.
4-4. 176.8 h, 31,249.9, 22.7 percent.
4-5. (a) 2.93, 1.04. (b) 2.16.
4-6. 23.45 MN/m^2, 16.8 MN/m^2, 2.59 MN/m^2, 3.44 MN/m^2, 14.7 percent. Specification requirements are satisfied since only 1 out of 50 samples has a strength less than 17.2 MN/m^2.
4-7. 23.82 MN/m^2, 2.6 MN/m^2, 10.9 percent.
4-8. $s = 2.835$ MN/m^2.
4-9. By adopting a class width of 1.4 MN/m^2, a cumulative frequency diagram gives a strength of 20.7 MN/m^2 exceeded by 90 percent of tests.
4-10. (a) 1073°C (b) 1197; 34.6°C; 3.225 percent.

Chapter 5

5-1. (a) 0.0867. (b) 0.0433. (c) 0.0625.
5-2. 1.575×10^{-12}.
5-3. $(\frac{3}{4})^4$.
5-4. (a) $\frac{11}{36}$. (b) $\frac{1}{3}$.
5-5. 3.6288×10^6.
5-6. $(3)^{14}$.
5-7. (a) 0.06. (b) 0.56. (c) 0.38.
5-8. (a) 0.56 (b) 0.8064
5-9. 0.72
5-10. No need to build new bridge.
5-11. The components function independently.
5-12. (a) $\frac{1}{12}$ (b) $\frac{1}{2}$
5-13. 2.333; 1.944
5-14. 0.8571
5-15. $\dfrac{(x^2 - 600^2)}{x^2}$; 0.36; 1200 h

Chapter 6

6-1. 0.000748 cm.
6-2. 4.57 N, 1.42.
6-3. (a) 16.2. (b) 2.64. (c) 16.2. (d) 1.62.
6-4. 0.43 cm.
6-5. 0.0437 kg, 0.0189 kg.

6-6. 60.06 s, 0.324 s, 0.09 s, 0.272 s, 0.54 percent.
6-7. (a) 3455 (b) 2.8 (c) 0.08 percent (d) 1.1 (e) 1.1

Chapter 7
7-1. (a) 0.349. (b) 0.387. (c) 0.930. (d) 0.264.
7-2. 0.091.
7-3. Consignment is suspect.
7-4. (a) 0.096. (b) 0.904. (c) 0.1, 0.315.
7-5. 0.294, 0.010.
7-7. (a) 0.579. (b) 0.209. (c) 0.194.
7-8. Hypothesis is rejected.
7-9. The bridge should be strengthened.
7-10. (a) 5 (b) New nozzle did not solve the problem.

Chapter 8
8-1. (a) Distribution is Poissonian.
(b) s from the given data is 10.06. s based on the Poisson distribution is 11.01.
8-2. 0.0039.
8-3. True.
8-4. Reject the consignment.
8-5. 0.647.
8-6. The failures occur randomly.
8-7. (a) 1.442 (b) 1.201 (c) 0.764.
8-8. yes.

Chapter 9
9-1. (a) 95.4 percent. (b) 36.8 percent. (c) 93.3 percent.
9-2. 6578 N, 1111 N.
9-3. (a) 0.1893. (b) 0.5773.
9-4. (a) 0.0055. (b) 0.7794. (c) 109.
9-5. 23805 pairs, 142193 pairs.
9-6. 2715 h, 199 h.
9-7. 0.0019 ohm^{-1}.
9-8. 13.

Chapter 10
10-1. (a) 20 $\mu\Omega$, 17.3 $(\mu\Omega)^2$. (b) 1.04 $\mu\Omega$. (c) (22.7, 17.3).
10-2. The numbers are not equally distributed.
10-3. (a) 42.4 percent. (b) 26.4 percent.
10-4. (a) 50 percent, (b) 43.2 percent, (c) 1.25 percent.
10-5. 32.
10-6. (a) 159,279 $\pm$ 0.026 cm, 10. (b) 4.56 percent (c) 4 or more.

10-7. $y = \frac{57}{3.262\sqrt{2\pi}} \exp[-(13.798 - x)^2/2(3.262)^2]$.

The histogram and fitted normal curve are in fair agreement.

10-8. $p(x) = 0.3989e^{-0.250x^2}$, $y = 12.45e^{-0.250x^2}$, 10.7 percent, 1.45 percent.
Good agreement is shown with the values previously calculated.

10-9. (a) $54.7^1 \pm 4.5^5$ (based on 10 readings taken at random).

(b) $y = \frac{64}{2.43\sqrt{2\pi}} \exp[-(54.13 - x)^2/2(2.43)^2]$.

(c) From a plot on normal probability paper, mean is 54.1 and s is 2.4. Good agreement.

10-10. $y = \frac{24.64}{\sqrt{2\pi}} e^{[-(23.28 - x)^2/2(3.45)^2]}$; 3.9 percent

10-11. 32.98 MN/m^2, 37.33 MN/m^2.

10-12. (a) h = 0.954 s/kg (b) 2.5 or 3; h = 1.335 s/kg.

Chapter 11

11-1. 5.92 is an outlier.

11-2. $\bar{x} = 3.55$.
(a) No outliers at 5 percent level of significance.
(b) 4.6 is an outlier; new mean is 3.3^4 cm.

11-3. (a) 8.92 should be discarded.
(b) 9.01 cm, original mean = 9.00 cm.

Chapter 12

12-1. χ^2 calculated is 10.316. Consignment is suspect at 1 percent level of significance.

12-2. χ^2 calculated is 0.946. Distribution does not differ significantly from Poisson distribution.

12-3. χ^2 calculated is 2.531. There is no justification in rejecting the goodness of fit.

12-4. χ^2 calculated is 8.552. Class B receives better instruction at 1 percent level of significance.

12-5. χ^2 calculated is 191. There is a significant difference between the machines, type A leading to "less trouble," at 0.1 percent level of significance.

12-6. χ^2 calculated is 11.492. There is a significant difference at the 0.1 percent level.

12-7. χ^2 calculated is 9.389. There is a significant difference at the 1 percent level.

12-8. χ^2 calculated is 5.429; (a) No significant difference at 5 percent level. (b) No significant difference at 1 percent level.

12-9. χ^2 calculated is 4.002. There is a significant difference at the 5 percent level.

Chapter 13

13-1. t calculated is 0.64. No significant difference at the 10 percent level of significance.

13-2. t calculated is 2.146. No significant difference at the 5 percent level of significance.

13-3. t calculated is 1.84. No significant difference at the 5 percent level of significance.

13-4. t calculated is 1.135. No significant difference.

13-5. t calculated is 2.236. No significant difference at the 5 percent level of significance.

13-6. t calculated is 3.345. Significant difference at the 1 percent level of significance.

13-7. t calculated is 1.88. No significant difference.

13-8. t calculated is 0.735. No evidence that angle is becoming smaller.

13-9. t calculated is 6.852. Significant difference at 0.1 percent level of significance.

13-10. t calculated is 3.02. Significant difference at the 5 percent level of significance.

13-11. By using normal probability we cannot say that group A and group B are significantly different. There is a difference between groups B and C.

13-12. t calculated is 0.39. No significant difference at 10 percent level of significance. Using the more discriminatory paired "t" test, t calculated is 3.436. Difference is significant at the 5 percent level of significance.

13-13. 1. No significant difference.
2. No significant difference.
3. There is significant difference.

Chapter 14

14-1. (a) σ_A^2 does not differ significantly from σ_B^2 at the 10 percent level.
(b) σ_A^2 does not differ significantly from $1.75\sigma_B^2$ at the 10 percent level.
(c) σ_A^2 is not significantly greater than σ_B^2 at the 5 percent level.

14-2. No significant difference.

14-3. No significant difference.

14-4. No significant difference.

14-5. No significant difference.

14-6. Significant difference at the 1 percent level of significance.

14-7. Significant difference between the two variabilities at the 1 percent level of significance.
14-8. No significant difference in standard deviations.
14-9. Significant difference at the 1 percent level of significance.
14-10. $v = 1823.6 \pm 64.2$ cm^3.
14-11. $s = 4.88$ m gives a better estimate of 'g'.
14-12. $\sigma_v = 25.133$ cm^3.
14-13. $\sigma_\delta = 0.00248$ mm.
14-14. 0.831 percent, 0.831 percent, "e" should be increased in accuracy.

Chapter 15

15-1. $y = 9.669 + 1.157x$
$x = -8.113 + 0.846y$
15-2. $y = 20.270 - (41.888/x)$.
15-3. (a) $y = -14.143 + 4.036x$.
(b) 20.16 ± 5.33.
15-4. (a) $y = 10 + 2x$.
(b) The hypothesis that $b = 0$ is not true at the 0.1 percent level.
(c) No significant difference.
(d) (1.47, 2.53).
15-5. $y = -313.52 + 6.95x$, 15.31, 103.5, 68.6, (69.2, 137.8).
15-6. 240.49, 243.62, 246.19.
15-7. $A = 82°15'35''$, $B = 110°37'45''$, $C = 66°24'35''$, $D = 100°42'05''$, $(A + B) = 192°53'20''$.
15-8. (a) $\log y = 1.602 + 0.161x$. (b) 211. (c) 565, 79.
15-9. $y = 115.5 - 21.7 \log x$.
15-10. (a) $y = 5.074 + 0.882x$. (b) $y = 5.128 + 0.746x$.
For (a) the value of $b = 0.882$ is significant at the 0.1 percent level.
The value of $a = 5.074$ is significant at the 10 percent level.
For (a) the slope is significantly different from 1.00 at the 10 percent level.
For (b) the value of $b = 0.746$ is significant at the 0.1 percent level.
The value of $a = 5.128$ is significant at the 1 percent level.
For (a) the slope is significantly different from 1.00 at the 10 percent level.
For (b) the slope is significantly different from 0.85 at the 1 percent level.
15-11. $y = x^{0.801}/4.78$.
15-12. 57°35'15'', 92°17'41'', 75°53'22''.
15-13. +10.705 m, −51.764 m, +32.849 m.

Chapter 16

16-1. 0.9152; the correlation is significant at the 1 percent level of significance.

16-2. r is 0.607. The correlation is significant at the 5 percent level but not at the 1 percent level of significance.

16-3. The corresponding population correlation does not differ from zero at either level of significance.

16-4. At least 46.

16-5. $y = 29.612 + 2.838x$, 0.996. The slope is not significant at the 5 percent level of significance.

Chapter 17

17-1. (a) $y = 83.077 - 0.317x_1 + 0.552x_2 + 0.176x_3$.

(b) $F = 3.26$; the null hypothesis that all the true partial regression coefficients are equal to zero cannot be rejected at the 5 percent level of significance.

(c) t_a is 7.489, a is significant.
t_{b_1} is 0.903, b_1 is not significant.
t_{b_2} is 2.462, b_2 is significant.
t_{b_3} is 0.329, b_3 is not significant.

(d) 0.742. Not significant at the 5 percent level of significance.

(e) $r_{yx_1} = -0.327$, $r_{yx_2} = 0.710$, $r_{yx_3} = 0.258$, $r_{x_1x_2} = -0.183$, $r_{x_1x_3} = 0.173$, $r_{x_2x_3} = 0.331$.

17-2. (a) $y = 94.213 - 0.333x_1 - 0.171x_2 + 0.971x_3$.

(b) 74.482.

(c) The null hypothesis that all partial regression coefficients are not significant cannot be rejected at the 5 percent level of significance.

(d) $t_a = 10.916$, a is significant.
$t_{b_1} = 0.753$, b_1 is not significant.
$t_{b_2} = 0.408$, b_2 is not significant.
$t_{b_3} = 2.143$, b_3 is not significant.

(e) 0.5865. Not significant at the 5 percent level of significance.

Chapter 18

18-1. Significant difference at the 1 percent level.

18-2. No significant difference.

18-3. No significant difference.

18-4. The type of terrain has no effect on runoff at the 5 percent level.

18-5. Significant difference at the 1 percent level.

18-6. No significant difference.

18-7. (b) No significant difference.
(c) No significant difference.

Chapter 19

19-1. $m = 2.43$ and $\gamma = 4.6 \times 10^{-7}$, 53 percent.

19-2. $m = 1.92$ and $\gamma = 1.035 \times 10^{-4}$.
χ^2 for the Weibull distribution is 10.97.
χ^2 for the normal distribution is 9.25.

19-3. (b) 4080 m^3/s, 2076 m^3/s.
(c) 4.2 years, 2.3 years.
(d) 7330 m^3/s, 8100 m^3/s, 9100 m^3/s.

19-4. $X = 800 - 94y$.

Chapter 20

20-1. Warning limits for mean chart: 3657 kN/m^2, 3363 kN/m^2.
Action limits for mean chart: 3742 kN/m^2, 3278 kN/m^2.
Warning limits for range chart: 706 kN/m^2, 144 kN/m^2.
Action limits for range chart: 920 kN/m^2, 62 kN/m^2.

20-2. Warning limits for mean chart: 18.50, 10.32.
Action limits for mean chart: 20.87, 7.95.
Warning limits for range chart: 13.30, 1.10.
Action limits for range chart: 18.33, 0.25.

20-3. (a) Not in control.
(b) Yes.

20-4. The control limits are (0, 5.72).
The probability limits are (0, 6.09).

20-5. (a) $s = 2.835$ MN/m^2.
(b) MWL: 26.31 MN/m^2, 21.33 MN/m^2.
MAL: 27.74 MN/m^2, 19.90 MN/m^2.
RWL: 11.94 MN/m^2, 2.44 MN/m^2.
RAL: 15.56 MN/m^2, 1.06 MN/m^2.
(c) All points are within the limits.

20-7. (a) MWL: 534.1 N, 495.9 N.
MAL: 545.2 N, 484.8 N.
RWL: 77.6 N, 11.7 N.
RAL: 103.7 N, 4.0 N.
(b) $s = 19.5$ N
(c) All welds meet the minimum specifications.

Chapter 21

21-1. (a) If the mean yield of 36 tests is greater than 10.1, reject the hypothesis that the processes are the same.
(b) Since there is no overlap (from a practical point of view) the required probability is zero.
(d) 38.

21-2. 6.

21-3. The effect of the different shifts is significant at the 1 percent level. There is a significant difference in the production figures during the four years at the 1 percent level.

21-4. The performance does not depend on the process at the 5 percent level. The source affects the performance at the 1 percent level.

21-5. (a) 15.46.
(b) 4.86.
(c) 16.16.
(d) No significant effect due to micrometers at the 1 percent level. No effect due to different gage blocks at the 1 percent level.
(e) There is no significant effect due to different engineers at the 5 percent level.

21-6. (a) $n = 25$ approximately.
(b) The hypothesis is rejected.

Index

D

N

Q

R

S

T

W

Y, Z